Impact Behavior and Pedestrian Protection of Automotive Laminated Windshield

Jun Xu · Yibing Li

Impact Behavior and Pedestrian Protection of Automotive Laminated Windshield

Theories, Experiments and Numerical Simulations

Science Press
Beijing

Springer

Jun Xu
Department of Automotive Engineering
Beihang University
Beijing, China

Yibing Li
Department of Automotive Engineering
Tsinghua University
Beijing, China

ISBN 978-981-13-2440-6 ISBN 978-981-13-2441-3 (eBook)
https://doi.org/10.1007/978-981-13-2441-3

Jointly published with Science Press, Beijing, China

The print edition is not for sale in China Mainland. Customers from China Mainland please order the print book from: Science Press, Beijing, China.

Library of Congress Control Number: 2018953591

This Springer imprint is published by the registered company Springer Nature Singapore Pte Ltd.
The registered company address is: 152 Beach Road, #21-01/04 Gateway East, Singapore 189721, Singapore

Preface

The fundamental objective of this book is very simple: to provide the reader (typically professors, research scientists, engineers, and graduate students) with a working knowledge of mechanical behaviors of laminated windshield. To achieve such a goal, I simply organize theories, experiments (including setups, testing procedures, and results), and numerical computation modeling into this little book in a step-by-step manner. This book starts with the general introduction of the glass and its physical and chemical properties. Then, manufacturing of automotive-laminated windshield is introduced. General mechanical behaviors as well as crack initiation behaviors are fully discussed, which provides readers a global picture with vivid analysis details. A very useful guideline for windshield design engineer would be the methodology of finite element modeling by following the examples in this book. Finally, pedestrian protection and accident reconstruction based on the cracking morphology of windshield are presented.

This book summarizes all my own research work in the past 10 years, along with representative research efforts from other research groups worldwide. I, by the exceptionally warm encouragement from my advisor Prof. Yibing Li (who is also the co-author of this book), initiated the study of laminated windshield when I was a graduate student at Tsinghua University. I can still remember the struggles and difficulties that I encountered during my research; and of course, I also cherish the joys that I had when some breakthroughs were made. Later, some of my fellow group members join in this topic including Ms. Jingjing Chen, Dr. Mengyi Zhu, Dr. Bohan Liu, and Dr. Yueting Sun. The research work advanced much faster with their talented input and nonstop working. Continuous financial supports from National Science Foundation of China expedite the research in a great extent. Illuminative academic supports from Prof. Xuefneg Yao and Prof. Dongyun Ge should also be mentioned. After I got my Ph.D. from Columbia University and joined Beihang University, I continued my research with the support from Jaguar Land Rover (JLR). The latest advance on this topic mainly focused on the numerical computation part. Dr. Bill Feng, as the project manager of JLR funding, gave us strong supports in vehicle testing.

The writing of this book would not have been possible without the help provided by my graduate students at Beihang University. At Beihang University, I was fortunate enough to meet and have a group of energetic and talented young students, who inspired me further toward the finish of this book. To all of them, my salute and gratitude. In particular, the extraordinary help from Ms. Yanting Zheng and Ms. Danping Xiong should never go unmentioned. I would like to express my thanks to the committee of the National Fund for Academic Publication in Science and Technology to extend their warm help for the publication of this book.

Last, but not the least, I thank my wife, Ms. Tingting Li and 5-year-old daughter Lilliana. They have been the most supportive persons in the world for such a "boring" research work and I could not have walked one step further without their love.

Beijing, China Jun Xu
March 2018

Contents

Chapter 1
Introduction

Glass, which possesses excellent material properties such as chemical stability, transparency, and high wear resistance, is a non-crystalline amorphous solid because of the large viscosity of liquid melt and because the liquid melt during cooling cannot spread out to crystallize completely [1, 2]. Glass products can be extensively popularized because the raw materials of glass, such as quartz and limestone, are distributed broadly and inexpensive [3]. Glass products are widely used in various aspects of life, including windows, decorations, and tableware [1].

1.1 Origin and Development of Glass

Glass was invented 5000 years ago by ancient Egyptians. Most academics believed that glass in China originated from Egypt, and this belief was debated when cultural relics of glass, such as colored glaze cups and bottles, were unearthed in China. The raw material of Chinese glass is different with that of western glass, and this difference shows that China was one of the first inventors of glass [3].

According to western civilization history dating back to 5000 years ago, a group of Phoenician sailors debarked from their ship and prepared their evening meal on a Mediterranean beach by using "natural soda" (sodium nitrate) as the fireplace. Then something interesting occurred. When the fire had died down, the sailors found something transparent and shining; it was formed by the chemical reaction of sodium nitrate and sand on the beach, and this was the earliest glass [4, 5].

Glass has a long history. After the Phoenicians discovered glass, the manufacture of glass gradually spread among the Mediterranean people, Egyptians, Greeks, and Romans. The Egyptians developed many non-transparent colored glass products and striped glass bottles, which can be found in museums today [4]. Figures 1.1 and 1.2 show a Roman cage cup from 4th century CE and Roman glass amphoriskoi from 1st to 2nd century AD, respectively.

J. Xu and Y. Li, *Impact Behavior and Pedestrian Protection of Automotive Laminated Windshield*, https://doi.org/10.1007/978-981-13-2441-3_1

Fig. 1.1 Roman cage cup from 4th century CE [1]

In the 1st century, ancient Romans developed transparent glass by increasing the temperature of furnaces and created various glass products with different shapes, such as vases and decorations, as shown in Fig. 1.3.

Afterward, glass manufacture prospered in Italy during the Middle Ages when Venice was at the height of its glory, and glass manufacture was regarded as a state secret. To monopolize glass manufacture, Venice relocated its glass workshops on the tiny island of Murano and kept the workers who knew the secrets of glass manufacture away from the external world and paid them high salaries [4, 5]. Figure 1.4 shows a glass cup called a Venetian goblet that was made in Italy in the early 19th century.

However, the secrets did not last for a long time because several workers escaped from the island. Glass manufacture gradually spread worldwide. In 1688, a French national named NAF developed plate glass that was larger than that manufactured by the Italians. However, the cost of the glass increased because of the manufacturing process.

Ordinary families had only a few glass products because glass was expensive products. Meanwhile, glass was widely used in churches and holy palaces, and expensive stained glass (Fig. 1.5) was used in places that deserved it. In 1828, a French worker named Robin invented the first machine for glass blowing. However, the machine was not promoted due to the low quality of its products [3]. Then in 1838, the tech-

Fig. 1.2 Roman glass amphoriskoi from 1st to 2nd century AD [1]

(a) Process of thermally shaping glass (b) Glass decorations

Fig. 1.3 Glass products [5]

nology of glass manufacture improved; its cost decreased, and glass became a regular product in daily life [5].

In 1903, the famous American automotive manufacturer, Ford, produced automotive glass through an auto-production line. The American glass industry developed rapidly, and America became the world's top glass producer.

The use of glass in China dates back to the Tang dynasty when glass called colored glaze emerged at a large scale and was regarded as a treasure because of its translucent

Fig. 1.4 Venetian goblet made in Italy in the early 19th century [1]

and brilliant appearance [5]. Figure 1.6 shows a colored-glaze handicap bottle from the Tang dynasty.

The wars in the period of Ming Qing influenced glass manufacture. However, the development of glass reached a period of high prosperity after the Qing dynasty when the manufacturing of glass recovered, and various categories of glass with high quality and different shapes emerged [6]. Figure 1.7 shows a Peking glass vase created in the Daoguang period.

After the founding of New China, the glass industry went through two stages, namely, recovery period after the ruin of the Qing dynasty and rapid development period from 1980 to the present. Plate, vacuum, and laminated glass were then developed [6].

1.2 Categories of Glass

Glass, one of the oldest known materials, is one of man's most useful products for different uses today [4]. It is widely used in architectural design (glass applied as building windows, as shown in Fig. 1.8), automotive industry (glass used as wind-

Fig. 1.5 Stained glass applied as windows in a church [5]

Fig. 1.6 Colored-glaze handicap bottle from the Tang dynasty [7]

Fig. 1.7 Peking glass vase from the Daoguang period [1]

shields, as shown in Fig. 1.9), military, and modern communication (glass used as fiber, as shown in Fig. 1.10) because of its numerous functions [8].

Glass has two main categories, namely, traditional glass (divided into flat and domestic glass) and new glass.

Plate glass, which is shown in Fig. 1.11, is a type of glass whose thickness is significantly smaller than its length and width. According to its usage, plate glass can be divided into glass for architectural buildings, safety glass, windshields for automobile, plane, and naval craft, and decorative glass. Original plate glass that is manufactured directly can be classified as float, roll, and so on. Float glass is produced by obtaining a molten glass solution floating on the surface of molten tin liquid and then shaping this solution as plate glass. The manufacturing process of roll glass uses a single or coupled metal roller to roll and extend the glass liquid into a glass band.

Domestic glass has abundant applications in our daily life, and examples include bottle glass, ware glass, instrument glass, optical glass, electronic glass, glass fiber, and glass composite materials (Fig. 1.12).

Meanwhile, new glass is used in optoelectronic information, biology and medicine, energy field, ecological environment and nanometer glass materials, as shown in Fig. 1.13. The non-oxide component systems of new glass are different from those of traditional glass because the former meets a specific property requirement [8].

This book focuses on the application of glass in architecture and automotive industries.

Fig. 1.8 Glass used in a building's windows [9]

Safety glass (Fig. 1.14), which is a type of glass manufactured through a special material and strength treatment, such as toughening, is mainly applied in automotive and consists of tempered and laminated glass [12].

The origin of safety glass in automotive dates back to the early 20th century when Ford and Benz invented automotive vehicles that were like carriages without windows and had only a hood. Windows (windshields) were needed to increase vehicle speed. Therefore, glass, mainly organic and normal inorganic, was first applied in the automotive industry in 1909. Subsequently, tempered and laminated glass took the place of organic and normal inorganic glass due to the low hardness and low wear resistance of the latter type of glass [12].

Fig. 1.9 Automotive windshield on a 1959 Edsel Corsair [10]

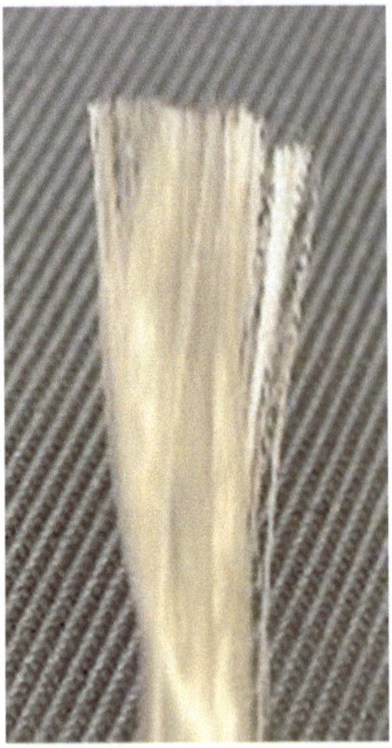

Fig. 1.10 Glass fiber [1]

Laminated glass, a type of safety glass, is basically a three-layer laminated composite structure that holds glass piles together by using an interlayer, such as polyvinyl butyral (PVB) or ethylenevinyl acetate (EVA) [13], as shown in Figs. 1.15 and 1.16. Laminated glass has extensive applications in architecture, structural parts, and automotive industry. PVB laminated glass is widely applied in aerospace, the military, and new high-technology industries due to its transparency, excellent adhesive property, thermal resistance, and high mechanical property.

Notably, the design of architectural glass is different from that of automotive glass despite their similar construction process. Architectural glass requires the inside plies

Fig. 1.11 Flat glass [11]

to remain unbroken when subjected to small missile impact. By contrast, automotive glass requires the glass and interlayer to fracture when subjected to impact by vehicle occupants for the purpose of safety protection [15].

1.3 Application and Related Research on Glass in Architecture

The categories of architectural glass are numerous, and they include plate, floating, tempered, laminated, heat reflection, and bullet-proof glass. Figure 1.17 shows architectural glass used as a building wall in a city. Laminated glass is also widely used as architectural glass [16].

The use of safety glass in architecture emerged early, and it developed rapidly as people became highly aware of the benefits of using safety glass in architecture. The safety glass applied in architecture not only meets architectural and aesthetic functions but also satisfies safety requirements, which are due to the possible occurrence of natural hazards, such as rainstorms and earthquakes [12]. When bombings occur in cities, most of the injuries result from flying and falling building window glass shards [18]. Thus, the glass used in architecture, such as safety glass, should be of high-quality must exhibit good performance. Figure 1.18 shows bulletproof glass. The explosion-proof property and impact resistance of architectural glass are research hotspots and have been investigated extensively due to the increasing attention on the safety problem of human–environmental survival [19].

For example, Hooper [16] studied the response of architectural glass under structural loading, and its bending behavior was investigated through experimental and theoretical studies. Results showed that the bending resistance of architectural laminated glass depends mainly on the thickness and shear modulus of the interlayer.

(a) Glass jars

(b) Wine bottle

(c) Beer bottle

Fig. 1.12 Domestic glass [11]

Flocker and Dharani [15] used a finite element wave code to model a low-velocity and small missile impact on laminated architectural glass to predict the fracture of internal plies subjected to windborne debris. They found that cone cracking prior to the penetration of external plies reduces the overall critical stress.

Flocker and Dharani [20] investigated the effect of the geometry of architectural glass plies on impact resistance under low-velocity, small missile impact via finite element analysis (FEA). The investigations showed that the geometry of glass plies significantly affect the impact resistance of architectural glass and that three-layered laminated glass provides better impact resistance than four-layered and seven-layered laminated glass.

(a) Low-iron solar glass

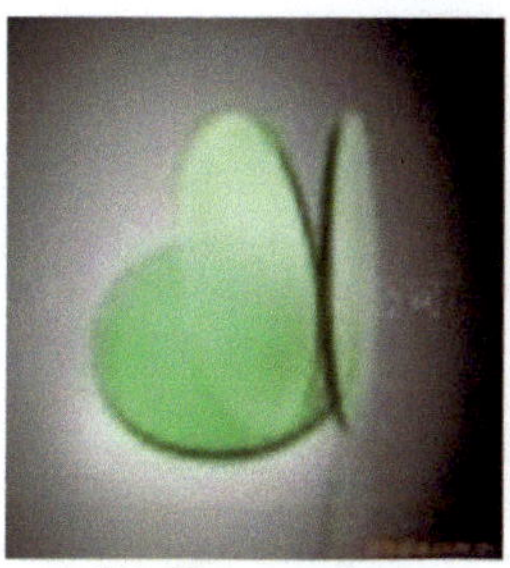

(b) Green glass

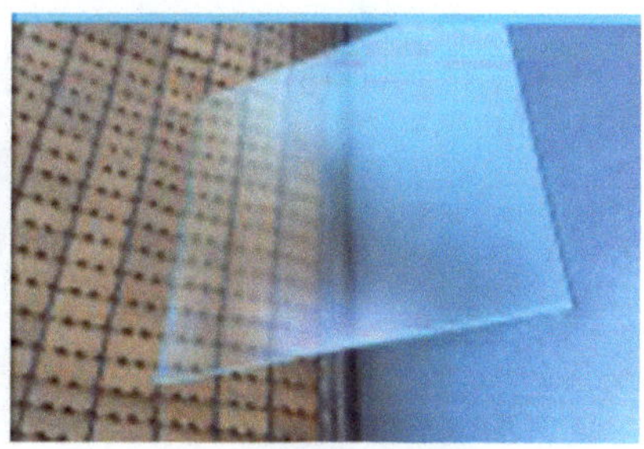

(c) Anti-glare glass

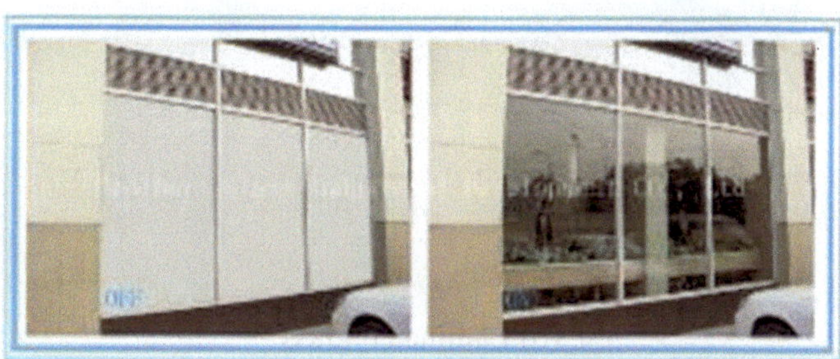

(d) Smart glass

Fig. 1.13 New glass [11]

To investigate the fragment behavior of architectural glass panels under blast loading, Ge et al. [21] conducted explosive tests to determine the projection velocity and splash distance of glass fragments and established the empirical expression of fragment flight trajectory, which can be used to estimate the hazard of broken glass. The theoretically predicted values of fragments fit the test results well.

(a) Tempered curved glass

(b) Laminated glass

Fig. 1.14 Safety glass [11]

Wei and Dharani [22] investigated a rectangular, laminated, architectural glazing subjected to blast loading through a simulation. They found that the negative phase of blast loading exerts a significant effect on the dynamic response of the glass. The mid-span panel deflection and maximum tensile stress caused by the negative pressure are nearly twice the values produced by the positive blast loading phase.

Aside from these aforementioned research, other studies, such as Refs. [23–26], have examined the impact resistance and blast impact behavior of architectural glass.

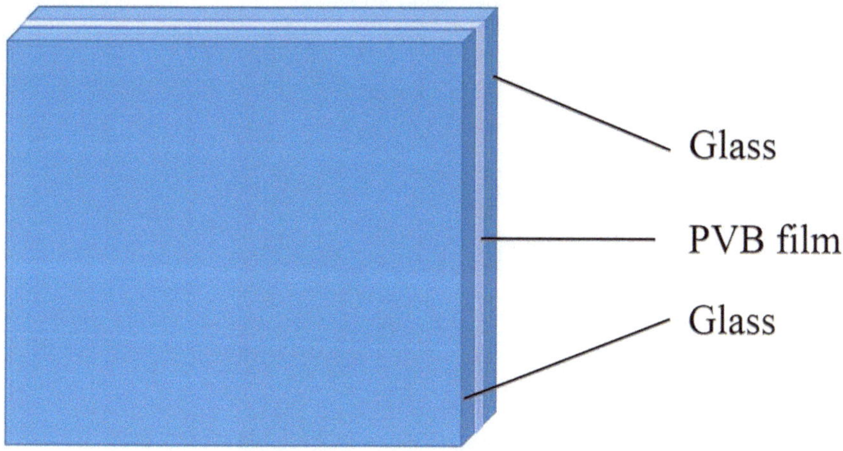

Fig. 1.15 Structure of laminated glass

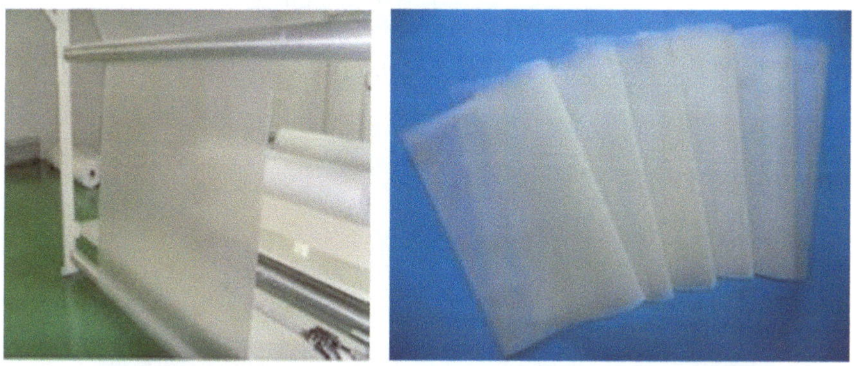

(a) Clear PVB films (b) Clear EVA films

Fig. 1.16 Interlayer materials of laminated glass [11]

Meanwhile, the development tendency of architectural glass could involve an increased number of functions while meeting safety requirements [12].

1.4 Application and Related Research on Glass in Automotive

The glass used in the windows of vehicles accounts for 3% of the total vehicle mass. Generally, the glass used in vehicles is mainly made of silicate glass, which consists

Fig. 1.17 Building wall installed with architectural glass [17]

Fig. 1.18 Bulletproof glass [11]

of sodium oxide, calcium oxide, silicon oxide (accounts for 70% of the total glass content), and magnesium [2].

Vehicle glass, which can be divided into laminated and tempered, is secondary glass processed from raw glass. The windshield mainly uses laminated glass. The front windshield commonly uses laminated tempered glass that is tempered by raw glass through heat and chilling treatments by cold wind [27].

However, vehicle windshields in the 1940s were still made of two pieces of plate tempered glass with a thickness of up to 6 mm. Until the early 1950s, Americans used curved tempered glass as the vehicle front windshield. Then, this type of windshield was quickly applied around the world because of its excellent quality [28].

In the 1960s, the single-piece bending windshield gradually took the place of the flat windshield in vehicles [2]. Figure 1.19 shows a plate glass used as a windshield on a 1952 DeSoto.

Automotive glass mainly uses PVB laminated glass. Laminated glass includes dry and wet glass classified according to the manufacturing method [29]. The manufacturing of laminated glass in foreign countries rapidly developed. American glass companies started producing laminated glass in 1910. In 1920, Muskat used cellulose ester as the glass interlayer. Nippon Sheet Glass Company developed three-layered laminated glass in 1922. America Ford established the glass industry and was the first to use laminated glass as a windshield in A-type vehicles. In 1938, America invented the PVB film, which was then widely applied in the automotive industry. Later on, Volvo used laminated glass as the front windshield of vehicles in 1944.

In China, National Construction Institute was the first to develop laminated glass. The development of laminated glass was limited by the late start of the development of the interlayer, and the PVB films used in top laminated glass needed to be imported from America, Japan, and other countries although China has been developing PVB films through the flowing method since 1965. The sales volume of PVB films reached 50,000 tons in 2004 in our country, whereas the corresponding value in other countries was 80% [29]. In the meantime, the production and consumption of laminated glass increased annually from 2001 to 2005, as shown in Tables 1.1 and 1.2. The production and consumption of laminated glass increased by 3.5 and 2 times in five years, respectively.

Laminated glass is a composite structure that consists of the interlayer PVB (polyvinyl butyral) and two layers of soda-lime silicate glass. Soda-lime silicate glass, which has a high alkali content (14 mol% Na_2O) [30], entails a compromise

Fig. 1.19 Flat windshield glass on a 1952 DeSoto [10]

Table 1.1 Production of laminated glass in China from 2001 to 2005 [29]

Item		2001	2002	2003	2004	2005
Total product (10^4 m^2)		561.15	828.28	1284.49	1712.63	1949.01
Output	Number (10^4 m^2)	184.97	359.61	556.98	903.68	1038.97
	Percent (%)	32.96	43.42	43.36	52.77	53.31

Table 1.2 Consumption of laminated glass in China from 2001 to 2005

Item		2001	2002	2003	2004	2005
Total consumption (10^4 m^2)		468.90	574.38	798.71	903.99	1103.52
Architecture	Number (10^4 m^2)	21.83	34.40	42.37	54.16	74.14
	Percent (%)	4.66	5.99	5.30	5.99	6.72
Automotive	Number (10^4 m^2)	399.75	506.93	712.14	800.55	969.67
	Percent (%)	85.25	88.26	89.16	88.56	87.87
Others	Number (10^4 m^2)	47.32	33.05	44.20	49.28	59.71
	Percent (%)	10.09	5.75	5.53	5.45	5.41

between the outstanding properties of pure silica and the cost of producing massive quantities of glass for different requirements [31]. The interlayer PVB has an excellent adhesion property, which contributes to the good pedestrian protection safety of laminated glass.

Many studies have been conducted on laminated glass. These studies used theoretical analyses, experiments, and simulations.

1.5 Significance of Research on Automotive Windshield Glass

The number of road traffic accidents has been increasing annually in recent years [32] because of the continuous increase in the number of vehicles in many countries. Pedestrian–vehicle impact accidents account for the largest proportion of road traffic accidents. The number of pedestrian deaths caused by road traffic accidents exceeds 0.2 million every year all over the world [33, 34]. The causes of road traffic accidents vary, and examples include road traffic conditions, driver errors, and weather. Therefore, pedestrian safety has been brought to public attention, and many researchers focused on improving pedestrian safety.

In pedestrian–vehicle accidents, the main cause of pedestrian deaths is head damage [35–37]. The vehicle structures that come into contact with the head during an accident include the windshield, the frame of the windshield, the A pillar, and the engine hood [38]. The windshield is the main contact structure that leads to severe damage of the head of pedestrians during pedestrian–vehicle accidents [39–41], and the contact proportion is 41.7% [38] more than the contact proportion of other vehicle structures.

Laminated glass is important in human protection and structural integrity due to its excellent energy-absorbing and adhesive capabilities [33]. It also helps reduce the injury of occupants in windshield-damaged accidents. Given that head–windshield impact contact frequently occurs during pedestrian–vehicle accidents, the study of the impact of laminated windshields, which are important to pedestrian safety protection [33], on pedestrians' heads is a hot research topic. It needs to be investigated to improve pedestrian protection and reduce head injuries.

The mechanical properties of laminated glass differ from those of single glass and the interlayer [33] because of its composite structure, which has become a research hotspot [30, 33, 42–44]. The mechanical behavior and crack characteristic of laminated glass are also important in understanding laminated glass.

Nonetheless, studies have not fully explored the mechanical behavior and crack characteristics of laminated glass that are vital to safety engineering and pedestrian protection due to the complexity caused by the composite material, i.e., polymer material sandwiched between two brittle materials. The mechanical behavior and crack characteristics of laminated glass provide important information on material property, accident reconstruction, and energy absorption [45], which play a role in the impact response because they contain the impact history subjected to foreign object impact [46]. Therefore, the mechanical behavior of laminated glass and macroscopic crack initiation and propagation need to be investigated to allow for improved design or prevention with laminated PVB glass as a structural component.

1.6 Summary

In this chapter, we introduce the origin and development of glass, which can be traced back to 5000 years ago in ancient Egypt. China is also one of the first inventors of glass, given that the use of glass in China dates back to the Tang dynasty. The categories of glass, namely, traditional glass (divided into two categories of flat and domestic) and new glass, are introduced in detail. We also present the application of glass in architecture and automotive fields. Studies related to glass are likewise introduced.

References

1. Wikipedia, Glass. https://en.wikipedia.org/wiki/Glass#cite_note-57
2. Q. Xiao, Automotive glass and its development prospects. Adv. Ceram. **31** (2010)
3. Y. Zhao, H. Yin, *Glass Technology* (Chemical Industry Publishing House, 2006)

4. B.D.A. Diehn, *History of Glass* (Ohio State University College of Engineering, 1941)
5. L. Jiu, The history of glass. China Constr. Metal Struct. **5** (2014)
6. C. Wang, S. Li, Y. Tao, X. Zhang, Glass development history and future trends. Glass (2010)
7. H. Jianwu, The influence of foreign culture on the jade stone and colored glaze of Tang dynasty. Collector **10**, 41–45 (2006)
8. M. Xu, Kinds and purposes of glass. Glass (2008)
9. Wikipedia, Flat glass. https://en.wikipedia.org/wiki/Flat_glass
10. Wikipedia, Windshield. https://en.wikipedia.org/wiki/Windshield
11. C.G. Network, Glass material. http://www.glassinchina.com/
12. X. Shi, *Safety Glass* (Chemical Industry Publishing House, 2006)
13. Wikipedia, Laminated glass. https://en.wikipedia.org/wiki/Laminated_glass
14. G. Glass, Product Center. http://www.globalstarglass.com/exhibit.asp
15. F.W. Flocker, L.R. Dharani, Modelling fracture in laminated architectural glass subject to low velocity impact. J. Mater. Sci. **32**, 2587–2594 (1997)
16. J.A. Hooper, On the bending of architectural laminated glass. Int. J. Mech. Sci. **15**, 309–323 (1973)
17. Wikipedia, Architectural glass and aluminum. https://en.wikipedia.org/wiki/Architectural_Glass_and_Aluminum
18. H.S. Norville, N. Harvill, E.J. Conrath, S. Shariat, S. Mallonee, Closure of "Glass-Related Injuries in Oklahoma City Bombing". J. Perform. Constructed Facil. **14** (2000)
19. S.P. Wang, X.N. Gao, Research on safe extent of architectural glazing subjected to explosion Jiangxi Sci. (2008)
20. F.W. Flocker, L.R. Dharani, Low velocity impact resistance of laminated architectural glass. J. Architectural Eng. 12–17 (1998)
21. J. Ge, G.Q. Li, S.W. Chen, Theoretical and experimental investigation on fragment behavior of architectural glass panel under blast loading. Eng. Fail. Anal. **26**, 293–303 (2012)
22. J. Wei, L.R. Dharani, Response of laminated architectural glazing subjected to blast loading. Int. J. Impact Eng. **32**, 2032–2047 (2006)
23. N°, Laminated architectural glass subjected to blast, impact loading. Am. Ceram. Soc. Bull. **84**, 42–44 (2005)
24. S. Zhao, L.R. Dharani, X. Liang, Analysis of damage in laminated architectural glazing subjected to blast loading. Adv. Struct. Eng. **11**, 129–134 (2008)
25. M.S. Shetty, J. Wei, L.R. Dharani, D.S. Stutts, Analysis of damage in laminated architectural glazing subjected to wind loading and windborne debris impact. Buildings **3**, 422–441 (2013)
26. M.S. Shetty, L.R. Dharani, J. Wei, D.S. Stutts, Failure probability of laminated architectural glazing due to combined loading of wind and debris impact. Eng. Fail. Anal. **36**, 226–242 (2014)
27. Y. Jia, Discussion on development trend of automotive glass. Glass (2010)
28. Z. Zhang, Research on automotive safety glasses development. *Tianjin Auto* (2005)
29. Z. Liu, S. Pang, *Energy-Saving Glass and Green Glass* (Chemical Industry Publishing House, 2009)
30. G.-I. Shim, S.-H. Kim, H.-W. Eom, D.-L. Ahn, J.-K. Park, S.-Y. Choi, Improvement in ballistic impact resistance of a transparent bulletproof material laminated with strengthened soda-lime silicate glass. Compos. B Eng. **77**, 169–178 (2015)
31. J.E. Shelby, *Introduction to Glass Science and Technology* (Springer, USA, 2005)
32. W.H. Organization, Global status report on road safety 2013: supporting a decade of action. World Health Organization (2013)
33. S. Chen, M. Zang, D. Wang, S. Yoshimura, T. Yamada, Numerical analysis of impact failure of automotive laminated glass: a review. Compos. B Eng. **122**, 47–60 (2017)
34. W.H. Organization, Pedestrian safety: a road safety manual for decision-makers and practitioners (2013)
35. WHO. WHO global status report on road safety 2013: supporting a decade of action. World Health Organ. (2013)

36. M. Yoshiyuki, Summary of IHRA pedestrian safety WG activities (2005)—proposed test methods to evaluate pedestrian protection afforded by passenger cars, in *Proceedings of the 19th International Technical Conference on the Enhanced Safety of Vehicles*, 2005
37. X.Q. Xu, B.H. Liu, Y. Wang, Y.B. Li, J. Xu, A numerical model on PVB laminated windshield subjected to headform low-speed impact. J. Phys: Conf. Ser. **451**, 012016 (2013)
38. D. Otte, Severity and mechanism of head impacts in car to pedestrian accidents, in *Proceeding of the 1999 International IRCOBI Conference on the Biomechanics of Impact*, 23–24 September 1999, Sitges, Spain, 1999
39. R. Fredriksson, E. Rosén, A. Kullgren, Priorities of pedestrian protection–a real-life study of severe injuries and car sources. Accid. Anal. Prev. **42**, 1672–1681 (2010)
40. J. Yao, J. Yang, D. Otte, Investigation of head injuries by reconstructions of real-world vehicle-versus-adult-pedestrian accidents. Saf. Sci. **46**, 1103–1114 (2008)
41. H. Zhao, G. Yang, F. Zhu, X. Jin, P. Begeman, Z. Yin, K.H. Yang, Z. Wang, An investigation on the head injuries of adult pedestrians by passenger cars in China. Traffic Inj. Prev. **14**, 712 (2013)
42. H.L. Ma, Z. Jia, K.T. Lau, J. Leng, D. Hui, Impact properties of glass fiber/epoxy composites at cryogenic environment. Compos. B **92**, 210–217 (2016)
43. L. Galuppi, G. Royer-Carfagni, Enhanced effective thickness of multi-layered laminated glass. Compos. B Eng. **64**, 202–213 (2014)
44. M.M. Ansari, A. Chakrabarti, M.A. Iqbal, An experimental and finite element investigation of the ballistic performance of laminated GFRP composite target. Compos. B Eng. **125**, 211–226 (2017)
45. J. Xu, Y. Li, X. Chen, Y. Yan, D. Ge, M. Zhu, B. Liu, Characteristics of windshield cracking upon low-speed impact: numerical simulation based on the extended finite element method. Comput. Mater. Sci. **48**, 582–588 (2010)
46. J. Xu, Y. Sun, B. Liu, M. Zhu, X. Yao, Y. Yan, Y. Li, X. Chen, Experimental and macroscopic investigation of dynamic crack patterns in PVB laminated glass sheets subject to light-weight impact. Eng. Fail. Anal. **18**, 1605–1612 (2011)

Chapter 2
Manufacturing of Automotive Laminated Windshields

Automotive glass includes tempered and laminated glass. Tempered glass is used for the side windows and backlite of vehicles, and laminated glass is mainly for the windshields [1] (Fig. 2.1).

Tempered glass is a type of float glass (Fig. 2.2) that is subjected to a temperature that is equal to the softening point of glass (i.e., approximately 680 °C) and cooled rapidly by cold air jets. Subjecting float glass to a tempered treatment increases the strength of this glass by four to five times [1], and this type of float glass is called tempered glass.

The aim of glass tempering is to create stress in the glass and a parabolic residual stress field that has tensile stresses in the core and compressive stresses on the surfaces of the glass [2].

With regard to the damage condition, tempered glass shatters into a series of small granular chunks with blunt edges rather than random, jagged shards when it breaks, which makes this type of glass safe; it also causes reduced injury during vehicle

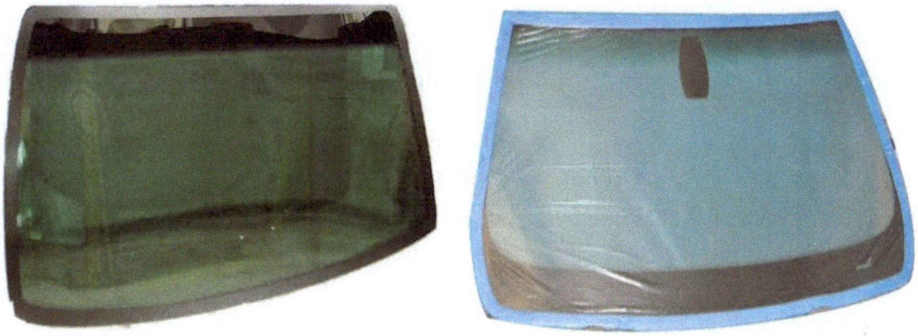

(a) Backlite (b) Front windshield

Fig. 2.1 Tempered and laminated glass applied in automotive as side windows and windshield, respectively [7]

J. Xu and Y. Li, *Impact Behavior and Pedestrian Protection of Automotive Laminated Windshield*, https://doi.org/10.1007/978-981-13-2441-3_2

Fig. 2.2 Float glass

Fig. 2.3 Broken tempered glass [3]

accidents [1, 3], so it is also referred to as safety glass [4]. Broken tempered glass is shown in Fig. 2.3.

The laminated glass used in automotive windshields is mainly PVB laminated glass composed of a PVB interlayer, which prevents the glass from breaking into large sharp pieces [5], sandwiched by two mono soda–lime glass pieces. When this glass breaks on impact, the shard remains stick together instead of shattering, as shown in Fig. 2.4, and produce a crack similar to a "spider web" in the windshield [6], as shown in Fig. 2.5.

Using PVB laminated glass as an automotive windshield presents the following advantages.

(1) Outstanding connecting properties of glass fragments

Fig. 2.4 Broken laminated glass [3]

Fig. 2.5 Automotive windshield with a crack under impact [5]

When the external impact is more powerful than the breaking strength of glass, glass, which is a typical brittle material, breaks into numerous fragments [8]. These fragments with different splattering speeds could result in different degrees of injury to human bodies, such as scratches and cuts, which could lead to permanent damage if the site of injury is the eyes.

(2) Good impact resistance

The impact resistance of common glass is lower than that of laminated glass because the entire structure of the former does not bear any external load when the glass breaks. Meanwhile, PVB laminated glass can efficiently improve the impact resistance of the windshield because of the PVB interlayer with a large damage strain that makes glass robust to loading after the glass breaks.

(3) Noise reduction capability

Common glass has a poor noise insulation effect, especially when vehicles run at a high speed. Related studies have demonstrated that PVB laminated glass can efficiently shield against noise with a frequency in the range of 125–4000 Hz [9] and can reduce the vibration and noise of glass due to the PVB sticky film.

(4) Prevention of thermal radiation

The absorption of thermal energy by PVB laminated glass is about 28%, which is less than the percentage for common glass [9]. PVB laminated glass prevents rapid increments in temperature inside cars. The ultraviolet radiation resistance of PVB laminated glass is better than that of common glass.

(5) Long-lasting durability

The durability of PVB laminated glass is better than that of common glass because of its outstanding mechanical property.

(6) Good visual effect

Compared with the visual effect of common glass, that of PVB laminated glass is sharper, that is, the scenery is clearer and the color contrast is higher. Image distortion is also small, which easily provides drivers with information outside the cars.

In the following text, we present the chemical components and manufacturing process of glass and the PVB interlayer.

2.1 Chemical Composition of the Constituent Material

2.1.1 Glass

The glass in laminated glass should possess excellent optical and mechanical properties, and the transmittance of the glass should be higher than 85%. Moreover, the quality of the raw glass of laminated glass applied as the front windshield must reach

Table 2.1 Chemical components of soda–lime

	$Ca(OH)_2$	H_2O	NaOH	KOH
Component (%)	75	20	3	1

Table 2.2 Chemical components of soda–lime silicate glass [15]

	SiO_2	Na_2O	MgO	CaO	Al_2O_3	Fe_2O_3
Component (wt%)	73	14	4	9	0.1	0.1

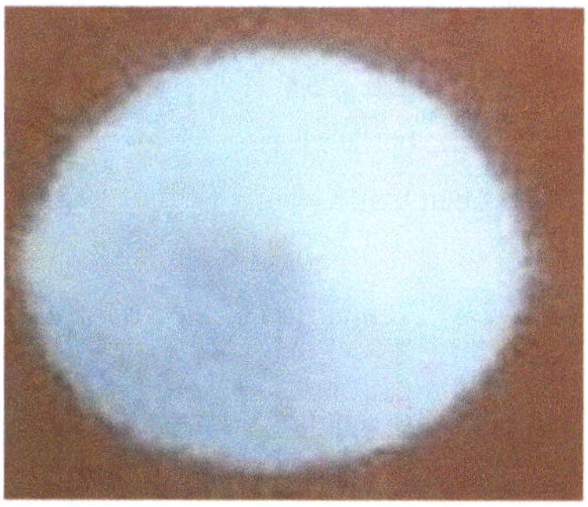

Fig. 2.6 One of the raw materials of glass: silica sand [7]

the level of automotive grade in the national standard for float glass GB/T 11614-1999 [10]. Raw and float glass should be of high quality, and glass with initial defects should be avoided.

The glass of laminated windshields is usually manufactured using soda–lime silicate glass [11], which has a high alkali content (4 mol% Na_2O) [12] and is a linear elastic material ($70 \leq E_g \leq 74$ GPa; $0.22 \leq \nu_g \leq 0.23$) whose failure does not involve plastic deformation [13]. The chemical components of soda–lime and soda–lime silicate glass are shown in Tables 2.1 and 2.2, respectively. Figure 2.6 shows silica sand that accounts for 60%–70% of the raw material of glass, and silica sand with high quality generally consists of a large amount of SiO_2 (about 99–99.8%) [14]. The impurities in glass are classified as harmless and harmful depending on whether they affect the glass quality or not. Fe_2O_3, Cr_2O_3, and TiO_2 are defined as harmful impurities because they can color glass. Table 2.3 shows the allowable contents of impurities in silica sand of different glass products.

2.1.2 PVB Interlayer

Given PVB's excellent properties, such as transparency, optical property, resistance to ultraviolet rays, high tensile strength, energy mitigation, and adhesion property to glass, metal, wood, and other fiber products, PVB is widely used in the manufacture of safety glass, coating products, and materials of electric products [16] and has become the most widely used interlayer of laminated glass due to its high adhesive property

Table 2.3 Allowable contents of impurities in silica sand of different glass products [14]

Glass	Allowable Fe_2O_3	Allowable Cr_2O_3	Allowable TiO_2
High-class glass	0.015	–	–
Optical glass	<0.01	<0.001	<0.05
Colorless vessel	<0.02	<0.001	<0.10
Polished glass	<0.03	<0.002	–
Window glass	<0.10–0.20	–	–
Electric bulb	<0.05	–	–
Chemical apparatus and pharmaceutical glass	<0.10	–	–
Half-white bottle glass	<0.30	–	–
Dark green bottle glass	<0.5	–	–

Table 2.4 Comparison of the mechanical properties of PVB and other materials [10]

Item	PVB	Glass	Stainless steel	High-strength fiber
Tensile elastic modulus (GPa)	0.04	70	200	124
Tensile strength (GPa)	32	50	1720	2760
Elongation at break (%)	250	0.071	2.0	2.5
Fracture energy (MJ/m)	55	0.018	17.2	34.5

with inorganic glass [10]. Furthermore, PVB is viscoelastic in nature, and its stiffness is sensitive to temperature and load duration [13]. Cellulose ester, polyester, PVB, and PU can be used in the interlayer of laminated glass. However, PVB is applied widely in the interlayer of laminated glass.

A comparison of the mechanical properties of PVB and other materials is shown in Table 2.4. PVB, which has a lower tensile elastic modulus and higher break elongation than other materials, is more suitable for use as the interlayer of laminated glass.

The PVB film used in laminated glass should meet the appearance quality requirements shown in Table 2.5.

PVB is a resin commonly used for applications that require strong binding, optical clarity, adhesion to many surfaces, toughness, and flexibility. PVB is formed from the reaction of polyvinyl alcohol and butyraldehyde [17].

With regard to composition and structure, PVB is prepared with polyvinyl alcohol (PVA) and butyraldehyde through condensation with a catalyst, and it presents white or yellow powder [16]. It is a random amorphous copolymer of vinyl butyral, vinyl alcohol, and vinyl acetate, and its general formula is shown in Fig. 2.7.

Table 2.5 Appearance quality requirements of PVB films [10]

Defect	High-class product	Qualified product
Color	White	White
Bubble	Not allowed	Allow two bubbles whose length is 1–2 mm under the circle range of ϕ300 mm
Impurity	Allow two, which should not be more than ϕ2 mm under the circle range of ϕ500 mm	Allow four, which should not be more than ϕ3 mm under the circle range of ϕ500 mm
Crack	Not allowed	Not allowed
Strip	No effect on usage, can depend on the suppliers and buyers	No effect on usage, can depend on the suppliers and buyers

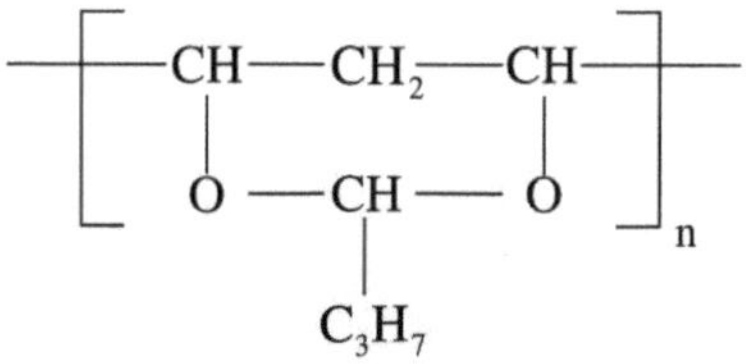

Fig. 2.7 General formula of PVB

2.2 Constituent Material Manufacturing: Theories and Technology

2.2.1 *Glass*

The glass produced in the industry is a type of soda–lime silica glass that is mainly formed by the mixture of silica (with a certain purity), limestone, and soda that go through a series of processing procedures as follows: melting under 1550–1600 °C, forming, and cooling [18].

Glass manufacturing involves four primary operations, namely, batching, melting, fining, and forming [2, 19]. The correct mix of the raw material is selected based on chemistry, purity, uniformity, and particle size [19], and the raw material is melted in different glass furnaces to produce different glass products. Fining aims to produce molten glass that is uniform in terms of composition and temperature and also bubble free [2].

The process of manufacturing plate glass is as follows [2]:

(a) The raw material is melted in a furnace.
(b) The glass is subjected to fining to eliminate bubbles.
(c) The melted glass is poured into a pool of molten tin. The glass floats under a nitrogen atmosphere to prevent corrosion of the tin bath. Then, the molten glass spreads out and forms a smooth flat sheet.

(a)

(b)

Fig. 2.8 Tempering furnace [20]

(d) The glass is cooled to avoid residual stresses within the glass in the annealing lehr.
(e) The glass is inspected, cut into typical sizes for different purposes, and then stored.

The manufacturing method for automotive glass generally adopts the float process, which is similar to the method mentioned above. The raw glass material is melted at 1500 °C, poured into a liquid suspended pool, and cooled at a temperature of 600 °C. Then, float glass is formed, and its upper surface is the liquid surface and the bottom surface is the liquid tin surface. After cooling, the glass can be cut into specific sizes for different proposes. This glass is prepared for the manufacturing of tempered and laminated glass [18]. Figure 2.8 shows the tempering furnace.

2.2.2 PVB Interlayer

The quality of laminated glass depends on the quality of the PVB interlayer [21]. The manufacturing methods of PVB include precipitation, one-shot, and dissolution methods and are classified according to the change in reactants in the process of the reaction.

The reaction of generating PVB is a type of condensation reaction. The theory of generating PVB is as follows [22]:

(a) One aldehyde molecule and one hydroxide radical react with each other and generate the hemiacetal, which is unstable.
(b) The hemiacetal reacts with another hydroxide radical to generate the stable acetal.

The three types of manufacturing methods of PVB are summarized below.

(1) Precipitation method

The precipitation method produces high-purity PVB resin. The method proceeds as follows [16]:

(a) PVA is dissolved in water at a temperature of 85–95 °C.
(b) The solution is filtered, and the PVB is removed without dissolution.
(c) The PVB is pressed into an acetal reaction kettle.
(d) Hydrochloric acid and butyraldehyde are added to the reaction kettle, and the reaction takes place.
(e) The PVB precipitate is washed, neutralized, and dried.
(f) The manufacturing of PVB is finished.

The condensation reaction is a key process in PVB manufacturing, and temperature, time, and concentration should be accurately controlled [16].

(2) One-shot method

The one-shot method, which is also called one-shot dissolution, proceeds as follows [16]:

(a) A methyl alcohol solution is prepared with polyvinyl acetate aggregated by vinyl acetate.
(b) Alcoholysis treatment is applied to the solution with hydrochloric acid.
(c) Butyraldehyde is added to the mixture solution of ethyl alcohol and acetic ether where PVA floats, and PVA acetalization occurs under the return temperature.
(d) As the reaction progresses, PVB is produced and dissolved in the solvent.
(e) Alkali is added to the solution to adjust the pH value to 6, and water is poured into produce the PVB precipitate.
(f) The PVB precipitate is washed, neutralized, and dried.
(g) The manufacturing of PVB is finished.

Notably, the acid in PVB production that results from the hydrolysis of polyvinyl acetate is difficult to wash [16].

(3) Dissolution method

The dissolution method, which is mainly used in laboratories for PVB manufacture, has several advantages, such as production of uniform PVB particles, high degree of acetalization, and steady reaction. The method proceeds as follows [16]:

(a) A methyl alcohol supernatant solution is produced with PVA.
(b) Hydrochloric acid and butyraldehyde are added to the solution for acetylation.
(c) As the reaction progresses, PVA is gradually dissolved and forms a uniform solution.
(d) Water is added to the solution, and PVB is dissolved out.
(e) The PVB precipitate is washed, neutralized, and dried.

2.3 Laminated Glass Manufacturing

Currently, two main manufacture methods of laminated glass are available, and these two are PVB film hot-press method (dry method) and grouting method (wet method). The latter is rarely used today.

2.3.1 Dry Method

The process of the dry or hot-press method is as follows [23]: washing and laminating the glass, pre-pressing and pressing in a high-pressure kettle, and inspection for freight containers (Fig. 2.9).

The main manufacturing process of laminated glass is discussed below.

(1) Cutting

Cutting is conducted by a numerical-control cutting grinding machine (Fig. 2.10). Appropriate cutting force is applied to cut the glass by conveying the glass to the cutting platform with a conveyer belt controlled by a servo motor. When the cutting is finished, the grinding begins. The grinding heads in both sides of the machine grind the four sides of the glass. Then, the conveyer belt transports the glass to a clean platform and carries out cleaning (Fig. 2.11).

(2) Laminating

Before laminating, PVB needs to be cut out into a suitable size that is generally larger than the glass size by using a cut-out machine or through manual work. The moisture content of PVB is inversely proportional to the connecting strength of PVB and

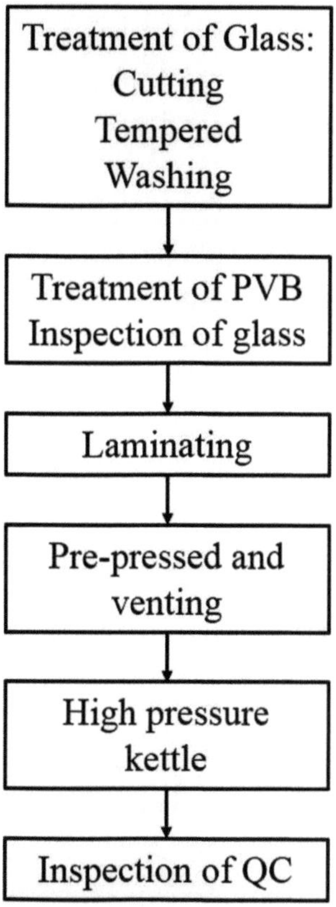

Fig. 2.9 Production flowchart for laminated glass [23]

glass because the water absorption of PVB is proportional to relative air humidity. Moreover, the mobility and contractility of PVB are proportional to temperature, and room temperature and humidity need to be controlled strictly in a laminating work room. Generally, temperature and humidity are controlled within the range of 18–24 °C and 18–28%, respectively [25].

During lamination, the glass surface that needs to be laminated should not have dust, and a vacuum cleaner can be used if needed. Two pieces of glass with a PVB film are laminated at a standard atmospheric pressure of 8–10 and a temperature of 100–150 °C by mechanical arms, as shown in Fig. 2.12.

(3) Pre-pressing

Pre-pressing of laminated glass, which is vital to the quality of laminated glass, aims to vent the air between the connection area of PVB and glass piles. Examples of this method include surface rolled pressing and vacuum approaches [23].

The surface rolled pressing approach has several advantages, such as manufacturing in a continuous and rapid manner and possessing a simple processing technology. The vacuum approach, which aims to prevent the premature sealing of glass edges,

Fig. 2.10 Transverse cutting machine [7]

Fig. 2.11 Edge polishing machine [24]

is widely used in the manufacturing of curved laminated glass, but the surface rolled processing approach cannot meet the manufacturing requirement. However, the vacuum approach is more complex than the surface rolled approach; in the former, the vacuum pressure condition must be larger than 10 kPa, the surrounding temperature must be in the range of 100–120 °C, and the surface temperature of glass should be in the range of 95–105 °C [23].

(4) High-pressure kettle processing

Fig. 2.12 Lamination of PVB laminated glass [26]

High-pressure kettle processing is crucial to the manufacture of laminated glass. The settings of temperature, pressure, and time are also vital to the quality of laminated glass. The heating and cooling of pressing glass must be at a low velocity [23].

(5) Edge cutting

After lamination, the laminated glass is conveyed to the edge-cutting work platform by a conveyer belt. Then, workers cut the redundant PVB films that extrude along with the glass edges.

However, cutting the laminated glass is difficult because of the existence of a plastic interlayer. An unsafe method of cutting laminated glass involves pouring a flammable liquid, such as denatured alcohol, into the crack to melt the interlayer and separate the two sides [5].

Nevertheless, several safe methods of cutting laminated glass (shown below) are recommended by the UK government's Health and Safety Executive [5].

(a) Special-purpose laminated cutting tables
(b) Vertically inclined saw frames
(c) Blowlamp or hot air blower
(d) High-pressure abrasive waterjet

The laminated glass should be checked for quality. Afterward, it can be packed.

In addition, large-size laminated glass needs to go through bending processing for design proposes. Bending processing is carried in a bending furnace, as shown in Fig. 2.13. Figure 2.14 shows a bending and tempering system controlled by a computer. Figure 2.16 shows performance windshields from Asahi India Glass, Limited

Fig. 2.13 Laminated bending furnace [20]

Fig. 2.14 Bending and tempering system for large-size glass [27]

(AIS). The boundary support of laminated glass installed in the vehicle uses a rubber sealing strip to seal the four edges of the laminated glass.

2.3.2 *Wet Method*

The wet method with its simple technology, minimal processing, and few pieces of required equipment can produce large and special products, such as laminated wired glass and fire-resistant glass [10]. The wet method is rarely used today. Here, we introduce it briefly. The process of the wet method is shown in Fig. 2.16.

Raw glass sheets are laminated after cutting, edging, and washing, and the sheets should be dry and clear on the surface. The lamination process in the wet method is different from that in the dry method, which involves the folding of glass sheets

Fig. 2.15 Performance windshields from Asahi India Glass, Limited [20]

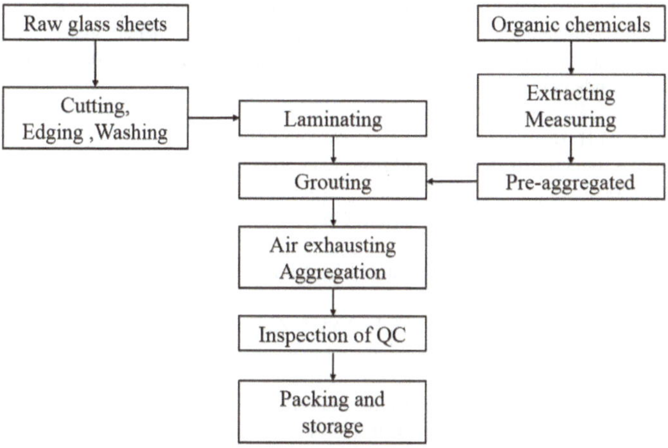

Fig. 2.16 Production flowchart of laminated glass in the wet method [10]

and PVB films. The process of lamination in the wet method requires the bonding of glass sheets with a cavity between them for grouting. During grouting, aggregated fluid is poured into the cavity, and PVB strips are utilized to seal the opening of the cavity [10].

2.4 Summary

In this chapter, we introduce automotive glass, which includes tempered and laminated glass. Tempered glass is for the side windows and backlite of vehicles, and laminated glass is mainly for the windshields. The composition of constituent materials (glass and PVB films) in laminated windshields and the manufacturing of glass and PVB films are then introduced. Manufacturing of automotive glass generally adopts the float process. The manufacturing methods of PVB include precipitation, one-shot, and dissolution methods classified according to the change in reactants in the process of the reaction. Finally, the manufacturing process of PVB laminated glass, which can be divided into dry and wet methods, is introduced in detail.

References

1. AIS, Tempered glass for sidelite & backlite. https://www.aisglass.com/products-services/automotive-glass/tempered-glass-sidelite-backlite
2. M. Fröling, *Strength Design Methods for Laminated Glass* (2011)
3. Wikipedia, Safety glass. https://en.wikipedia.org/wiki/Safety_glass
4. M. Haldimann, Structural use of glass. Int. Assoc. Bridge Struct. Eng. (2006)
5. Wikipedia, Laminated glass. https://en.wikipedia.org/wiki/Laminated_glass
6. AIS, Laminated windshields. https://www.aisglass.com/products-services/automotive-glass/laminated-windshields
7. C.G. Network, Processed Glass. http://www.glassinchina.com/products/ProcessedGlass1/
8. W. Holand, G.H. Beall, *Glass Ceramic Technology* (2012)
9. T. Yoshioka, Functional laminated glazing with advanced PVB technology. AutoTechnology **4**, 62–65 (2004)
10. Z. Liu, S. Pang, *Energy-Saving Glass and Green Glass* (Chemical Industry Publishing House, 2009)
11. G. Castori, E. Speranzini, Structural analysis of failure behavior of laminated glass. Compos. Part B Eng. (2017)
12. G.-I. Shim, S.-H. Kim, H.-W. Eom, D.-L. Ahn, J.-K. Park, S.-Y. Choi, Improvement in ballistic impact resistance of a transparent bulletproof material laminated with strengthened soda-lime silicate glass. Compos. B Eng. **77**, 169–178 (2015)
13. M. Overend, C. Butchart, H. Lambert, M. Prassas, The mechanical performance of laminated hybrid-glass units. Compos. Struct. **110**, 163–173 (2014)
14. Y. Zhao, H. Yin, *Glass Technology* (Chemical Industry Publishing House, 2006)
15. J.E. Shelby, *Introduction to Glass Science and Technology* (Springer, USA, 2005)
16. L. Hong, The research on synthetic process of high viscosity PVB resin for safe glass, Zhejiang University of Technology (2011)
17. Wikipedia, Polyvinyl butyral. http://en.wikipedia.org/wiki/Polyvinyl_butyral (2009)
18. Q. Xiao, Automotive glass and its development prospects. Adv. Ceram. **31** (2010)
19. E.L. Bourhis, *Glass: Mechanics and Technology* (2007)
20. AIS, Automotive glass. https://www.aisglass.com/products-services/automotive-glass
21. H.G. Feng, Process formulation design of PVB film for laminated safety-glass. China Plastics Industry (2005)
22. Q. Song, Y. Ma, R. Zhao, Q. Zhang, Q. Luo, W. Jiang, New process of manufacturing polyvinyl butyral by suspension method. China Adhes. (2012)
23. S. Diewen, Discussion on technology of laminated glass. Glass **42** (2015)

24. C.G. Network, Glass Machine & Equipment. http://www.glassinchina.com/products/GlassMachineEquipment3/
25. X.W. Jun, Production of printed curvature laminated glass for front windshield of car. Glass **31**, 53–56 (2004)
26. S.G. LLC, Brief introduction to PVB laminated glass manufacturing process. http://www.aisglass.com/pvb_laminated.asp
27. AIS, Technical support. https://www.aisglass.com/node/91#tab-1

Chapter 3
Mechanical Behavior and Constitutive Modeling of Laminated Glass

3.1 Mechanical Behavior of Laminated Glass

3.1.1 Test Method

(1) Tensile test

Tension test is widely used as a tension mechanical parameter in the measurement of metallic or composite materials. The sample in a tension test is usually dog-bone shaped according to GB/T 228-2002 (*Tensile test of metal materials at room temperature*) to make the deformation of the middle part of the material average and uniform. The elastic modulus, E_m, of the sample is calculated with the axial stress obtained by a transverse strain gauge or an extensometer fixed on the sample. The Poisson's ratio, μ, of the sample is also obtained with a transverse strain gauge, which can also modify the transverse strain effect. Strain values ε_{1i} and ε_{2i} are obtained with axial and transverse strain gauges. The true strain values are expressed as ε_1 and ε_2 because of the transverse effect of the strain gauge. The difference between the indicated and true strain can be described as

$$\varepsilon_1 = (1 - \nu_0 H)(\varepsilon_{1i} - H\varepsilon_{2i}), \tag{3.1}$$

$$\varepsilon_2 = (1 - \nu_0 H)(\varepsilon_{2i} - H\varepsilon_{1i}), \tag{3.2}$$

where H is the lateral effect coefficient of the strain gauge, which is the ratio of the transverse to vertical sensitivity of the strain gauges. ν_0 is the Poisson's ratio of the material for calibrating the sensitivity coefficient of the strain gauge.

The elastic modulus and Poisson's ratio of a test sample can be described as

$$E_m = \frac{P}{A_m \varepsilon_1}, \nu_m = -\frac{\varepsilon_2}{\varepsilon_1}. \tag{3.3}$$

J. Xu and Y. Li, *Impact Behavior and Pedestrian Protection of Automotive Laminated Windshield*, https://doi.org/10.1007/978-981-13-2441-3_3

(2) Compression test

Yield strength and compressive strength under unidirectional compression can be obtained through a compression test. According to the rules of GB/T 7341-2005 (*Tensile test of metal materials at room temperature*), the following should be satisfied.

(a) The design of the shape and size of a sample should guarantee that the sample part in the scale distance is in uniform and unidirectional compression conditions during the test.
(b) The deformation measured by an extensometer should be equal to the deformation in the scale distance on the axis of the sample.
(c) The end of the sample should not be deformed before the test.

Moreover, the length of the sample should meet the design requirement. Buckling destruction could occur in the sample before being deformed under compression if its length is too large.

(3) Bending test

Bending is different from tension and compression. The upper portion of the neutral layer of the material is in a compression condition, whereas the lower portion is in a tension condition. Common bending tests include three- and four-point bending, as shown in Fig. 3.1.

The load–deflection curve of a sample can be obtained by using a strain gauge to measure the deformation of the sample during uniform loading. Then, the bending elastic modulus E_y and bending strength σ_y of the sample can be calculated.

Three-point bending test:

$$E_y = \frac{\Delta P l^3}{4bt^3 \Delta f}, \sigma_y = \frac{3P_{\max} l}{2bt^2} \tag{3.4}$$

Four-point bending test:

$$E_y = \frac{\Delta P l^3}{8bt^3 \Delta f}, \sigma_y = \frac{3P_{\max} l}{4bt^2} \tag{3.5}$$

3.1.2 Experimental Analysis of Laminated Glass

The mechanical behavior of PVB laminated glass is complex, especially under dynamic loading, due to the combination of the rate-dependent effect of PVB and the brittle-polymer composite structure. Therefore, experimental data are important for the establishment of constitution and FE models for further investigation of the mechanical behavior of PVB laminated glass under static and dynamic loadings.

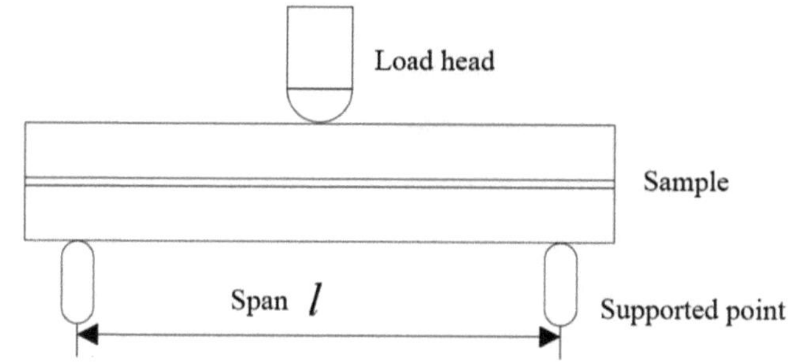

(a) Schematic of three-point bending mechanical experiment

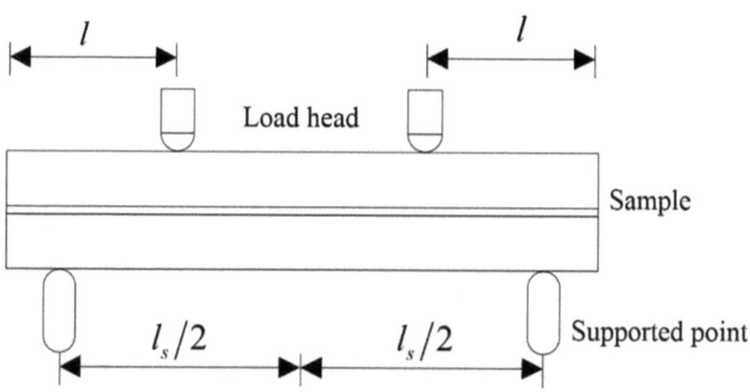

(b) Schematic of four-point bending mechanical experiment

Fig. 3.1 Schematic of static bending mechanical experiment

Typically, when studying the mechanical responses of composite materials, experiments can be categorized into three regimes, namely, quasi-static, low impact speed, and high impact speed, in terms of strain rates [1].

3.1.2.1 Quasi-static Loading Experiment

We adopted a compression experiment rather than a bending test because of the difficulties in conducting a quasi-static tension experiment with a slippery surface and because the windshield undergoes compression subject to human head impact.

(1) Experimental setup

A circular flake of PVB laminated glass with a dimension of $u = 10$ mm was used as the testing specimen. A 0.76 mm thick PVB interlayer was sandwiched between two pieces of 2 mm thick glass, as shown in Fig. 3.2. Glass, as a typical rate-independent brittle material, has a density of $\rho = 2500\ \text{kg/m}^3$, Young's modulus of $E = 70$ GPa, and Poisson's ratio of $m = 0.22$ [2]. The PVB laminated glass was fabricated by placing

(a) Illustration of the test specimen before the test

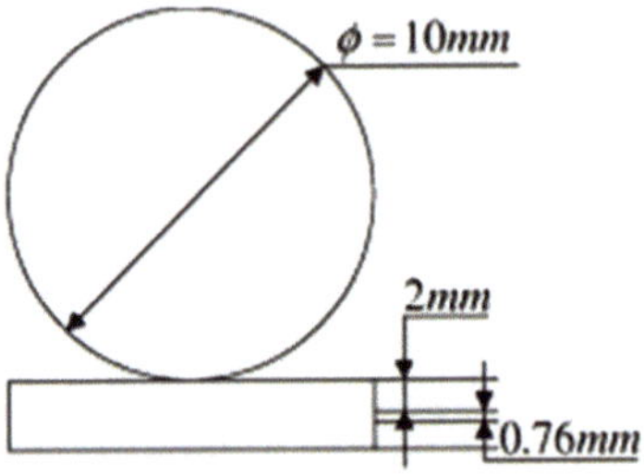

(b) Detailed dimension of the specimen

Fig. 3.2 Illustrations and dimensions of the specimen used in quasi-static and dynamic impact experiments

two glass sheets in an autoclave and implementing compression with a 0.76 mm thick PVB interlayer under a specific cycle of pressure (10 bar) and heat (120 °C). Thus, no slide occurs among the three layers [3]. To make the experiment highly applicable to engineering, the thickness of the specimen used was similar to that in most passenger cars. In the experiment, the specimen was placed in a cylinder container that has a steel pin with an internal diameter of $u = 38$ mm for compression. The experiments were conducted with a material testing system (MTS). Load was applied perpendicularly through the pin to the sample. A clip-gage-type extensometer was installed between the gaps to measure the displacement of the pin as a function of load during the experiment, as shown in Fig. 3.3. The loading rates were 1, 0.1, and 0.01 mm/min corresponding to the strain rates of $\dot{\varepsilon} \approx 1 \times 10^{-3}\, s^{-1}$, $\dot{\varepsilon} \approx 1 \times 10^{-4}\, s^{-1}$, and $\dot{\varepsilon} \approx 1 \times 10^{-5}\, s^{-1}$ on the upper and lower sides of the specimen. The strain rates were selected to measure the quasi-static mechanical property of PVB laminated glass. During the tests, the unloading threshold value was set to 80% of the maximum loading value; thus, the complete load–displacement history was recorded. The temperature and relative humidity during the experiments remained in the range of 25–26 °C and 40–45%, respectively.

(2) Experiment results

The failure process of a composite subject to quasi-static compression loading usually involves a sequential damage accumulation process [1]. We set the unloading threshold value to 80% of the maximum load value due to the major load decrease that occurred during the experiment when a major crack was observed. Thus, we defined the stress and strain at this moment as major failure onset (MFO) stress and strain. Figure 3.4 shows the “load decrease” phenomenon and the morphology of the cracked samples.

Figure 3.5 presents three stress–strain relation curves under the three loading rates. The stress–strain curves show “nonlinearity” in the mechanical behavior of PVB laminated glass, and this “nonlinearity” was probably caused by the nonlinear mechanical responses of the PVB interlayer and progressive micro-crack growth in the sample. As the loading rate increased, the MFO strain increased while the MFO

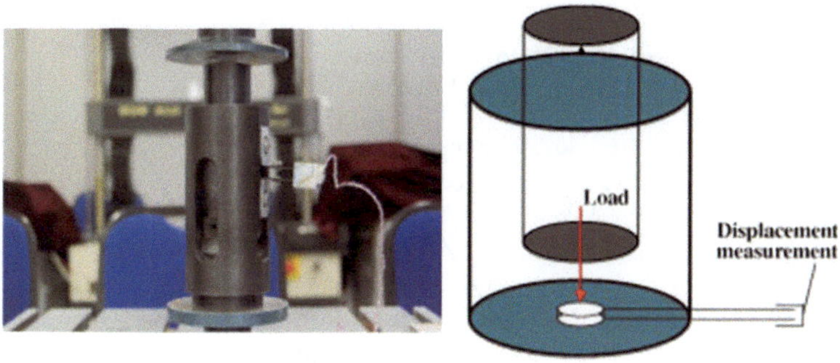

(a) Compression experiment setup (b) Principles of compression

Fig. 3.3 Quasi-static compression experimental setup in MTS

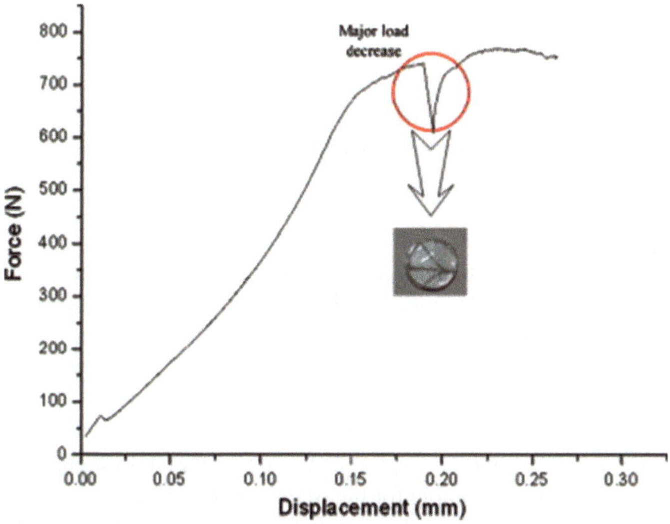

Fig. 3.4 Major load decrease occurrence with the onset of a major crack

stress remained nearly the same. The primary reasons are as follows: (i) in a situation with an extremely low strain rate (quasi-static), the outer glass panel plays a critical role in the mechanical response and (ii) glass is a rate-independent material, whereas PVB is a rate-dependent one.

3.1.2.2 Dynamic Split Hopkinson Pressure Bar Compression Experiment

Charpy and drop-weight tests are good options for low impact speeds, whereas SHPB or Kolsky's apparatus is suitable for high strain rates; the latter is a popular experi-

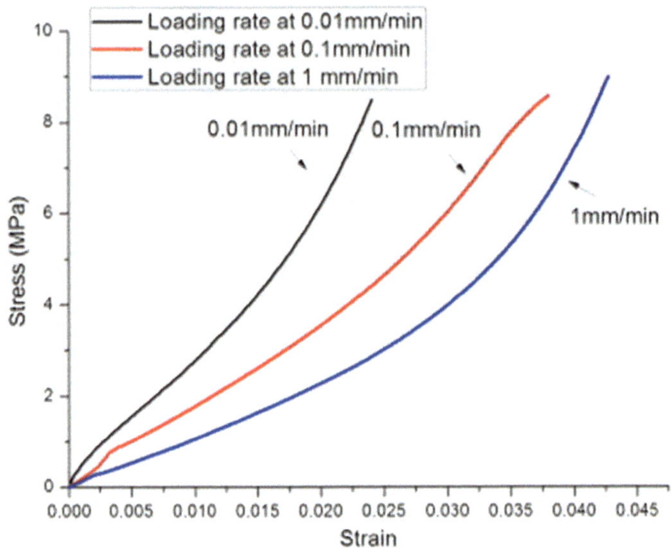

Fig. 3.5 Stress–strain curves obtained from quasi-static compression experiments under three different loading rates, i.e., 1, 0.1, and 0.01 mm/min

mental technique involving pattern analysis [4, 5]. Here, 1D dynamic compression experiments were conducted to measure the mechanical behavior of PVB laminated glass at different loading rates.

(1) SHPB setup

To overcome the disadvantage of MTS in low strain rate loading, we considered the advantage of the SHPB method. Experiments with strain rates ranging from $10^2\ s^{-1}$ to $10^3\ s^{-1}$ are particularly suitable for the SHPB method [5].

A typical SHPB system includes input, output, and striker bars, illustrated in Fig. 3.6. A gas gun controlled by gas pressure is used to propel the striker bar to hit the input bar. Two strain gauges are located on the surfaces of the input and output bars. Signals from the gauges are written down by the data recorder. Duralumin (hard aluminum) bar was selected in this study because glass has a very high Young's modulus ($E_{glass} = 70$ GPa). Details about the physical parameters of the bars and strain gauge locations are listed in Table 3.1. The geometries and dimensions of the specimen tested in the dynamic experiments were similar to those used in the quasi-static experiments.

Experiments were conducted at different strain rates, i.e., 2000 s^{-1}, 4000 s^{-1}, and 6000 s^{-1}, by applying different gas pressures on the striker bar. For each SHPB experiment, we directly measured impact velocity, incident strain wave signal $\varepsilon_i(x_i, t)$, reflected strain signal $\varepsilon_r(x_i, t)$, and transmitted strain signal $\varepsilon_t(x_i, t)$. According to wave propagation theory in solids, dynamic stress $\sigma_s(t)$, dynamic strain $\varepsilon_s(t)$, and strain rate $\dot{\varepsilon}_s(t)$ at any time during the loading–unloading process in the experiment are [6]

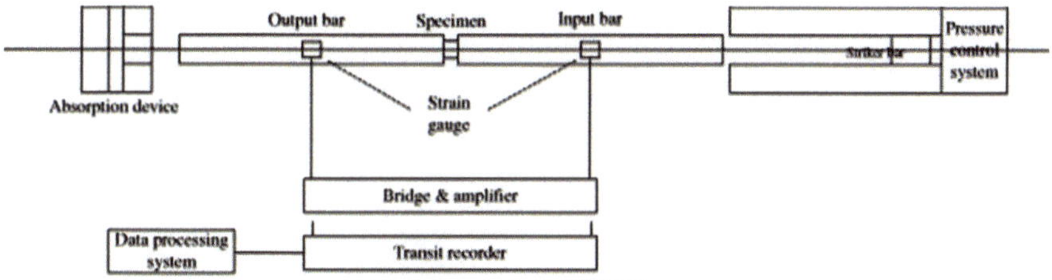

(a) Schematic of SHPB

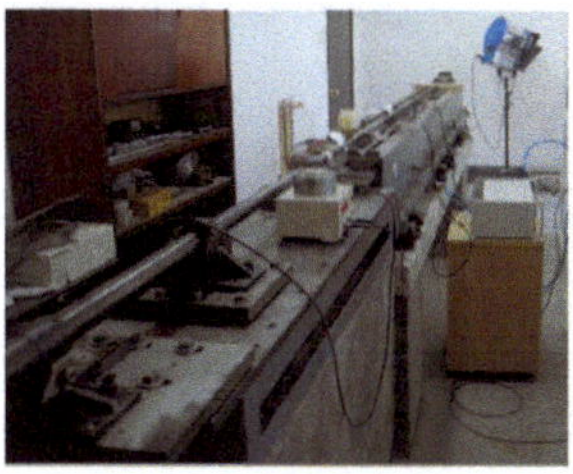

(b) Real-world SHPB system platform

Fig. 3.6 SHPB setup for dynamic testing of PVB laminated glass

Table 3.1 Specifications of the duralumin SHPB bars

	Strike bar	Input bar	Output bar
Length L (m)	0.3/0.2	1.002	1.002
Diameter φ (mm)	14.5	14.5	14.5
Longitudinal wave speed C_0 (m/s)	5159	5159	5159
Young's modulus (GPa)	206	206	206
Mass density ρ (kg/m^3)	7740	7740	7740
Strain gauge location (distance from gauge to specimen) D (m)	N/A	0.532	0.328

$$\sigma_s(t) = \frac{E_b A_b}{2A_s}\left[\varepsilon_i(\chi_i) + \varepsilon_r(\chi_i) + \varepsilon_t(\chi_o)\right], \tag{3.6}$$

$$\varepsilon_s(t) = -\frac{C_0}{l_s}\int_0^t \left[\varepsilon_i(\chi_i) - \varepsilon_r(\chi_i) - \varepsilon_t(\chi_o)\right]\mathrm{d}t \tag{3.7}$$

$$\dot{\varepsilon}_s(t) = -\frac{C_0}{l_s}[\varepsilon_i(\chi_i) - \varepsilon_r(\chi_i) - \varepsilon_t(\chi_o)], \tag{3.8}$$

where A_b and A_s are the cross-section areas of the bars (three bars have the same section areas) and the specimen, respectively; l_s is the original length of the specimen; C_0 is the longitudinal elastic wave velocity of the bars (where $C_0 = 5159$ m/s); and E_b is the Young's modulus of the bars (where $E_b = 206$ GPa).

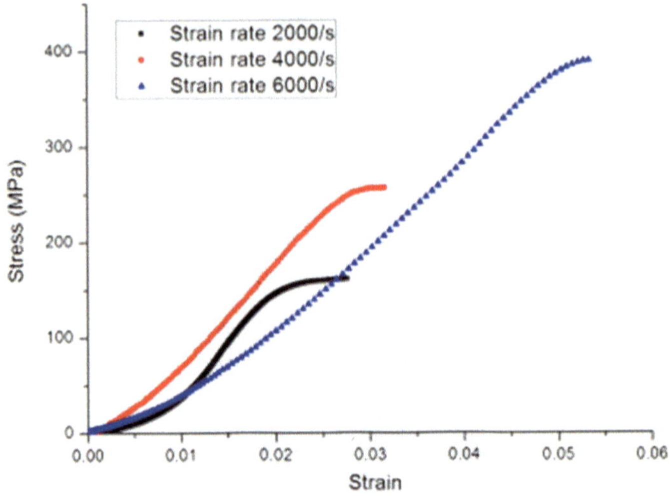

Fig. 3.7 Stress–strain curves obtained from dynamic SHPB experiments at strain rates of 2000/s, 4000/s, and 6000/s

The specimen was not subjected to an additional constraint when it was placed between the input and output bars to minimize the inertia effect. Temperature and humidity were similar to those in the quasi-static compression tests.

(2) Dynamic experimental results

Figure 3.7 shows the SHPB test results under the three strain rates. At a high strain rate, the MFO stress and MFO strain increased with the increase in the strain rate. The MFO stresses were 178, 257, and 286 MPa at strain rates of 2000 s^{-1}, 4000 s^{-1}, and 6000 s^{-1}, respectively. The values were approximately 15–30 times higher than those in the quasi-static situation, indicating that PVB laminated glass is a strong rate-dependent material, especially at high strain rates. Glass is usually considered a rate-independent material [2], so the PVB interlayer plays a dominant role after the failure of glass in a high strain rate situation. Minimal difference was observed in MFO strain domains in quasi-static and dynamic cases because the total thickness of glass layer accounted for over 84% of the total thickness. Thus, we no longer considered the relatively larger displacement in the dynamic cases than in the static scenarios. In the high strain rate situation, the testing samples turned into tiny fragments or powders after the experiments because of the iteration of the stress wave propagation in and between bars. Hence, we could not analyze the damage by referring to the debris. However, similar to the quasi-static case, we found that the stress–strain curve exhibited nonlinearity, which could also be fit via exponent-like functions. In the next section, we present a constitutive model describing quasi-static and dynamic stress–strain behaviors.

3.2 Constitutive Model of Laminated Glass

3.2.1 Johnson–Cook Constitutive Model

To recommend suitable constitutive relations for future numerical simulations, we adopted a phenomenological model due to its convenience and because only a few variables need to be calibrated. After examining the global stress–strain relation, we found that the constitutive behavior is exponent-like for static and dynamic scenarios. The Johnson–Cook (JC) model, one of the most extensively used constitutive models for describing material behavior, is also mathematically an exponent-like expression form. Considering that the JC model is commonly used to describe the constitutive behavior of metal, we used thc JC model as a basic mathematic model (without considering any physical context) to establish the constitutive relation of PVB laminated windshield. The data illustrated in Figs. 3.5, 3.6 and 3.7 are stress–strain relations without consideration of unloading, and we discuss the unloading in another part (i.e., the point with the largest strain and stress is the failure/ultimate point).

The JC model can be expressed as the following equation if the temperature effect is disregarded [7].

$$\sigma = A + B(\varepsilon_p)^n, \tag{3.9}$$

where σ is the yield stress, ε_p is the equivalent strain, and A, B, and n are material constants are obtained from experiments.

The strain–stress curve obtained from the experiments at a strain rate of $\dot{\varepsilon} \approx 1 \times 10^{-5}\ \mathrm{s}^{-1}$ was selected as the reference curve to determine the coefficients of A, B, and n in the first section of Eq. (3.9). In our specimen, stress–strain began from the origin, i.e., point (0, 0), so $A = 0$. With the least squares method, the stress–strain relation can be fitted as

$$\sigma = 606.6(\varepsilon_p)^{1.167}\left(1 + C\ln(\dot{\varepsilon}_p^*)\right), \tag{3.10}$$

where σ has a unit of MPa. Thus, $B = 606.6$. $\dot{\varepsilon}_p^* = \frac{\dot{\varepsilon}_p}{\dot{\varepsilon}_0} \cdot \dot{\varepsilon}_p$ and $\dot{\varepsilon}_0$ are the strain rate and the reference strain rate, respectively. C is the material constants obtained from the experiments. Figure 3.8 demonstrates the fitting model and experiment data.

To effectively describe the rate-dependent nonlinearity of the stress–strain curve of PVB laminated glass, an extra term should be added to Eq. (3.10) [8]. Thus we incorporated the function $\int(\dot{\varepsilon}_p)$ into Eq. (3.7) and established a constitutive model in the form of

$$\sigma = 606.6(\varepsilon_p)^{1.167}\int \dot{\varepsilon}_p. \tag{3.11}$$

We selected a polynomial function to describe the nonlinearity variation in initial stresses at different strain rates. Thus,

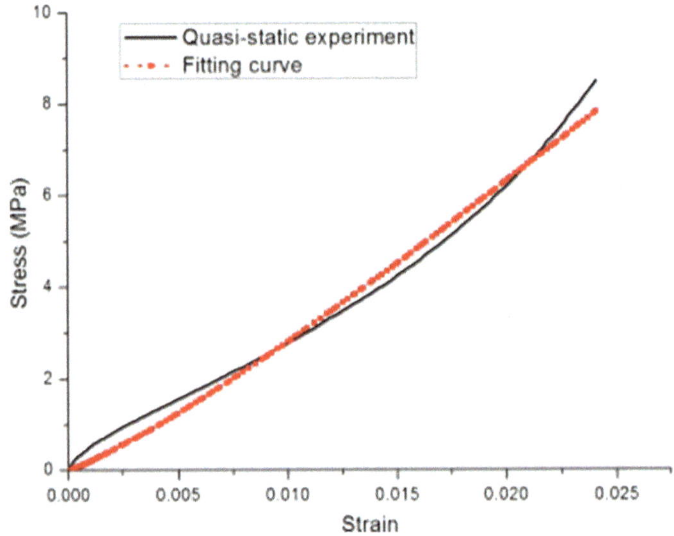

Fig. 3.8 Experimental data at a strain rate of $\dot{\varepsilon} \approx 1 \times 10^{-5}\ \mathrm{s}^{-1}$ used to fit and determine coefficients in the first terms

$$\int (\dot{\varepsilon}_0) = \left(\frac{A_i}{A}\right) = D\left(\ln \frac{\dot{\varepsilon}_p}{\dot{\varepsilon}_0}\right)^2 + E\left(\ln \frac{\dot{\varepsilon}_p}{\dot{\varepsilon}_0}\right) + 1, \tag{3.12}$$

where A_i is the initial stress at each strain rate.

We used experimental data at strain rates of $10^{-5}\ \mathrm{s}^{-1}$, $10^{-4}\ \mathrm{s}^{-1}$, $10^{-3}\ \mathrm{s}^{-1}$, $2000\ \mathrm{s}^{-1}$, $4000\ \mathrm{s}^{-1}$, and $6000\ \mathrm{s}^{-1}$ to fit the exponent function for $\int (\dot{\varepsilon}_p)$. The constants are best fit to be $D = -0.0436$, $E = 0.109$. The fitted curve is shown in Fig. 3.9.

To show the validity and accuracy of the constants, experimental data at a strain rate of $\dot{\varepsilon} \approx 4000\ \mathrm{s}^{-1}$ are verified in Fig. 3.10, and good agreement was observed between the experimental data and model prediction results, thereby showing the accuracy of the proposed model.

Thus, a constitutive relation model for PVB laminated glass is recommended as follows:

$$\sigma = 606.6(\varepsilon_p)^{1.167}\left(-0.0436\left(\ln \frac{\dot{\varepsilon}_p}{10^{-5}}\right)^2 + 0.109\left(\ln\left(\frac{\dot{\varepsilon}_p}{10^{-5}}\right) + 1\right)\right). \tag{3.13}$$

Equation (3.13) can be further extended to predict the dynamic fracture of PVB laminated glass because the experimental data used to establish the constitutive model are about the fracture of the testing sample. Therefore, we provided an additional equation to determine the relationship between fracture strain and strain rate by fitting the failure strains from different strain rates as follows:

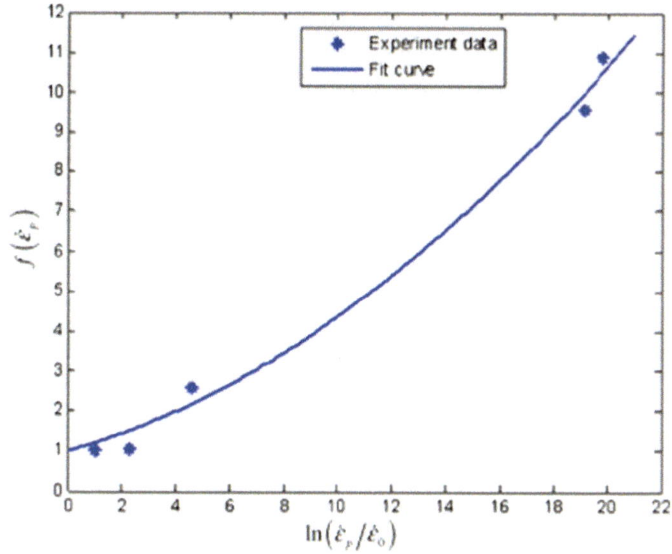

Fig. 3.9 Polynomial curve fitting for $f(\dot{\varepsilon}_p)$

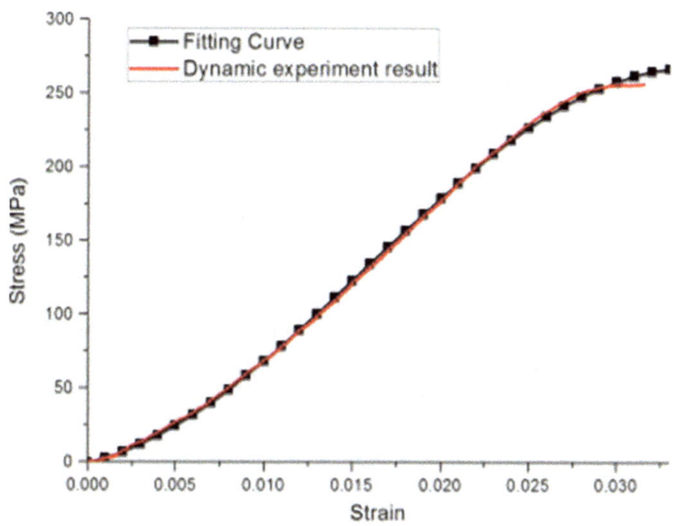

Fig. 3.10 Verification of the constitutive model by experimental data at $\dot{\varepsilon} \approx 4000\ \mathrm{s}^{-1}$ in dynamic SHPB test

$$\varepsilon_p = 14.8 - 3.06\dot{\varepsilon}_p^2 + 0.20\dot{\varepsilon}_p^4 - 0.00421\dot{\varepsilon}_p^6. \quad (3.14)$$

Equations (3.13) and (3.14) can be used in combination to easily determine the dynamic fracture of PVB laminated glass for science and engineering applications in various impact situations (strain rates).

3.2.2 *Damage-Modified Nonlinear Viscoelastic Constitutive Relation Model*

The constitutive relation of a laminated composite material can effectively reflect the material's mechanical behavior. To study the crack in a laminated composite material, the constitutive relations of laminated glass should be established.

Cracks on the PVB windshield in pedestrian–vehicle traffic accidents have not been studied thoroughly. In this research, a damage-modified nonlinear viscoelastic constitutive equation was first developed based on the updated Lagrangian approach. Second, after describing the impact between the pedestrian head and PVB laminated windshield, an FEA model with proper material parameters verified by a classical example was established. Lastly, the internal stresses of the cracks were studied. The Poisson's ratio, distance from the crack edge to the impacted point, and impact velocity were also parametrically investigated.

The PVB windshields used in the automotive industry are laminated composite materials that comprise a PVB interlayer sandwiched by two soda–lime glass sheets. The impact between the pedestrian head and vehicle PVB windshield can be regarded as a scenario where a composite plate is subjected to a concentrated force by a spherical indenter on its top surface, as shown in Fig. 3.11. The composite plate consists of $n = 2$ individual brittle soda–lime glass layers with thickness t_g and modulus E_g and sandwiched with a PVB interlayer of thickness t_p and modulus E_p.

By studying the constitutive relations of a PVB windshield, the crack in the PVB windshield can be analyzed thoroughly. Lili et al. [9, 10] studied the mechanical behavior of polymer by performing dynamic impact and quasi-static tensile experiments. According to the experimental results, the behavior of the polymer can be described as the following constitutive relation:

$$\sigma = \sigma_e + \sigma_{t_1} + \sigma_{t_2}, \tag{3.15}$$

where σ_e stands for the stress of the damage material. σ_{t_1} and σ_{t_2} denote the tensile stress of first and second Maxwell body, respectively.

$$\sigma_e = \sigma_m \left[1 - \exp\left(-\sum_{t=1}^{n} \frac{(m\varepsilon)^t}{t} \right) \right], \tag{3.16}$$

$$\sigma_{t_1} = E_1 \int_0^t \dot{\varepsilon}(\tau) \exp\left(-\frac{t-\tau}{\theta_1} \right) d\tau, \tag{3.17}$$

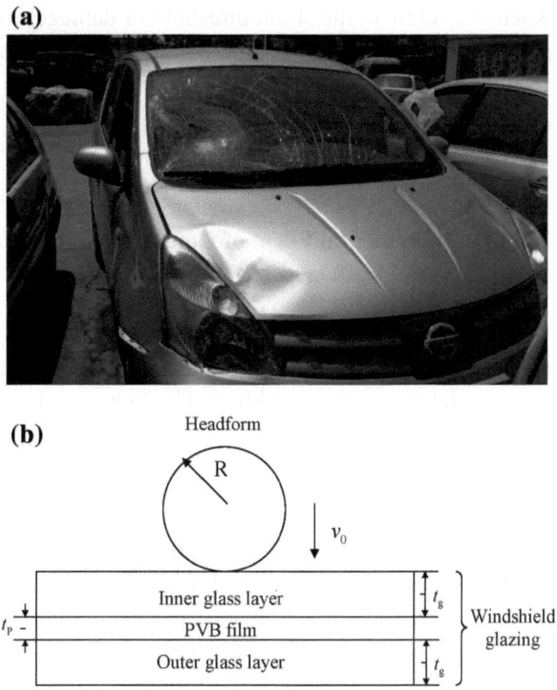

Fig. 3.11 **a** PVB windshield crack morphology in a real-world pedestrian–vehicle accident and **b** schematic of a sphere pedestrian head form impacting a windshield glazing

$$\sigma_{t_2} = E_2 t \int_0^t \dot{\varepsilon}(\tau) \exp\left(-\frac{t-\tau}{\theta_2}\right) d\tau, \tag{3.18}$$

where m, τ, E_1, and E_2, are non-dimensional relative initial elastic modulus, relaxation time as defined in Maxwell body, and elastic constants of the first and second Maxwell body, respectively. θ_1 and θ_2 are also constants denoting the relaxation times of the first and second Maxwell body. ε, $\dot{\varepsilon}$ stand for stress, strain respectively.

Nevertheless, these constitutive relations are only valid when strain is less than 7% [10]. They cannot express the relation completely. To further utilize these relations, Fenghua et al. [11] suggested the following equation to describe constitutive relations.

$$\sigma = (1 - D)\sigma_e + \sigma_{t_1} + \sigma_{t_2}, \tag{3.19}$$

where D is a "damage" variable and is defined as

$$D = \begin{cases} 0 & \varepsilon \le \varepsilon_{th} \\ \dot{D}_0 \dot{\varepsilon}^{\delta-1}(\varepsilon - \varepsilon_{th}) & \varepsilon > \varepsilon_{th} \end{cases}, \tag{3.20}$$

where ε_{th} refers to the strain threshold of damage evolution. $\dot{D}_0$ and δ are material parameters determined by experiments. Then, a dynamic damage criterion is put forward as follows:

$$D_f = \dot{D}_0 \dot{\varepsilon}^{\delta - 1} (\varepsilon - \varepsilon_{th}). \tag{3.21}$$

3.2.2.1 Incremental Form of Constitutive Relations When the Material Is not Damaged

When employing constitutive relations, several conditions, such as whether the strain is larger than the threshold strain, must be considered. Thus, we need an equivalent strain to determine this condition. The following is the equivalent.

$$\varepsilon_{eq} = \frac{\sqrt{2}}{3} \sqrt{(\varepsilon_1 - \varepsilon_2)^2 + (\varepsilon_2 - \varepsilon_3)^2 + (\varepsilon_3 - \varepsilon_1)^2 + \frac{3}{2}\left(\gamma_{12}^2 + \gamma_{23}^2 + \gamma_{31}^2\right)^2}, \tag{3.22}$$

where subscripts 1, 2, and 3 denote the directions of principal stresses.

On the basis of the updated Lagrangian virtual work method, the stress–strain relations in a material that undergoes large deformation with disregard of physical nonlinear properties can be expressed as [12]

$$\left\{S_{ij}\right\} = \int_0^t E(t - \tau)[A] \frac{\partial \{e_{kl}\}}{\partial \tau} d\tau, \tag{3.23}$$

where $\left\{S_{ij}\right\}$ and $\{e_{kl}\}$ are Kirchhoff stress tensors and Green strain tensors, respectively.

$$[A] = \frac{1 - v}{(1 + v)(1 - 2v)} \cdot \begin{bmatrix} 1 & \frac{v}{(1-v)} & \frac{v}{(1-v)} & 0 & 0 & 0 \\ \frac{v}{(1-v)} & 1 & \frac{v}{(1-v)} & 0 & 0 & 0 \\ \frac{v}{(1-v)} & \frac{v}{(1-v)} & 1 & 0 & 0 & 0 \\ 0 & 0 & 0 & \frac{1-2v}{2(1-v)} & 0 & 0 \\ 0 & 0 & 0 & 0 & \frac{1-2v}{2(1-v)} & 0 \\ 0 & 0 & 0 & 0 & 0 & \frac{1-2v}{2(1-v)} \end{bmatrix} \tag{3.24}$$

Similarly, given that a PVB windshield is a 3D structure, Eqs. (3.15)–(3.20) can be rewritten as [13–15]

$$S_{ij} = S_{ij,e} + S_{ij,t_1} + S_{ij,t_2}, \tag{3.25}$$

where

$$S_{ij,e} = \sigma_m \left\{ 1 - \exp\left(-mAE_{kl} - \sum_{t=1}^{n} \frac{(mE_{ij})^t}{t} \right) \right\}, \tag{3.26}$$

$$S_{ij,t_1} = E_1 \int_0^t A \frac{\partial E_{kl}}{\partial \tau} \exp\left(-\frac{t-\tau}{\theta_1} \right) d\tau, \tag{3.27}$$

$$S_{ij,t_2} = E_2 \int_0^t A \frac{\partial E_{kl}}{\partial \tau} \exp\left(-\frac{t-\tau}{\theta_2} \right) d\tau, \tag{3.28}$$

The incremental form of constitutive relations can be obtained based on Boltzmann superposition [16]. Implementation in commercial FEA software also requires the incremental form of the stress tensor of Eqs. (3.26)–(3.28) to be rewritten. Thus, the incremental form must be established first. According to the proof of the chain rule, Eq. (3.26) can be rewritten as follows:

$$dS_{ij,e} = \sigma_m \exp\left(-mAE_{kl} - \sum_{t=2}^{n} \frac{(mE_{ij})^t}{t} \right) \cdot \left[mA \frac{\partial E_{kl}}{\partial \tau} + \sum_{t=2}^{n} \frac{(mE_{ij})^{t-1}}{t} \frac{\partial E_{kl}}{\partial \tau} \right] \cdot \Delta t \tag{3.29}$$

Δt being short enough yields [16]

$$\frac{\partial E_{kl}}{\partial \tau} = \frac{\partial E_{kl}}{\partial t} = \frac{\partial E_{kl}}{\Delta \tau}. \tag{3.30}$$

Thus Eq. (3.27) becomes

$$\Delta S_{ij,e} = \sigma_m \exp\left(-mAE_{kl} - \sum_{t=2}^{n} \frac{(mE_{ij})^t}{t} \right) \cdot \left[mA\Delta E_{kl} + \sum_{t=2}^{n} \frac{(mE_{ij})^{t-1}}{t} \Delta E_{kl} \right] \tag{3.31}$$

Similarly, viscoelastic stress tensors for low and high strain rates at t are obtained as

$$\Delta S_{ij,t_1} = E_1\theta_1 A\frac{1-\exp\left(-\frac{\Delta t}{\theta_1}\right)}{\Delta t}\Delta E_{kl}\,,\\ -\left(1-\exp\left(-\frac{\Delta t}{\theta_1}\right)^t\right)S_{ij,t_1} \tag{3.32}$$

$$\Delta S_{ij,t_2} = E_2\theta_2 A\frac{1-\exp\left(-\frac{\Delta t}{\theta_2}\right)}{\Delta t}\Delta E_{kl}\\ -\left(1-\exp\left(-\frac{\Delta t}{\theta_2}\right)^t\right)S_{ij,t_2} \tag{3.33}$$

Therefore, the incremental form of a stress tensor during one unit time step Δt can be obtained as

$$\begin{aligned}\Delta S_{ij,n+1} &= \Delta S_{ij,e} + \Delta S_{ij,t_1} + \Delta S_{ij,t_2}\\ &= \sigma_m \exp\left(-mAE_{kl} - \sum_{t=2}^{n}\frac{(mE_{ij})^t}{t}\right)\cdot\left[mA\Delta E_{kl} + \sum_{t=2}^{n}\frac{(mE_{ij})^{t-1}}{t}\Delta E_{kl}\right]\\ &+ E_1\theta_1 A\frac{1-\exp\left(-\frac{\Delta t}{\theta_1}\right)}{\Delta t}\Delta E_{kl} - \left(1-\exp\left(-\frac{\Delta t}{\theta_1}\right)^t\right)S_{ij,t_1}\\ &+ E_2\theta_2 A\frac{1-\exp\left(-\frac{\Delta t}{\theta_2}\right)}{\Delta t}\Delta E_{kl} - \left(1-\exp\left(-\frac{\Delta t}{\theta_2}\right)^t\right)S_{ij,t_2}.\end{aligned} \tag{3.34}$$

The stress tensor can be calculated as follows:

$$S_{ij,n+1} = S_{ij,n} + \Delta S_{ij,n+1} \quad \varepsilon < \varepsilon_{th}\,, \tag{3.35}$$

where $t_{n+1} = t_n + \Delta t$.

3.2.2.2 Incremental Form of Constitutive Relations When the Material Is Damaged

When the material is not damaged, the incremental form of constitutive relations can be described with the function relation above. However, when the strain is larger than the threshold strain, the material enters the "damage" phase. The stress tensor in the damage phase at time t_{n+1} can be concluded as

$$\{S_{ij}\}_{n+1} = \{S_{ij}\}_n + \{\Delta S_{ij}\}_{n+1}, \tag{3.36}$$

and Eq. (3.35) can be rewritten as

$$S_{ij,n+1} = S_{ij,n} + \Delta S_{ij,n+1}. \tag{3.37}$$

Therefore,

$$\begin{aligned}S_{ij,n+1} &= S_{ij,n} + \Delta S_{ij,n+1} \\ &= \frac{1}{1-D_{n+1}} S^d_{ij,n+1} + 1^d - \frac{1}{1-D_n} S^d_{ij,n} \\ &= \frac{1}{1-D_{n+1}} \Delta S^d_{ij,n+1} + \frac{\Delta D_{n+1} S^d_{ij,n}}{(1-D_{n+1})(1-D_n)},\end{aligned} \tag{3.38}$$

where $\Delta D_{n+1} = D_{n+1} - D_n$ is the increase in the damage variable.

Rewriting Eq. (3.37) yields

$$\begin{aligned}\Delta S^d_{ij,n+1} &= (1-D_{n+1})\Delta S_{ij,n+1} - \frac{\Delta D_{n+1} S^d_{ij,n}}{(1-D_n)} \\ &= (1-D_{n+1})\Delta S_{ij,n+1} - \Delta D_{n+1} \Delta S_{ij,n}.\end{aligned} \tag{3.39}$$

Similarly, the stress tensor under damage can be calculated as

$$S^d_{ij,n+1} = S^d_{ij,n} + \Delta S^d_{ij,n+1} \quad \varepsilon \geq \varepsilon_{th}\,. \tag{3.40}$$

3.2.2.3 Verification of Constitutive Relations

The constitutive relations of a PVB windshield were obtained initially. The accuracy of these relations should be studied next. To verify the constitutive relations, we selected a classical example. A thin plate with an infinite length and $2b = 12$ mm width was subjected to uniform load $q(t) = 1$ kN m. A small crack existed in the center with length $2a = 1$ mm (Fig. 3.12). The material properties are listed in Table 3.2.

Numerical results were obtained by implementing the constitutive relations in FEA software and adding the command of material failure. Comparison of the numerical and theoretical results showed that the constitutive relations agree with the theoretical solutions well, as shown in Fig. 3.13.

Table 3.2 Material parameters of the plate

σ_m/MPa	n	m	E_1/MPa	θ_1/s	E_2/MPa	θ_2 /μs	ε_{th}	δ	D_0	D_f
1100	4	19.9	949	13.8	3981	67.4	0.055	1.22	0.863	0.0797

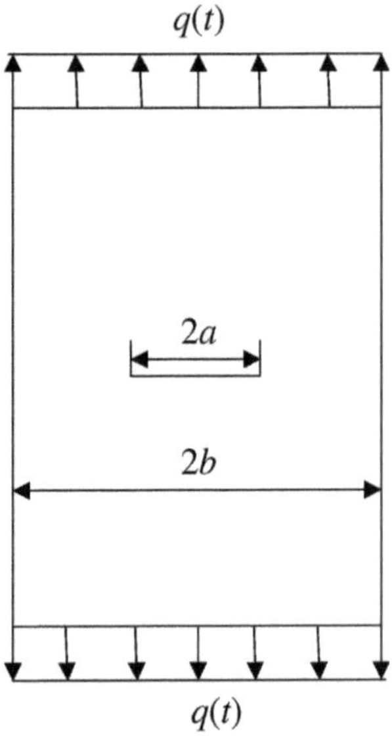

Fig. 3.12 Thin plate with a center crack under uniform load

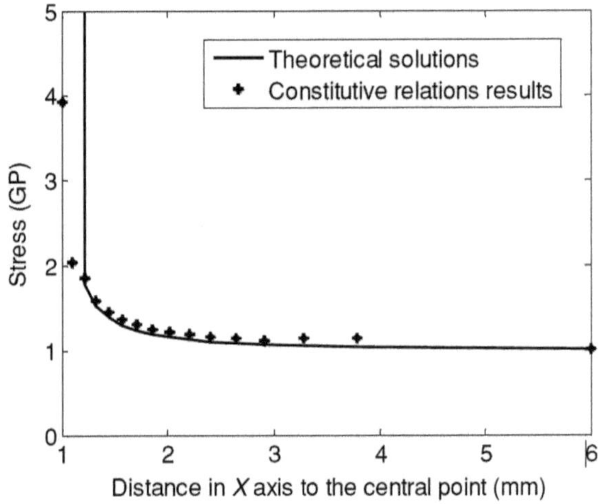

Fig. 3.13 Comparison of theoretical and FEA results

3.2.2.4 PVB Windshield FEA Modeling Based on Constitutive Relations of PVB Laminated Glass

Verification of the constitutive relations with a classical example indicated that the relations possess sufficient accuracy. Then, FEA modeling of a PVB windshield was performed.

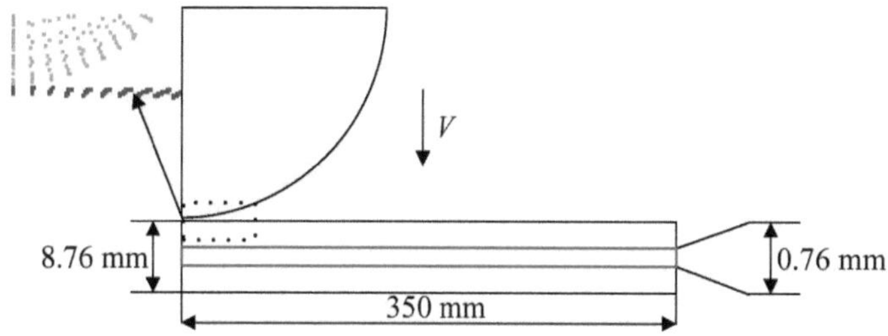

Fig. 3.14 Simplified physical model of impact between head and windshield

(1) Contact analysis

A non-conforming two-body contact model of a PVB windshield loaded by a head-form (sphere) is illustrated in Fig. 3.14. The thickness of the PVB windshield is 8.76 mm, and its length is 350 mm. The thickness of the interlayer PVB is 0.76. Shell elements were used to mesh the two bodies in contact. Hourglass control and automatic surface-to-surface contact with a dynamic friction coefficient of 0.1 were employed in LS-Dyna. Initial nodal gaps between the two contact surfaces were prescribed. The contact elements were not activated until the sphere penetrated into the composite glass for the purpose of saving computation time.

An extra command that allows material failure was added in the input file of LS-Dyna. Then, the abovementioned constitutive relations with a damage variable were fully employed in FEA. The accuracy of the contact FEA model was examined with the classical Hertzian pressure calculation method. An illustration of classical Hertzian theory is shown in Fig. 3.15 [17]. δ and a_0 stand for the mutual approach between two bodies and their contact radius.

The deviations between FEA and theoretical results were all within 1.5%. The comparison shown in Fig. 3.16 demonstrates the sufficient accuracy of the FEA model to stimulate the impact between pedestrian head and PVB windshield.

(2) Material model

PVB laminated glass was modeled according to the following principles: if the glass does not fail, the composite is treated as a shell; otherwise, the interlayer PVB film acts as a membrane. A Belytschko–Tsay shell element for the glass material and a membrane element for the interlayer were employed [18, 19]. Both elements were fully integrated in simulations so that no hourglass modes should be expected [20].

Glass and PVB film can be regarded as isotropic materials according to their physical and material properties. Therefore, we assumed that each layer of the composite windshield is isotropic with respect to its material symmetry lines and obeys Hooke's law. We also assumed that due to the influence of dynamic loading, no sliding occurs between two contact layers. Thus, we considered the composite windshield as having integrity. An equivalent elastic modulus and Poisson's ratio are introduced as follows [21]:

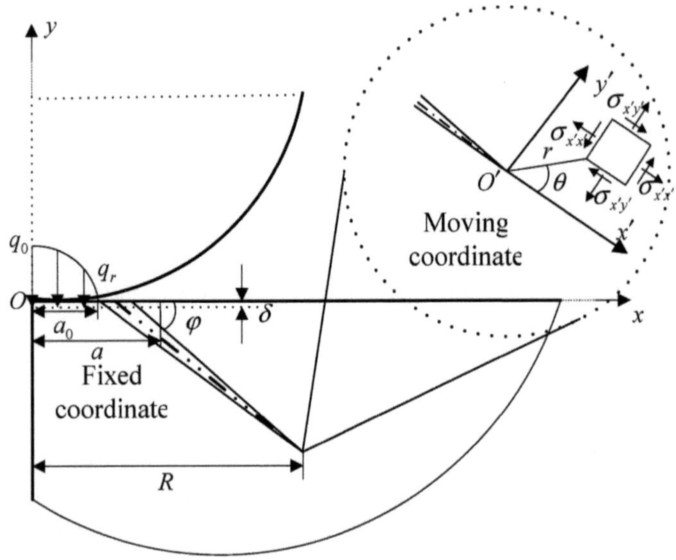

Fig. 3.15 Simplified crack model based on Hertzian pressure with moving coordinate at the crack tip

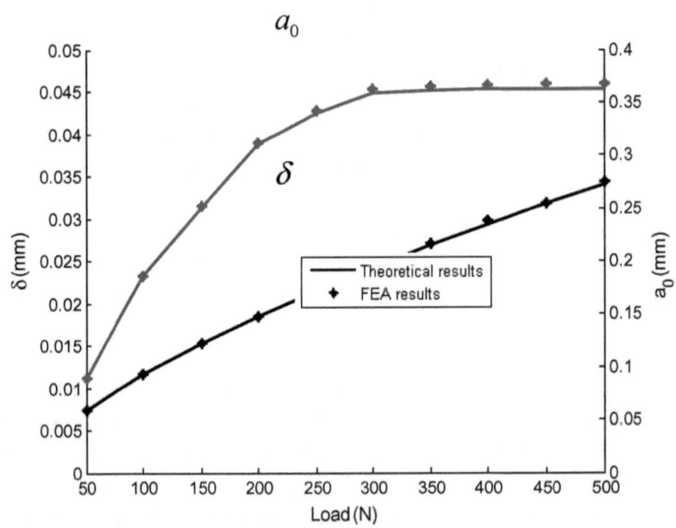

Fig. 3.16 Comparison of FEA and theoretical results

$$\bar{E} = \frac{2t_g E_g + t_p E_p}{2t_g + t_p}, \tag{3.41}$$

$$\bar{\nu} = \frac{2t_g \nu_g + t_p \nu_p}{2t_g + t_p}, \tag{3.42}$$

Table 3.3 Parameters used in constitutive relations and FEA

Components	Parameters and values
Headform	$E = 65\,\text{GPa},\ \rho = 1412\text{kg}/\text{m}^3,\ v = 0.22$
Glass	$E = 74\,\text{GPa},\ \rho = 2500\text{kg}/\text{m}^3,\ v = 0.25,\ t_g = 2\,\text{mm}$
PVB film	$k = 20\,\text{GPa},\ \rho = 1100\,\text{kg}/\text{m}^3,\ t_p = 0.76\,\text{mm}$
Windshield dimension	Panel dimensions (a × b): 1320 mm × 630 mm, with a small curvature in x and z directions in coordinates
Other configuration	Initial velocity: $v_0 = 10\,\text{m/s}$ with no gravity field

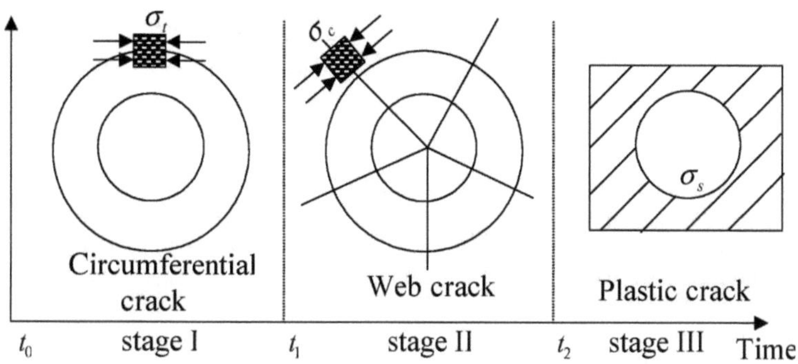

Fig. 3.17 Diagram of different crack-growing stages

where $\bar{E}$ and $\bar{v}$ are the equivalent Young's modulus and equivalent Poisson's ratio of windshield glazing, respectively.

The material parameters are listed in Table 3.3.

(3) Results and discussions

1) Internal stress analysis

The cracks on the PVB windshield are demonstrations of internal and external stresses. Thus, the FEA model above was used to analyze the internal stresses. Only by investigating the internal stresses can we determine the cracks on the windshields effectively.

During the dynamic crack evolution shown in FEA, the stages of damage evolution can be sketched (Fig. 3.17). The three types of cracks are circumferential, web, and plastic cracks. Circumferential cracks are mainly caused by pressure stress, and web cracks are mainly caused by tensor stress. Plastic cracks occur in the last stage in which the windshield deforms plastically.

In Fig. 3.18, three types of internal stresses, namely, tensile stress σ_t, compressive stress σ_c, and shear stress σ_s, are plotted against impact time. The value of cracks can be determined. Lawn and Wilshaw [22] stated that in addition to the Hertzian cone crack and the median vent crack, an inelastic deformation zone also exists, where shear and hydrostatic compression are

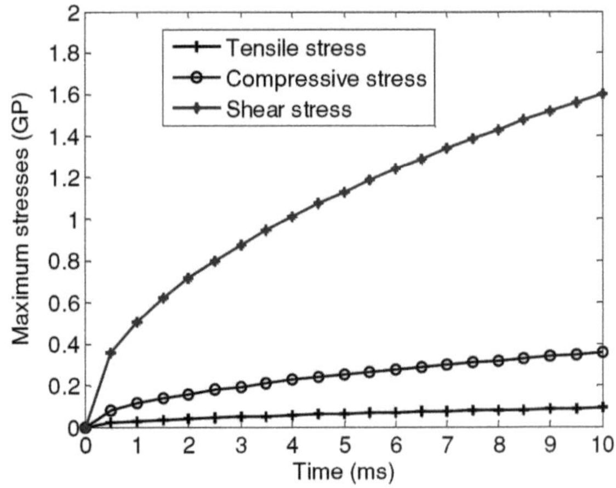

Fig. 3.18 Curves of three types of maximum internal stress in PVB windshield, namely, tensile, compressive, and shear stress, plotted against impact time

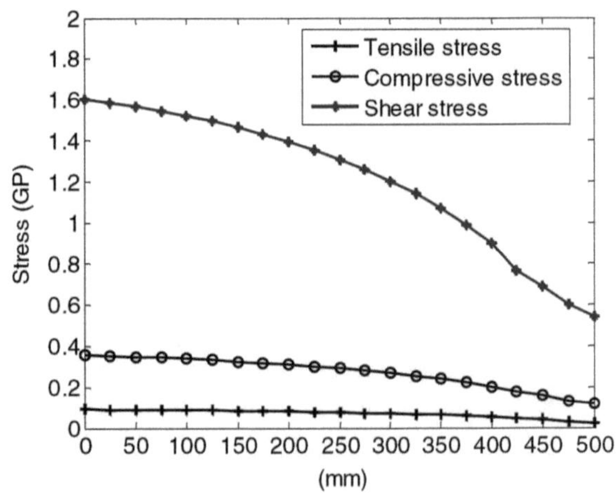

Fig. 3.19 Curves of three types of maximum internal stress in PVB windshield, namely, tensile, compressive, and shear stress, plotted against the increased distances from the central point

the highest. Similarly, Fig. 3.18 shows that shear stress is the highest among the three stresses, causing the windshield to deform plastically.

2) Internal stress variation among different points on the windshield plate
The internal stresses of different points on the windshield plate are different, and crack information can prove this. Cracks weaken as the radius increases for circumferential and radial cracks. As shown in Fig. 3.19, the maximum stresses appeared at the point impacted by the pedestrian head. The stresses decreased as the distances from the central point increased. The decrease rates of the three stresses were nearly the same according to the computational results.
3) Effects of Poisson's ratio on crack angle

Table 3.4 Comparison of the results of two models in terms of the effect of Poisson's ratio on crack angle

Poisson's ratio	0.20	0.25	0.30	0.35	0.40
SIF analysis	12.4	15.3	19.6	24.5	27.7
FEA	11.8	15.1	18.9	23.1	26.2
Differences (%)	5.20	1.32	3.70	6.10	5.80

Fig. 3.20 Effect of initial impact velocity on stress

Poisson's ratio is important to cone crack angle. A parametric study was thus performed to determine the effects on crack angle. Poisson's ratio ranged from 0.2 to 0.4 with a step width of 0.05. Table 3.4 shows a comparison of the results of FEA and stress intensity factor (SIF) analysis. The maximum difference between the results was below 7%. With the increase in Poisson's ratio, the conical crack angle also increased, although the increase pattern remains unknown.

4) Effects of impact velocity
Vehicle speed is a critical factor in pedestrian–vehicle traffic accidents. The extents and degrees of injuries largely depend on the impact velocity between the pedestrian and vehicle. Therefore, studies on cracks on windshields under different head impact velocities would provide an essential foundation in pedestrian protection. According to census data from the National Traffic Accident Database of Tsinghua University (NTADTU), in most cases, vehicles impact pedestrians at a speed of about 40–70 km/h. Thus, we varied the impact velocity of the head from 0 to 20 m/s and made impact velocity V_0 an independent variable. We selected the maximum radius of circumferential cracks R_c, the maximum length of radial cracks L_r, and the maximum radius of plastically deformed area R_p as dependent variables. These three dependent variables are illustrated and defined in Fig. 3.20.

As shown in Fig. 3.21, impact velocity significantly influenced the three selected variables. At an impact velocity of 20 m/s, R_c reached up to 642 mm, which is much

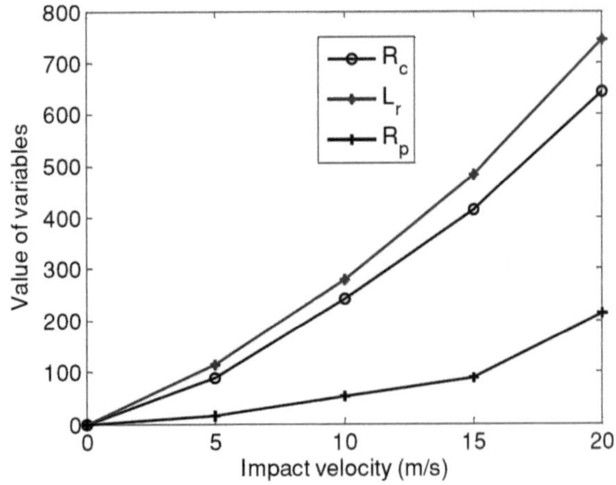

Fig. 3.21 Effect of initial impact velocity on cracks

larger than $R_c = 89$ mm under the impact velocity of 5 m/s. Vehicle velocity plays an important role in pedestrian head injury and windshield damage. In addition, inferring the extent of injuries of pedestrian head and impact velocity of the vehicle by using R_c, L_r, and R_p is reasonable.

3.2.3 *Viscoelasticity Discussion of PVB*

3.2.3.1 SHPB Setup

The SHPB setup was similar to that in the test in Sect. 3.1.2.2. The impact loading test of PVB was conducted at five different strain rates, i.e., 700/s, 1200/s, 2200/s, 3500/s, and 4500/s. The difference from the test in Sect. 3.1.2.2 is that the samples were placed in a diamond shape on the surface of the bar, and impact loading was conducted by the impact bar without the input bar, which is unsuitable because PVB is a polymer material with low material stiffness and low Young's modulus (about $E \approx 50$ MPa). Moreover, semiconductor strain gauges were used instead of traditional metal-made ones. The modified SHPB configuration is shown in Fig. 3.22.

The traditional PVB interlayer used for automotive was made into a circular flake-shaped specimen with thickness 4×0.76 mm $= 3.04$ mm and diameter $\emptyset =$ 3.90 mm, as shown in Fig. 3.22. The static mechanical properties of PVB are listed in Table 3.3. The specifications of the bars are listed in Table 3.5. The reason the four specimens were placed in a diamond shape (shown in Fig. 3.22) was to reduce the inertia and friction effect and thereby satisfy the 1D wave propagation assumption as best as possible. However, in a situation with a large strain rate, specimens

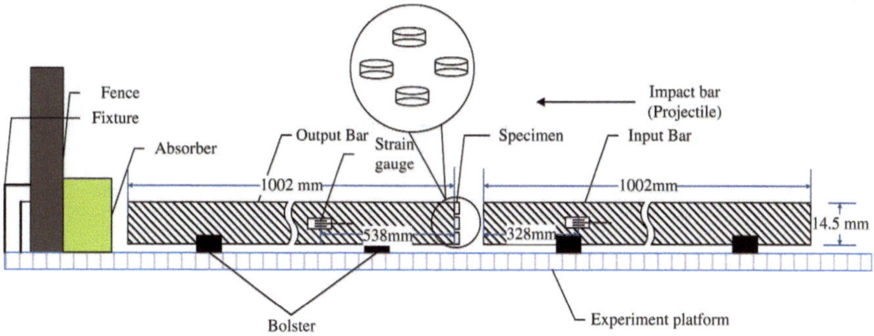

Fig. 3.22 SHPB configuration used in the experiments. All dimensions are in millimeters

Table 3.5 Quasi-static mechanical and geometrical properties of PVB

Specimen diameter Ø (mm)	Density ρ (kg/m^3)	Thickness t_p (mm)	Bulk modulus K (GPa)	Poisson's ratio ν
12	1100 [24]	0.76	20 [25]	0.49 [24]

may touch one another, which would slightly affect the accuracy of the experiment. Stress dispersion is also a notable problem in the SHPB method. Thus, we used the dispersion correction method mentioned in Ref. [23] to make necessary corrections in the output data.

3.2.4 Dynamic Experimental Results

(1) Validation of experimental results

The experimental data must be correct when determining accurate constitutive relations of PVB under impact loading. Two aspects need to be examined for the validity of the data. The first one is to check the stability of input velocities of the impact bar in the same experiment set, and the second one is to check the deviation of four experimental results at the same strain rate.

In our experiments, we observed that the impact velocities of the impact bar were well controlled. The deviations of the maximum and minimum velocities were 1.81, 5.42, 6.74, 4.99, and 3.96% at strain rates of 700/s, 1200/s, 2200/s, 3500/s, and 4500/s, respectively. Thus, the input factor was verified. We put four experimental results (stress–strain curves) at the same strain rate together in Fig. 3.23. Minor differences (less than 6% oscillation in the data) were found in the comparisons at all strain rates, indicating the good accuracy and reliability of the experiment results. Hence, we ended the experimental data validation process and proceeded to data analysis.

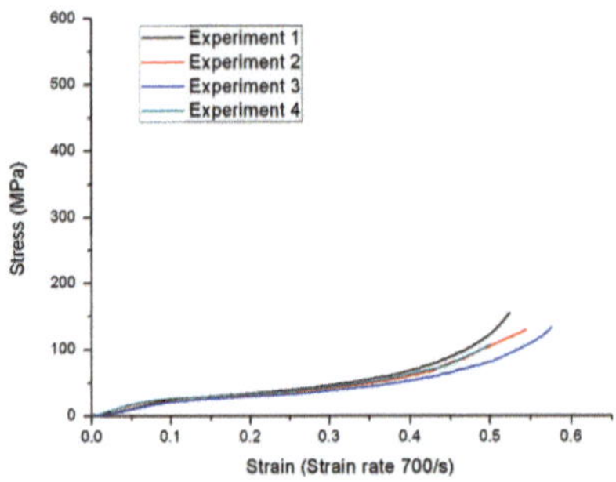

(a) Four experiments conducted at a strain rate of 700/s

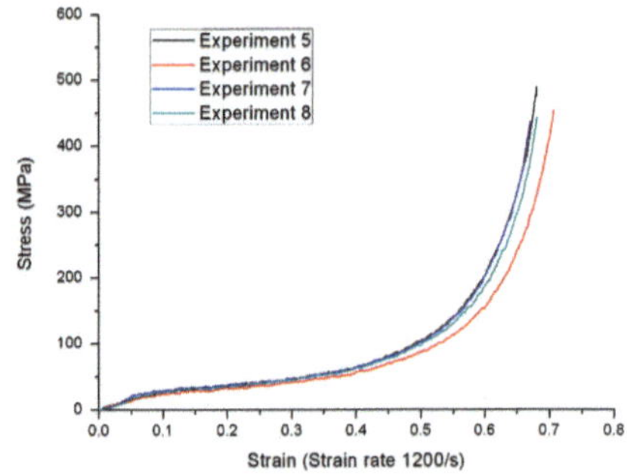

(b) Four experiments conducted at a strain rate of 1200/s

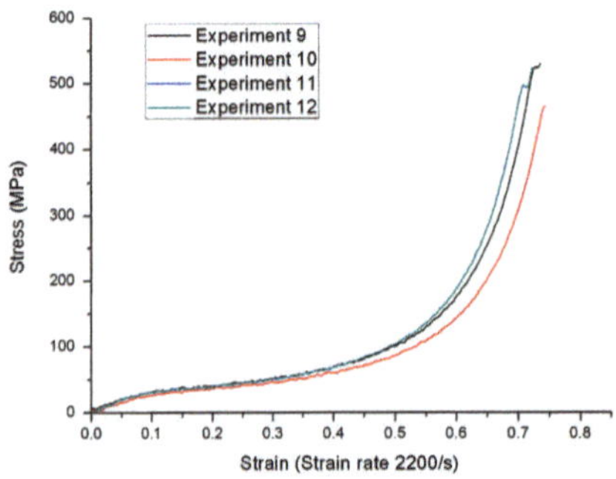

(c) Four experiments conducted at a strain rate of 2200/s

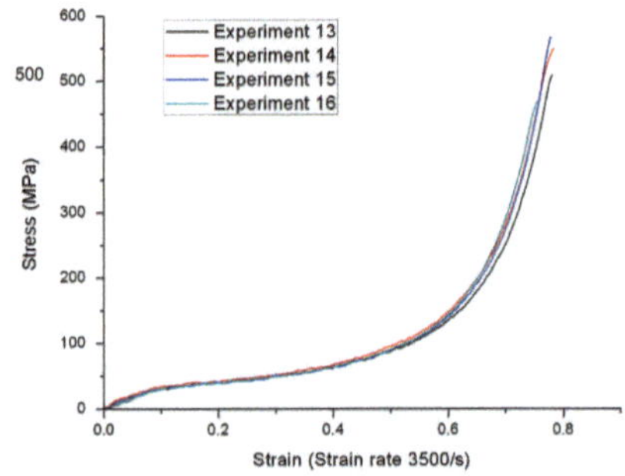

(d) Four experiments conducted at a strain rate of 3500/s

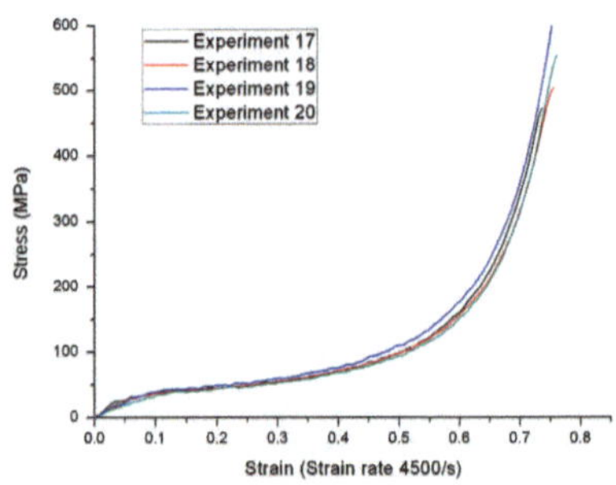

(e) Four experiments conducted at a strain rate of 4500/s

Fig. 3.23 Dynamic response of PVB material at each strain rate in four repeated experiments

(2) Results analysis

Typical stress–strain curves at every strain rate were selected (to see the curves clearly) and combined, as shown in Fig. 3.24a. In the experiments, we discovered that the PVB specimens were smashed into powder, which proves that PVB was completely damaged, although the extent of smashing was slightly different at the strain rate of 700/s and at strain rates above 1200/s. Therefore, we conclude that at all strain rates, the loading time was sufficiently long to push the specimen into the "stiff stage." The stress–strain curves obtained at strain rates above 1200/s were similar, and we could not ascertain if a "rate-dependent effect" existed among the different strain rates. The failure stress of PVB at the strain rate of 700/s was much lower than those at the other strain rates.

Figure 3.24b indicates that the "rate-dependent effect" did not exist among all the strain rates, i.e., the stress increased under the same strain as the strain rate increased (stresses at strain rates 3500/s and 2200/s were almost the same, and the average stress at 4500/s was about 1.5 times that at 700/s). We considered the strain rate under 1000/s as relatively "low speed" impact. Thus, we can reasonably say that PVB behavior depends between "low speed" impact and "high speed" impact (further discussion is provided in the following section).

We let failure energy be $W = \int_0^{\varepsilon} \sigma \, d\varepsilon$; thus, the energy absorbed by a specimen at each strain rate was calculated by integrating the area below the stress–strain curve. The unit area energy absorbed (uAEA, defined as W/A where A is the area of the specimen) and the unit volume energy absorbed (uVEA, defined as W/V where V is the volume of the specimen) are two important indicators and are depicted at different strain rates in Fig. 3.25. The energy absorption capability of PVB in the high-speed impact condition is about three times of that in the low-speed scenario in terms of uAEA and uVEA.

We divided PVB deformation under compression impact into two Stages I and II (shown in Fig. 3.24). In the first stage, the PVB material deforms mildly under compression loading without increasing the stress significantly (gentle stress increase) because of its low density (which means many dislocations exist in the microstructure of the material). From the perspective of the microstructure of the material, the energy carried by loading applies not only to the old dislocation movement but also to a large number of new dislocations produced by jarring or working the material sufficiently [26]. Strain hardening occurs in this stage, and the PVB material is compacted. Therefore, we call this stage the "compaction stage." PVB then deforms much harder as stress increases rapidly in the second stage, which can be regarded as a "stress hardening" phenomenon. Thus, we call this stage the "hardening stage." In this stage, the largest deformation (strain) of the specimen only accounts for about 30% of the total deformation (failure strain), and the stress increases by about four times that in the compact stage when the stage boarder is set to $\varepsilon = 0.5$. From the microstructure point of view, Meijer [27] stated that the deformation process in amorphous polymers is dominated by molecular motions on a segment scale. Thus, entanglement of a molecule plays a critical role in the second stage. As compaction continues, the entanglement among molecules becomes increasingly significant and thereby causes the "stress hardening" effect.

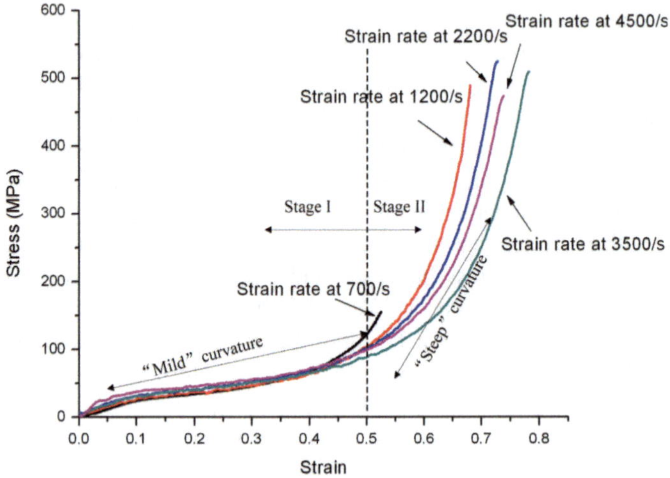

(a) Global stress–strain behavior of PVB under various strain rates

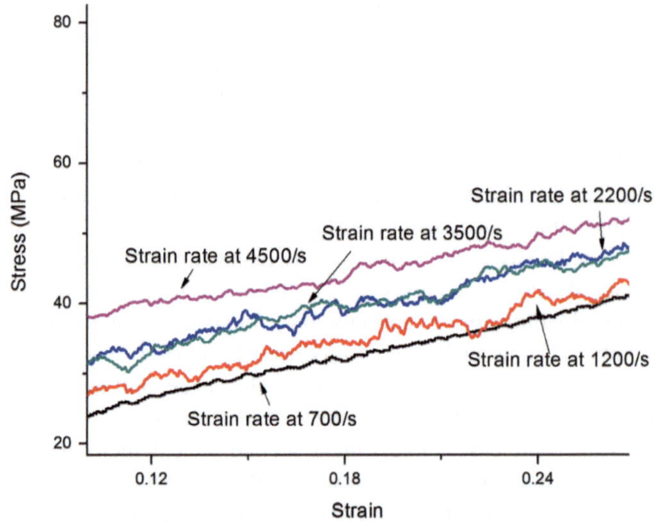

(b) Local stress–strain behavior in Stage I (zoomed in view of the initial stage in (a))

Fig. 3.24 Stress–strain curves of PVB material under different strain rates

3.2.5 Results for the Constitutive Response of PVB

The shapes of the stress–strain curves at all strain rates are similar to the curves described in the work of Yamashita et al. [28], Urayama et al. [29], and Kluppel

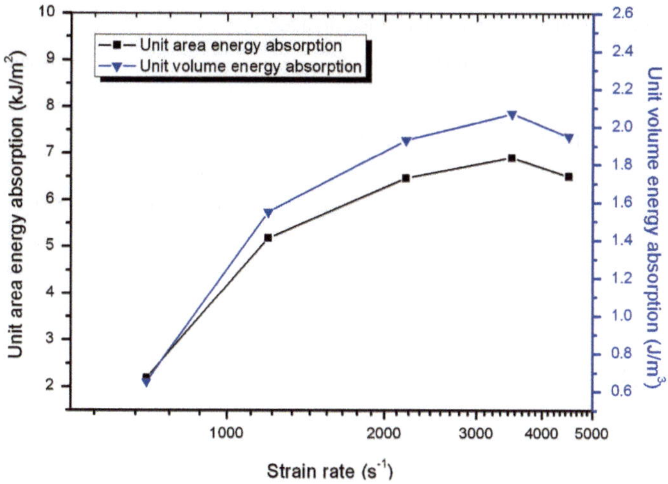

Fig. 3.25 Unit area and volume energy absorption at different strain rates in Log10set

et al. [30], indicating that PVB behaves as a rubber-like material in dynamic loading. Thus, two classical rubber-like constitutive models were adopted and compared in the current work. High- and low-speed impact scenarios were considered.

(1) Mooney–Rivlin theory and data fitting model

$$\sigma^e = -p_e\mathrm{I} + \alpha_1\mathrm{B} + \alpha_2\mathrm{B}\cdot\mathrm{B}, \tag{3.43}$$

where σ^e is the Cauchy stress tensor, p_e is pressure, $\alpha_1 = 2(\partial W/\partial \mathrm{I}_1 + \mathrm{I}_1\partial W/\partial \mathrm{I}_2)$, and $\alpha_2 = -2\partial W/\partial \mathrm{I}_2$. W represents the strain energy potential, which is assumed to be in a polynomial series. $\mathrm{I}_1 = \mathrm{tr}(B)$, $\mathrm{I}_2 = \left[\mathrm{I}_1^2 - \mathrm{tr}\left(B^2\right)\right]/2$, and $\mathrm{I}_3 = \det(\mathrm{B})$. Based on Brown's [31] discussion of the number of unknown parameters, strain energy with three terms in polynomial series can be given as

$$W = A_1(\mathrm{I}_1 - 3) + A_2(\mathrm{I}_2 - 3) + A_3(\mathrm{I}_1 - 3)(\mathrm{I}_2 - 3), \tag{3.44}$$

where A_1, A_2, and A_3 are determined through a 1D test.

In a uniaxial loading case where the stretch ratio is denoted by λ and with the definition of gradient F and left Cauchy–Green deformation tensor B [32], the invariants should be

$$\mathrm{I}_1 = \lambda^2 + 2\lambda^{-1}, \mathrm{I}_2 = \lambda^{-2} + 2\lambda. \tag{3.45}$$

Therefore, by substituting Eq. (3.45) into Eqs. (3.47) and (3.43), the constitutive relation in 1D loading condition is written as

Table 3.6 Empirical material parameters fitted in models

Strain rate (s^{-1})	Mooney–Rivlin parameters			Ogden parameter
	A_1	A_2	A_3	α_j
700	106.9032	−140.7256	−26.8505	11.6758
1200	257.0411	−317.8733	−64.7156	11.6747
2200	271.2663	−338.1145	−65.4526	10.91
3500	249.2075	−315.1729	−55.9025	11.5329
4500	295.2272	−371.3073	−67.1798	11.2809

$$\sigma_{11}^{e} = 2\lambda\left(1-\lambda^{-3}\right)\left(A_1\lambda + A_2 + A_3\left(\lambda^2+2\lambda^{-1}-3+\lambda\left(\lambda^{-2}+2\lambda-3\right)\right)\right). \tag{3.46}$$

With the relation $\lambda = 1 + \varepsilon_{11}$ and in consideration of the 1D compression test we conducted, the constitutive relation can be rewritten as

$$\begin{aligned}\sigma = {} & 2(1+\varepsilon)\left(1-\frac{1}{(1+\varepsilon)^3}\right)(A_1(1+\varepsilon)+A_2 \\ & +3A_3\left((1+\varepsilon)^2-(1+\varepsilon)-1+\frac{1}{(1+\varepsilon)}\right)\Bigg).\end{aligned} \tag{3.47}$$

After fitting, we observed good correlation between the experimental data and the suggested model in Fig. 3.26. All the parameters used are listed in Table 3.6.

(2) Ogden theory and data fitting

Ogden [33] proposed another strain energy formulation to describe rubber-like polymers.

$$W = \sum_{i=1}^{3}\sum_{j=1}^{n}\frac{u_j}{\alpha_j}\left(\lambda_i^{*a_j}-1\right) + \mathrm{K}(\mathrm{J}-1-\mathrm{InJ}), \tag{3.48}$$

where a_j is a non-integer, $\mathrm{J} = \lambda_1\lambda_2\lambda_3$; $\lambda_i^* = \lambda_i \mathrm{J}^{-1/3}$. Using $\partial \mathrm{J}/\partial\lambda_i = \lambda_j\lambda_k$,$\mathrm{J} = 1$, and $\lambda_i^* = \lambda_i$, the constitutive model takes the form [32, 33]

$$\sigma_i = \sum_{j=1}^{n} u_j\left(\lambda^{\alpha_j} + \lambda^{-\frac{\alpha_j}{2}}\right). \tag{3.49}$$

Likewise, we let $\lambda = 1 + \varepsilon$ and rewrite Eq. (3.49) to obtain

$$\sigma = \sum_{j=1}^{n} u_j\left((1+\varepsilon)^{\alpha_j} + (1+\varepsilon)^{-\frac{\alpha_j}{2}}\right), \tag{3.50}$$

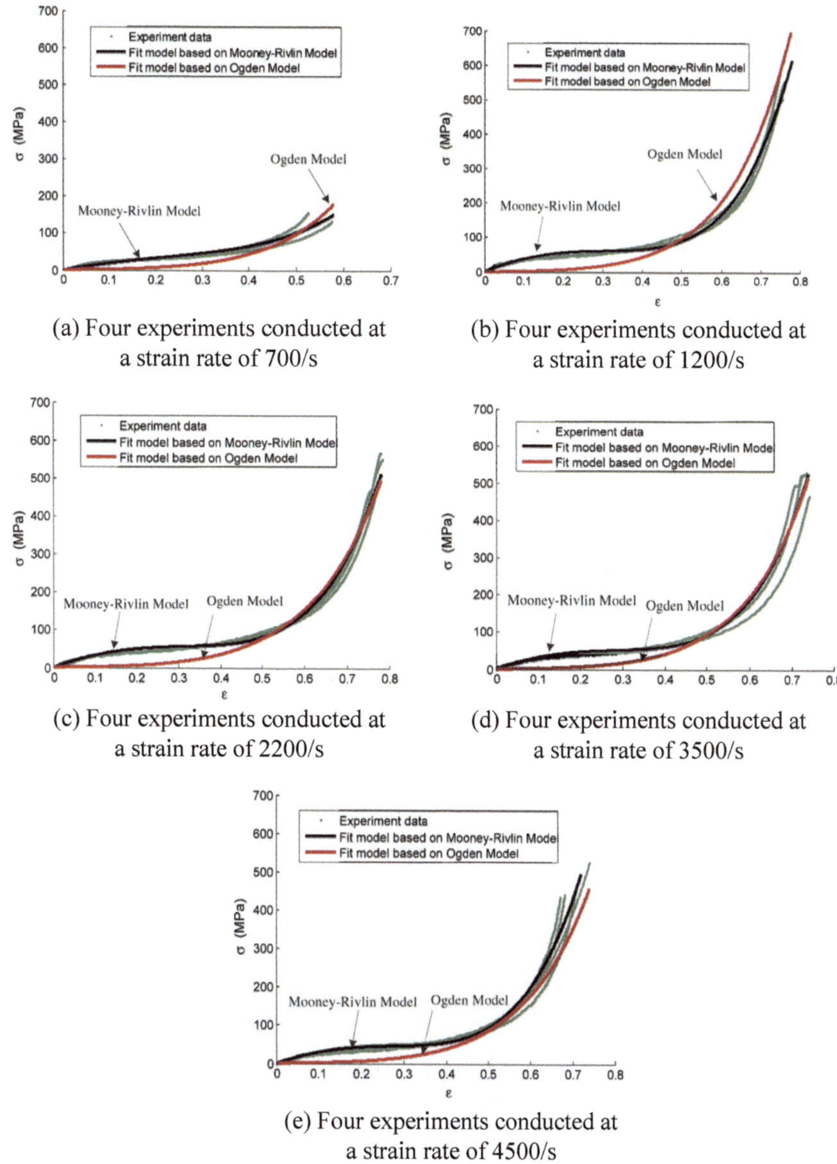

(a) Four experiments conducted at a strain rate of 700/s

(b) Four experiments conducted at a strain rate of 1200/s

(c) Four experiments conducted at a strain rate of 2200/s

(d) Four experiments conducted at a strain rate of 3500/s

(e) Four experiments conducted at a strain rate of 4500/s

Fig. 3.26 Comparison of experiment data and the proposed model

where u_j and a_j are empirical material parameters. For isotropic elasticity, $\sum_{j=1}^{n} u_j \cdot a_j = 2G$, with G being the shear modulus. $G = 5$ MPa [34]. Thus, Eq. (3.50) can be further simplified as

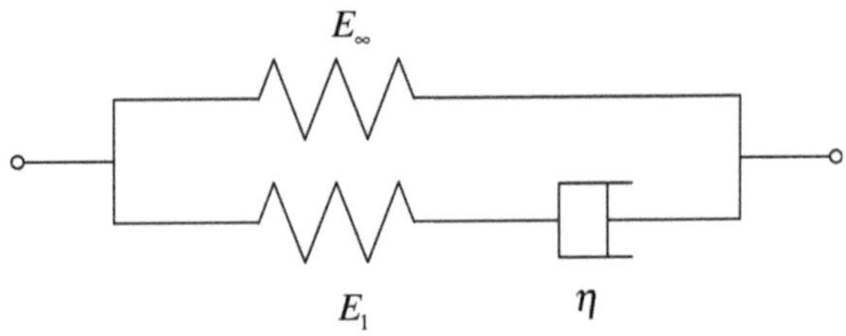

Fig. 3.27 Standard linear solid model

$$\sigma = \frac{2G}{\alpha_j}\left((1+\varepsilon)^{\alpha_j} + (1+\varepsilon)^{-\frac{\alpha_j}{2}}\right) = \frac{10}{\alpha_j}\left((1+\varepsilon)^{\alpha_j} + (1+\varepsilon)^{-\frac{\alpha_j}{2}}\right). \tag{3.51}$$

After fitting the experiment data based on the Ogden model, we obtained the best fitted curves shown in Fig. 3.26. All the parameters are listed in Table 3.6.

The comparison illustrated in Fig. 3.26 shows that the Mooney–Rivlin model performs better than the Ogden model according to the experimental data of the PVB material, especially in the starting and ending stages of deformation. Therefore, we conclude that the Mooney–Rivlin model is more suitable for describing the dynamic constitutive behavior of PVB materials than the Ogden model.

3.2.6 *Viscoelasticity*

Polymer materials usually have a property called "viscoelasticity." Viscoelasticity in mechanical behavior may be illustrated by the combined usage of springs and dashpots that represent elastic and viscous factors, respectively. The standard linear solid (SLS) model, which is based on Maxwell and Kelvin–Voigt basic models, is widely used. Figure 3.27 shows the SLS model.

The modulus in SLS is defined as

$$E(t) = E_\infty + E_0 e^{-\frac{\tau}{t}}, \tag{3.52}$$

where E_∞ and E_0 are the infinite time and initial modulus, respectively. τ is the relaxation time, and t is the total time.

According to the Boltzmann superposition integral, the stress–strain relation for the viscoelasticity part is

$$\sigma_v = \int_0^t \left(E_\infty + E_1\left(e^{-\frac{(t-\xi)}{\eta}}\right)\right)\frac{\mathrm{d}\varepsilon}{\mathrm{d}t}\mathrm{d}\xi. \tag{3.53}$$

To highlight the undetermined material parameter, Eq. (3.53) is converted to

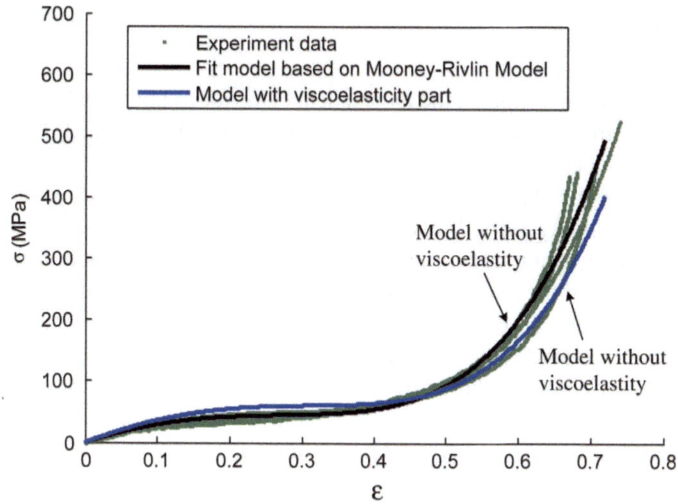

Fig. 3.28 Comparison of models with and without the viscoelasticity effect

$$\sigma_v = \int_0^t \left(B_1 + B_2\left(e^{-\frac{(t-\xi)}{B_3}}\right)\right)\dot{\varepsilon}\mathrm{d}\xi. \tag{3.54}$$

Thus, the total stress should be

$$\begin{aligned}\sigma_t &= \sigma + \sigma_v \\ &= 2(1+\varepsilon)\left(1 - \frac{1}{(1+\varepsilon)^3}\right)(A_1(1+\varepsilon) + A_2 + 3A_3) \\ &\left((1+\varepsilon)^2 - (1+\varepsilon) - 1 + \frac{1}{(1+\varepsilon)}\right) + \int_0^t \left(B_1 + B_2\left(e^{-\frac{(t-\xi)}{B_3}}\right)\right)\dot{\varepsilon}\mathrm{d}\xi.\end{aligned} \tag{3.55}$$

Data fitting showed that a small difference exists between the results of Eq. (3.47) (without viscoelasticity) and Eq. (3.53) (with viscoelasticity). Figure 3.28 shows a comparison by regarding the curves at a strain rate of 4500/s as an example. Hence, the Mooney–Rivlin fitting model well describes the constitutive relation behavior of PVB materials. Meanwhile, viscoelasticity can be neglected in further research of the material without losing much accuracy.

3.3 Summary

In this chapter, we investigated the mechanical behavior of laminated glass by quasi-static and dynamic test analyses, and constitutive modeling of such behavior was studied. The viscoelasticity of PVB was also examined.

First, we introduced the test methods for laminated glass, including tensile, compression, and bending tests. For experimental analyses of PVB laminated glass, quasi-static compression experiment based on MTS is investigated. The stress–strain curves showed the "nonlinearity" characteristic in the mechanical behavior of PVB laminated glass. As the loading rate increased, the MFO strain increased while the MFO stress remained nearly the same. Then, dynamic SHPB compression experiments were carried out at three different strain rates, i.e., 2000 s^{-1}, 4000 s^{-1}, and 6000 s^{-1} to investigate the mechanical behavior of PVB laminated glass. We found that the stress–strain curve also exhibited nonlinearity, which could also be fit via exponent-like functions. The MFO stress values were about 15–30 times higher than those in the quasi-static situation, indicating that PVB laminated glass is a strong rate-dependent material, especially at a high strain rate. Minimal difference was found in MFO strain domains in quasi-static and dynamic cases because the total thickness of the glass layer accounted for over 84% of the total thickness.

Then, we proposed a constitutive Johnson–Cook model to describe quasi-static and dynamic stress–strain behaviors. The parameters in the model were fit via a data fitting method. The results revealed excellent agreement between the suggested model and experimental data. The experimental data and rate-dependent constitutive model can be employed to numerically analyze human injury, vehicle crashworthiness, and accident reconstruction, which are interesting and attractive topics in the automotive industry.

For further discussion of the constitutive relations of PVB laminated glass, a damage-modified, nonlinear, viscoelastic, constitutive relation model of a PVB laminated windshield was recommended based on the updated Lagrangian method to study the crack evolution on a windshield plate subject to impact by a pedestrian head. The constitutive relations with a damage variable were implemented in FEA software to simulate the pedestrian head impact with the vehicle windshield after verification with a classical example and the classical Hertzian pressure calculation model. On the basis of the FE model above, the internal stress of PVB laminated glass was analyzed. Shear, compressive, and tensile stresses were the main causes of plastic deformation, radial cracks, and circumferential cracks, respectively, in the laminated glass subject to impact. Parametric study based on the FE model showed that an increase in Poisson's ratio led to an increase in crack angle. Impact velocity exerted a great influence of crack formation and pedestrian head injury.

Finally, the viscoelasticity of PVB was examined through an SHPB impact experiment with modified semiconductor strain gauges at five different strain rates, i.e., 700/s, 1200/s, 2200/s, 3500/s, and 4500/s. Four repeated experiments were conducted at each strain rate to ensure the validity of the output data. The stress–strain curves revealed the existence of a "rate-dependent effect" among the strain rates. The energy absorption capability of PVB in the high-speed impact condition was about three times of that in the low-speed scenario in terms of unit area and volume energy absorbed. Then, the typical constitutive behavior of PVB material was divided into compact stage and hardening stage. The strain hardening effect dominated in the first stage, whereas stress hardening dominated in the other stage. The constitutive behaviors differed at low- and high-speed impact, which caused PVB to absorb three times more energy in the high strain rate scenario than in the low strain rate case. On the contrary, PVB behaved the same at the strain rate above 1200/s. Mooney–Rivlin and Ogden models were employed to fit the stress–strain data obtained from the experiment because the curves were similar to that of the rubber-like material studied previously. Data fitting models based on the Mooney–Rivlin model were established and found to describe the dynamic behavior appropriately. The strain rate effect was further studied, and the results showed that the PVB material does behave rate-dependently under different strain rates. Comparison of the failure strengths at quasi-static, low strain rate, and high strain rate conditions showed that SIF is $\beta = 4$, which can be employed in future computations. Through data fitting, we discovered that minimal difference exists between results from constitutive forms with and without viscoelasticity. Thus, viscoelasticity can be disregarded in further research of the material without losing much accuracy. The viscoelasticity discussion on PVB provides important results on the mechanical constitutive response of PVB materials under dynamic impact situations along with prospective applications in automotive impact crashworthiness and pedestrian protection research.

References

1. M.G. Stout, D.A. Koss, C. Liu, J. Idasetima, Damage development in carbon/epoxy laminates under quasi-static and dynamic loading. Compos. Sci. Technol. **59**, 2339–2350 (1999)
2. W. Holand, G.H. Beall, Glass Ceramic Technology, (2012)
3. S. Zhao, L.R. Dharani, X. Liang, L. Chai, S.D. Barbat, Crack initiation in laminated automotive glazing subjected to simulated head impact. Int. J. Crashworthiness **10**, 229–236 (2005)
4. J.F. Bell, C. Truesdell, The experimental foundations of solid mechanics. Int. J. Mach. Tools Manuf **49**, 1234–1237 (2009)
5. J.E. Field, S.M. Walley, W.G. Proud, H.T. Goldrein, C.R. Siviour, Review of experimental techniques for high rate deformation and shock studies. Int. J. Impact Eng **30**, 725–775 (2004)
6. M. Xu, L. Wang, M. Xu, L. Wang, A new method for studying the dynamic response and damage evolution of polymers at high strain rates. Mech. Mater. **38**, 68–75 (2006)
7. G.R. Johnson, W.H. Cook, A constitutive model and data for metals subjected to large strains, high strain rates and high temperatures, (1983)
8. H. Yu, Y. Guo, X. Lai, Rate-dependent behavior and constitutive model of DP600 steel at strain rate from 10^{-4} to 10^{3} s^{-1}. Mater. Des. **30**, 2501–2505 (2009)
9. L.L. Wang, H.W. Lai, Z.J. Sun, X.U. Ming-Qiao, D.J. Huang, S.Q. Shi, Dynamic Deformation and Fracture of Polymers Taking Account of Damage Evolution. J. Ningbo Univ. **16**(6), 372–380 (2003)
10. W. Lili, Z. Xixiong, S. Shaoqiu, An impact dynamics investigation on some problems in bird strike on windshields of high speed aircrafts. Acta Aeronautica Et Astronautica Sinica **12**, B27 (1991)

11. F. Zhou, L. Wang, S. Hu, A damage-modified nonlinear visco-elastic constitutive relation and failure criterion of PMMA at high strain-rates. Combust. Explos. Shock Waves **12**, 333–342 (1992)
12. U. Fischer, Washizu, K., Variational Methods in Elasticity and Plasticity, 3rd edn. Oxford-New York, Pergamon Press (1982). XV, 630 S., £ 40.00. US $ 100.00. ISBN 0 08 026723 8, Zamm Journal of Applied Mathematics & Mechanics Zeitschrift Für Angewandte Mathematik Und Mechanik, 64 (1984) 70–71
13. S.R. Swanson, L.W. Christensen, A constitutive formulation for high-elongation propellants. J. Spacecraft Rockets **20**, 559–566 (2015)
14. X. Wang, Z. Feng, F. Wang, Z. Yue, Dynamic response analysis of bird strike on aircraft windshield based on damage-modified nonlinear viscoelastic constitutive relation. China Aviation J (English Version) **20**, 511–517 (2007)
15. Y.P. Shen, Y.H. Chen, Y.F. Pen, The finite element method of viscoelastic large deformation plane problem with kirchhoff stress tensors-green strain tensors constitutive relation. Acta Mechanica Solida Sinica, (1987) pp. 87–91
16. Z. Feng, On the Viscoelastic Finite Element Method, (1995)
17. Y. Cao, G. Zhang, Multiple surface crack initiation and growth in glass-ceramics loaded by a sphere. Eng. Fract. Mech. **67**, 277–292 (2000)
18. T.B. Belytschko, C.S. Tsay, Explicit algorithms for nonlinear dynamics of shells. Comput. Methods Appl. Mech. Eng. **43**, 223 (1981)
19. T. Belytschko, J.I. Lin, T. Chen-Shyh, Explicit algorithms for the nonlinear dynamics of shells. Comput. Methods Appl. Mech. Eng. **42**, 225–251 (1984)
20. M. Timmel, S. Kolling, P. Osterrieder, P.A.D. Bois, A finite element model for impact simulation with laminated glass. Int. J. Impact Eng **34**, 1465–1478 (2007)
21. S. Zhao, L.R. Dharani, L. Chai, S.D. Barbat, Analysis of damage in laminated automotive glazing subjected to simulated head impact. Eng. Fail. Anal. **13**, 582–597 (2006)
22. B. Lawn, R. Wilshaw, Review: indentation fracture: principles and applications. J. Mat. Sci. **10**, 1049–1081 (1975)
23. H. Zhao, Material behaviour characterisation using SHPB techniques, tests and simulations. Comput. Struct. **81**, 1301–1310 (2003)
24. S. Mukherjee, A. Chawla, P. Mahajan, D. Mohan, N.S. Mane, M. Singh, M. Sakurai, Y. Tamura, Modelling of head impact on laminated glass windshields, in *Proceedings of the 2000 International Ircobi Conference on the Biomechanics of Impact*, Montpellier, France, 20–22 Sept 2000
25. P.A.D. Bois, S. Kolling, W. Fassnacht, Modelling of safety glass for crash simulation. Comput. Mater. Sci. **28**, 675–683 (2003)
26. H. Wf, *Mechanical behavior of material New York* (Cambridge University Press, USA, 2010)
27. E.H.M. Han, L.E. Govaert, Mechanical performance of polymer systems: the relation between structure and properties. Prog. Polym. Sci. **30**, 915–938 (2005)
28. Y. Yamashita, A. Tanaka, Mechanical property of rubber reinforced with rice husk charcoal, In: PARD, STM, 2002
29. K. Urayama, T. Kawamura, S. Kohjiya, Multiaxial deformations of end-linked poly(dimethylsiloxane) networks. 4. Further assessment of the slip-link model for chain-entanglement effect on rubber elasticity. J. Chem. Phys. **118**, 5658–5664 (2003)
30. M. Kluppel, J. Schramm, A generalized tube model of rubber elasticity and stress softening of filler reinforced elastomer system. Macromol. React. Eng. **9**, 42–54 (2000)
31. B. Rp, *Physical testing of rubber*, 3rd edn. (Chapman & Hall, London, 1996)
32. LS-DYNA user manual, Livermore Software Technology Corporation, 2007
33. R.W. Ogden, Large deformation isotropic elasticity—on the correlation of theory and experiment for incompressible rubberlike solids. Proc. Royal Soc. London **326**, 565–584 (1972)
34. ECE/TRANS/180/Add.9, Global technical regulation No. 9 in, 2008

Chapter 4
Crack Initiation and Propagation in Laminated Glass upon Impact

The quasi-static mechanical behavior of PVB laminated glass has been widely investigated experimentally by starting from a fundamental point [1]. Unlike the quasi-static cracking problem, dynamic fracture is more complex and challenging, but only a few studies have focused on the dynamic fracture of PVB laminated glass, except for several that performed numerical simulations based on a probabilistic damage mathematical model [2], continuum damage mechanics [3–6], explicit finite element method [7–9], and extended finite element method [10], which often disregard the important PVB layer. Most of them did not involve explicit crack growth and pattern analysis. Thus, a dynamic experimental investigation is urgently required.

Investigation of dynamic crack patterns in PVB laminated glass may serve as a first step in closing this gap by exploring the dynamic impact fracture of PVB laminated windshields, which may also help satisfy pressing practical needs in engineering design and material selection.

4.1 Experimental Setup of Dynamic Out-of-Plane Loading

The impact system setup is sketched in Fig. 4.1. In the setup, a weight block can slide freely along the two standing poles, and a tip at the end of the drop weight is used to hit the force direction converter on the testing sample (Fig. 4.2). Converting the vertical impact direction into a horizontal one is necessary to facilitate film capture.

High-speed impact photography was employed to capture the dynamic crack growth on the plate in situ. The entire system consisted of a high-voltage charging controller, a multi-spark box, two field lenses, and an array camera with films (Fig. 4.2). In addition, a multi-spark high-speed camera was adopted to provide sufficient spot light source. The multi-spark box provides 16 independent spot lights triggered by the high-voltage charging controller by a preset adjustable interval time

J. Xu and Y. Li, *Impact Behavior and Pedestrian Protection of Automotive Laminated Windshield*, https://doi.org/10.1007/978-981-13-2441-3_4

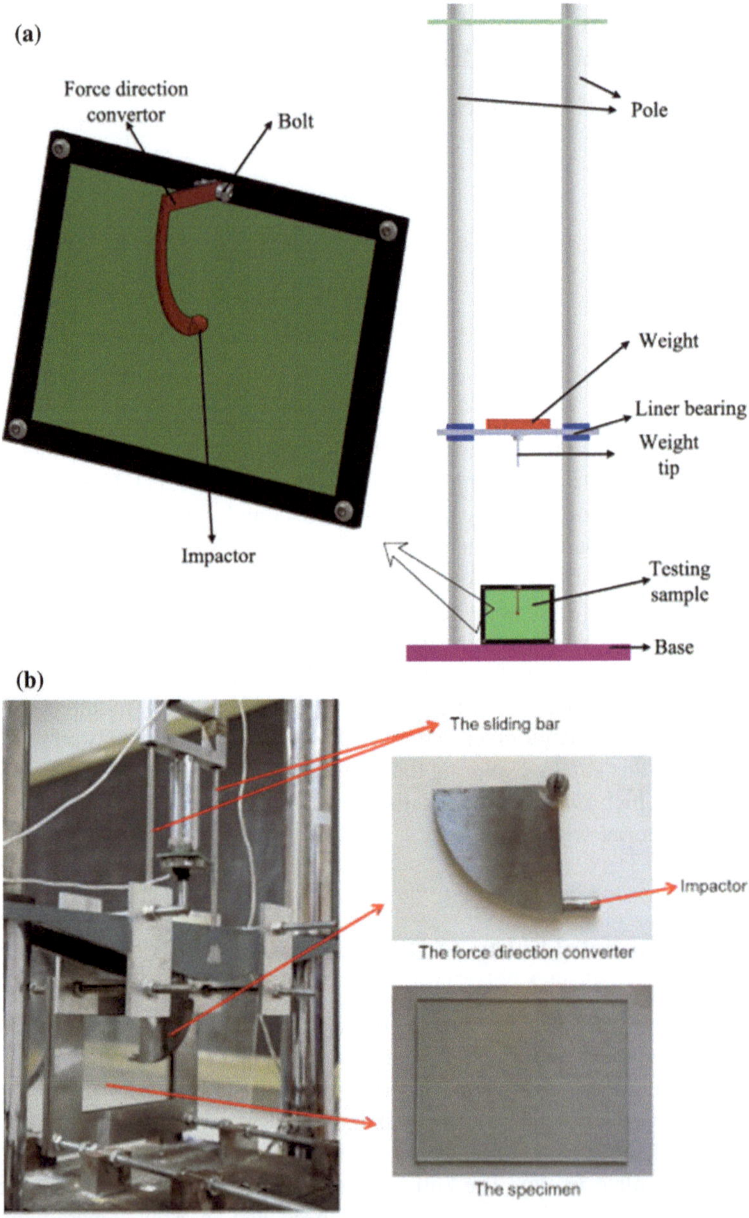

Fig. 4.1 Schematic of the drop-weight tower experiment setup. The testing sample is illustrated along with the force direction convertor

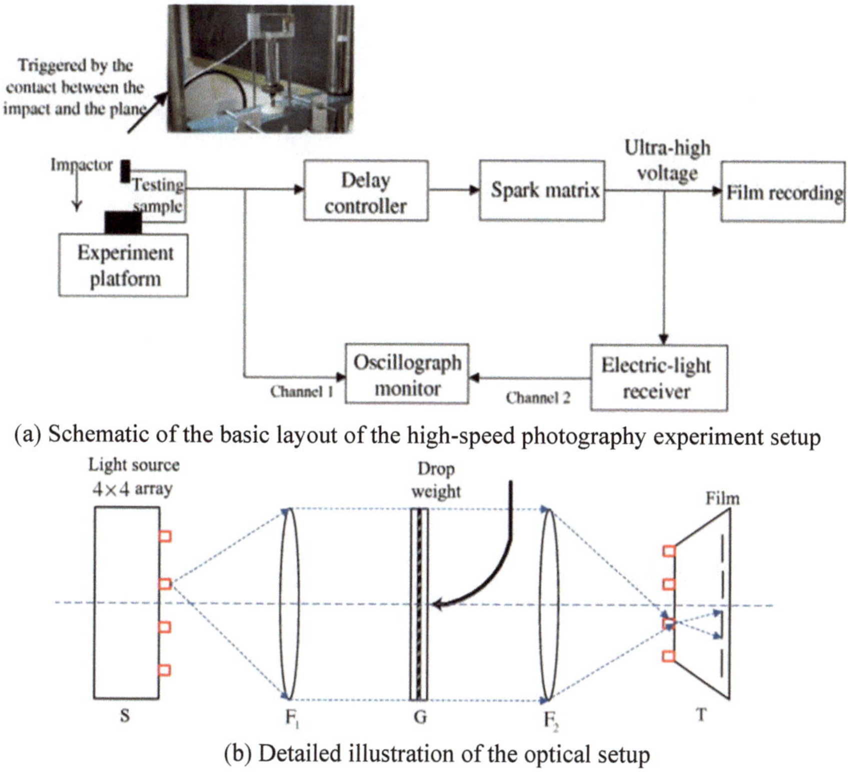

Fig. 4.2 Schematic of the entire kit of the experiment setup

from 1 to 9999 μs whose charging voltage can reach up to 30 kV. Meanwhile, a 4 by 4 array camera with 16 films was set correspondingly on the other side to receive the light and record crack image information.

To ensure that the images recorded on the films are in the time series of the ignition of the spot lights, two field lenses were used for the optical path (Fig. 4.2b). Moreover, a delay controller was introduced to control the time interval between each spot light. To determine the exact ignition time for each spot light, an electric light receiver and an oscillograph monitor were used. The electric light receiver was employed to collect electric light signals and send them to the oscillograph monitor. Reference control experiments on impact induced fracture were carried out using pure glass sheets, and the results are in good agreement with those in literature [11, 12] (Fig. 4.3). The current test system is thus validated.

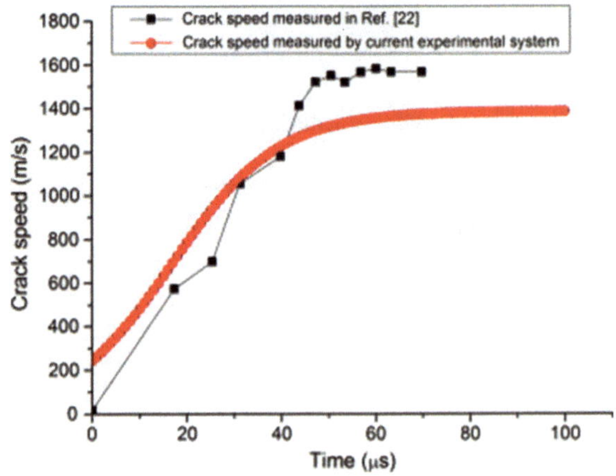

Fig. 4.3 Comparison of the crack velocity propagation speed of pure glass between experiment results extracted from the current study and Ref. [11]. The external impact energy is 1 J in the current experiment and 0.2 J in the experiment in Ref. [11]

4.2 Preliminarily Experiment to Investigate Radial Crack Propagation

4.2.1 Experiment Condition

In the preliminarily impact fracture experiment to investigate the radial crack propagation behavior of PVB laminated glass, the minimum mass of the drop weight was 0.5 kg, and the maximum height of the drop weight was 1.84 m (corresponding to the maximum impact velocity of 6 m/s). The mass and height of the drop weight were also varied in this experimental investigation. Prior to impact, the drop weight was suspended and fixed through an electromagnet. The temperature and relative humidity during the experiments remained in the range of 20–21 and 20–25%, respectively.

4.2.2 Specimen Preparation

The standard PVB layer has a Young's modulus $E_{PVB} = 100$ MPa, Poisson's ratio $\upsilon_{PVB} = 0.48$, and mass density $\rho_{PVB} = 870$ kg/m^3 [13]. To approximate the automotive industry, the PVB thickness is $t_{PVB} = 0.76$ mm for passenger cars. In addition to the middle PVB layer, for the upper and lower soda–lime glass sheets in laminated glass, the thickness was $t_{glass} = 2$ mm, with $\rho_{glass} = 2500$ kg/m^3, $E_{glass} = 70$ GPa, and $\upsilon_{glass} = 0.22$. They were bonded together at 10 bar and 120 °C following standard procedures, and the effective in-plane dimension was 200 mm × 150 mm.

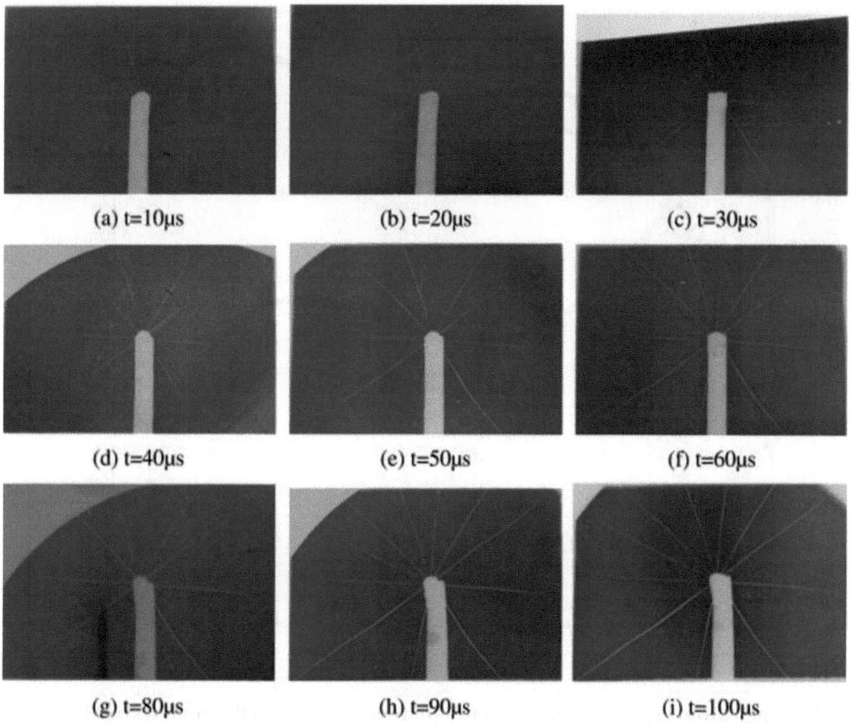

Fig. 4.4 Snapshots of crack propagation with 1 kg of drop weight and 1 m of drop height

4.2.3 Experimental Results and Discussion

(1) Crack morphology

A series of experiments under different loading conditions, i.e., with different heights and masses of the drop weight, were performed. Figure 4.4 shows typical dynamic crack propagation snapshots of the PVB laminated glass sheet in response to 1 kg of weight dropped from a height of 1 m. Extensive radial fracture was observed with no appearance of a circumferential crack, which is consistent with the condition in numerical simulation that when a foreign object impacts a large windshield, a radial crack always appears before a circumferential crack [10].

Approximately 10 radial cracks were observed on the composite sheet, and the development and evolution of each crack differed due to the inevitable stochastic flaw in glass. The average crack length was employed to discuss the behavior of radial crack propagation. By using a nonlinear fitting method on the experimental points obtained from the experiments, we obtained the crack tip position in time history. Then, the averaged time evolution history of radial crack tip velocity and acceleration was obtained from the time history of the radial position of the crack tip, as shown in Fig. 4.5.

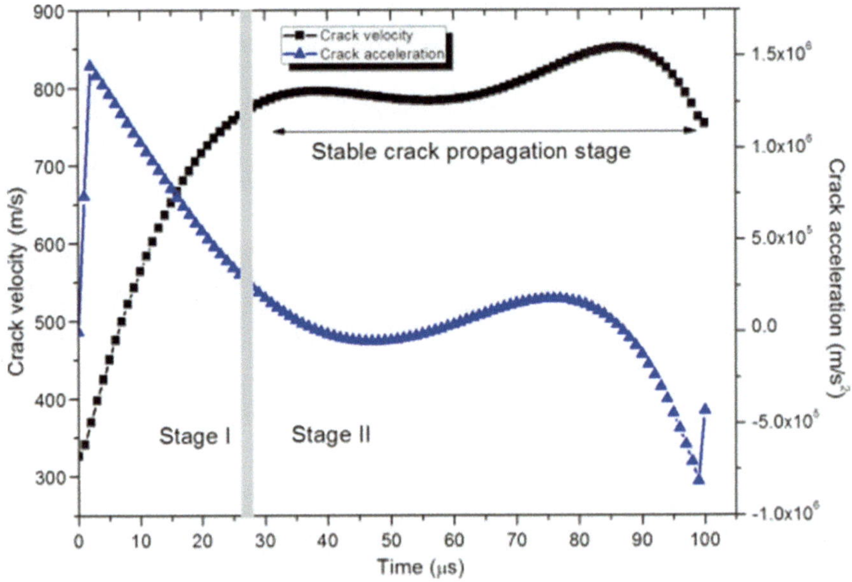

Fig. 4.5 Averaged crack tip velocity and acceleration in time history for the case in Fig. 4.2

The measured crack velocity initially increased and then fluctuated slightly, which is in accordance with the experiment results of the cracking speeds of pure soda–lime glass at various cracking modes in Refs. [11, 12] and with previous numerical simulation results of several brittle materials [14, 15] and composites [16, 17].

As proposed by Ravi-Chandar and Knauss [18], a micro-crack zone exists near the vicinity of the crack tip, and this zone impedes the further propagation of cracks. The spacing between the initial cracks is extremely small within the impact area, and such an intensively damaged zone may reduce the cracking speed substantially. With the advancement of cracks, the retardation effect is significantly reduced until it reaches an almost constant influence when the crack speed becomes steady in a later stage. Thus, crack evolution may be qualitatively divided into two stages. The cracking speed keeps rising in stage I, whereas the crack speed becomes stable but exhibits some fluctuation in stage II. The transition occurs roughly just in time (further discussion will be provided later). The crack propagation acceleration in stage I decreases until stage II begins; that is, as the stress in the crack tip accumulates and develops, the crack tip is able to break through the energy barrier surrounding the material to reach equilibrium.

In this example, the averaged steady-state cracking speed during stage II is approximately $\bar{v}_{PVB} \approx 811$ m/s. Meanwhile, the stable cracking speed of polymethyl methacrylate (PMMA) is $\bar{v}_{PMMA} \approx 600$ m/s [12] and that of tempered soda–lime glass is $\bar{v}_{GLASS} \approx 1470$ m/s [11] (Fig. 4.6 provides a detailed comparison). Although the impact energies in Ref. [13] and in this experiment are slightly different, the comparison in Fig. 4.6 is still valid. PVB laminated glass may be regarded a composite

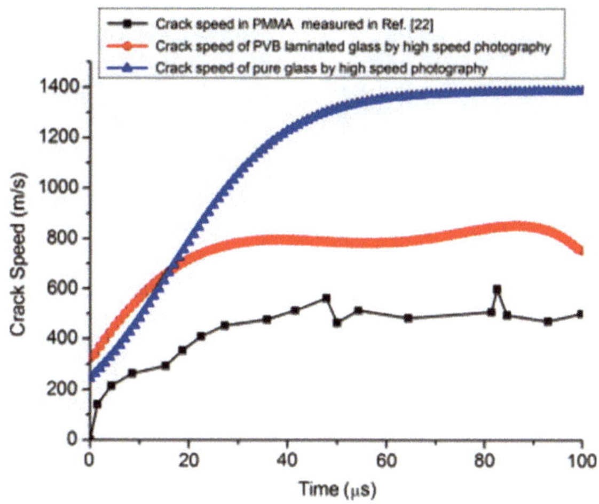

Fig. 4.6 Comparison of crack velocity propagation speeds among pure glass, PVB laminated glass, and PMMA. The data for PMMA are extracted from Ref. [11]

material whose overall effective "fracture toughness" is between that of PMMA and pure glass. Figure 4.6 also indicates that the general variation trends of crack speeds in two different materials are almost the same. Rayleigh wave speed v_R of the glass sheet can be used to normalize the crack speed (because cracking occurs in the glass sheet), i.e., $v_R = 3098$ m/s. Thus, the stable cracking speed of PVB laminated glass is about 0.26 v_R in this particular case, which is lower than the typically measured maximum speed of about 0.4–0.5 v_R in brittle materials. The PVB interlayer slows down the crack speed, which makes the material and brittle-polymer sandwiched structure candidates for energy absorption/shielding.

(2) Different driving mechanisms of in-plane cracking on two glass layers

Figure 4.7 shows a comparison of typical crack propagation on supported and loaded glass layers at the loading speed with PVB thickness $h = 0.76$ m. Here, time $t = 0$ μs is used to denote crack tip initiation for radial cracks on the supported glass layer. The radial cracks (Fig. 4.7a–d) initiated first on the supported glass layer. Then, radial cracks on the "loaded glass layer" were generated after about $t = 600$ μs, as shown in Fig. 4.7f. The cracks on the two glass layers were completely overlapped although they propagated at different times.

The stress-wave-dominated inertial effects during impact loading on PVB laminated glass play an important role in crack propagation [19, 20]. The plate is sufficiently small to disregard bending deformation during impact due to the high modulus of the glass material and the fixed mode of the specimen.

For a convenient analysis, a schematic of a layer system consisting of two glass plates bonded by a PVB interlayer is shown in Fig. 4.8. Impact occurs on the surface of the loaded glass layer of the PVB laminated glass, and the compressive waves

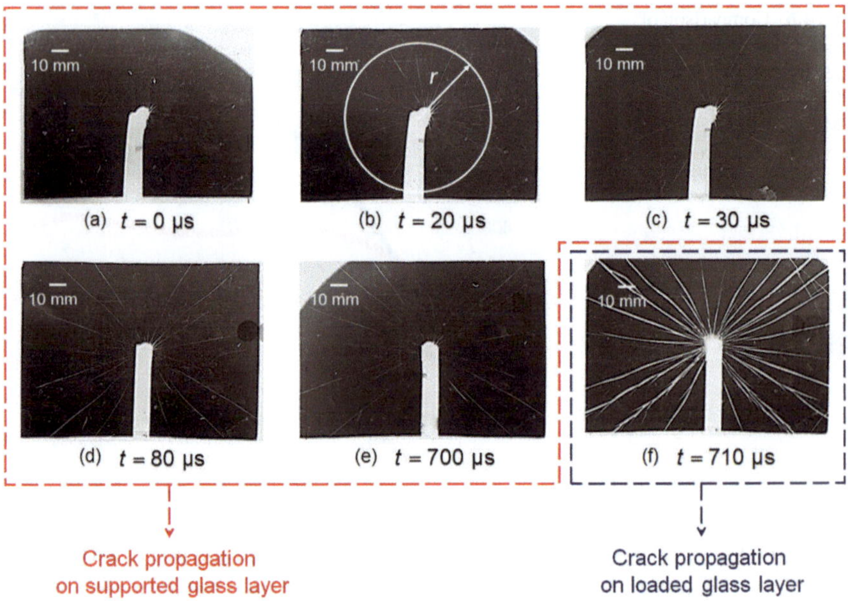

Fig. 4.7 Selected sequence of images depicting radial crack growth on two glass layers

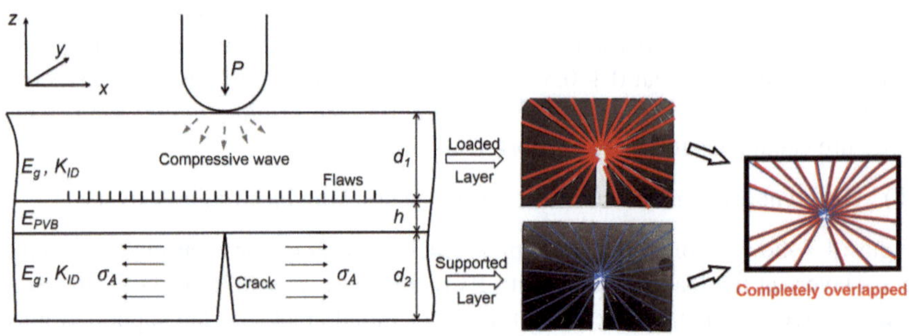

Fig. 4.8 Schematic of the impact process in a layer system consisting of two glass plates bonded by a PVB interlayer. The cracks propagate into the loaded glass layer by re-initiation from a surface flaw

caused by the impact travel down the depth direction (Fig. 4.8). Owing to its viscoelasticity, the PVB layer plays an important role in reducing the amplitude and speed of waves that eventually pass through by internal friction [21] compared with the glass layer. The compressive waves reflect from the boundaries as tensile waves, which continuously arrive at the supported plate; thus, tensile-dominated cracks initiate when the limit of dynamic fracture toughness KID of the material is reached [22]. Considering that a glass plate is suitable for compressive stress waves but prone to fail in tensile stress waves. Radial cracks initiate first on the supported glass layer.

Table 4.1 Parametric experiments in terms of drop weight and height variations

Experiment No.	Drop weight (kg)	Drop height (m)	Equivalent velocity (m/s)	Equivalent impact energy (J)
1	2	1	4.129	19.62
2	1	1.4	5.241	13.734
3	1	1	4.429	9.81
4	1	0.3	3.431	5.886
5	0.5	1	4.429	4.905

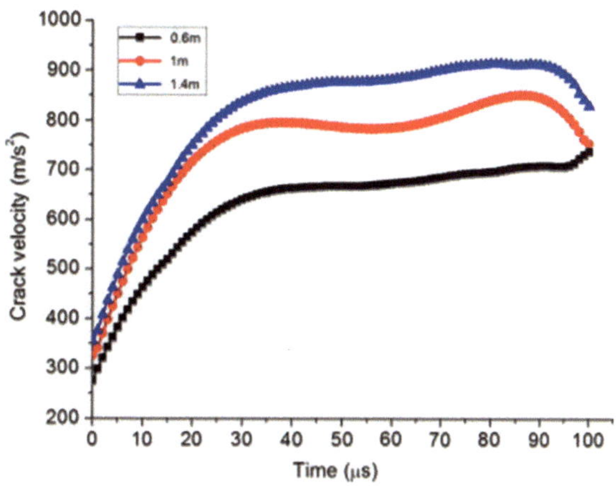

Fig. 4.9 Different crack tip velocity histories under different drop heights (impact speed)

(3) Effect of drop weight and height

Next, we investigated the influence of drop weight and height on crack propagation. The drop mass ranged from 0.5 to 2 kg, and the height varied from 0.6 to 1.4 m. Table 4.1 shows a detailed combination of the parametric experiments.

Figures 4.9 and 4.10 show that with the increase in drop weight and height, the basic trend of cracks remained almost the same. Meanwhile, if the value would increase, large mass and high impact speed will cause rapid crack propagation. Zhang et al. [23] investigated the effect of loading rate on crack velocities in concrete and found that with the increase in loading rate, crack velocity increases proportionally. In quasi-static indentation, radial and lateral cracks in ceramic also increase in length under a large indentation load.

These conclusions are qualitatively consistent with our experimental findings. Particularly, $\overline{v}_{PVB}$ increased to 17.70 and 12.21% when the impact speed changed from 3.43 to 4.43 and 5.24 m/s, respectively; $\overline{v}_{PVB}$ increased to 26.72 and 14.92% when the drop weight was increased from 0.5 to 1 and 2 kg, respectively.

Furthermore, if we combine the effect of impact mass and velocity into the term of impact energy E and normalize it by a reference potential energy $E_n = 9.81$ J, then the normalized steady-state cracking speed, i.e., $\overline{v}_{PVB}/v_R$, can be fitted as

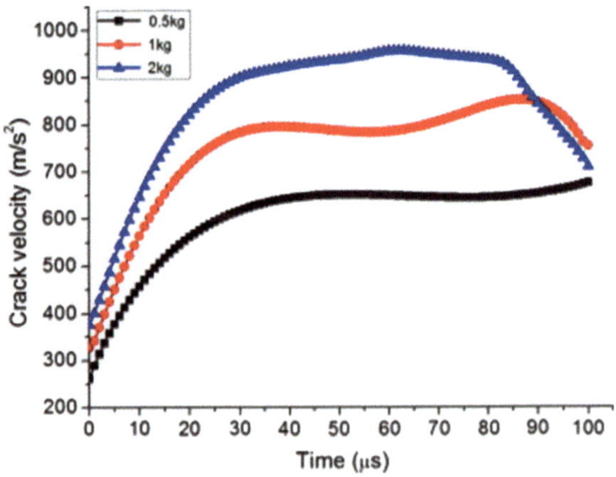

Fig. 4.10 Different crack tip velocity histories under different drop masses

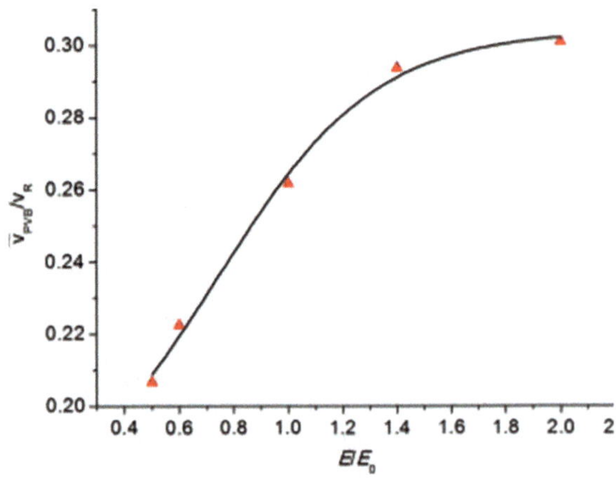

Fig. 4.11 Relationship between steady-state crack propagation velocity and external impact energy

$$\overline{v}_{PVB}/v_R = 0.167 \pm 0.137/(1 \pm 10\exp(1.50(0.739 - E/E_0)) \tag{4.1}$$

for the range of parameters in this study.

Figure 4.11 shows that the stable crack velocity is unlikely to increase indefinitely as the external impact energy increases, and it has a theoretical limit, i.e., Rayleigh wave speed v_R, but does not reach it.

Regarding the transition time between stages I and II, the general trend is that the smaller the external energy is, the faster the transition time is (although the difference is relatively small and still within the same order of magnitude). That is, if the impact energy is high, crack speed easily achieves the steady state. The detailed physical mechanism of such a variation in transition time will be explored in the future.

4.3 Another Experiment Investigating Radial and Circular Cracks

4.3.1 Experiment Condition

In another experiment to investigate radial and circular crack propagation in PVB laminated glass, a weight block with mass of 2 kg was evaluated to a certain height and then released to create impact loading. The height of the drop weight adopted in our work varied from 300 to 900 mm. An impactor was fixed at the tip of the force direction converter to control the contact shape and area. Different shapes may be designed for the impactor, and in this study, the impactor shape was designed to be cylindrical and the top was hemispherical. The diameter of the cylindrical impactor was 10 mm, and the total length of the impactor was 25 mm. The mass of the impactor (including the force direction converter) was 0.545 kg. In addition, a force sensor with sufficient accuracy and frequency was attached below the drop weight to record force–time history, and the signals recorded by the force sensor were sent to the oscillograph monitor to record the force–time history.

4.3.2 Specimen Preparation

The plate consisted of a PVB interlayer sandwiched by two brittle glass sheets. For the PVB interlayer, we set Young's modulus $E_p = 0.1$ GPa, short time shear modulus $G_0 = 0.33$ GPa, long time shear modulus $G_1 = 0.69$ MPa, bulk modulus $K = 20$ GPa, Poison's ratio $\nu_p = 0.49$, density $\rho_p = 1100$ kg/m^3, and decay factor $\beta = 12.6$ s^{-1}. For the glass, we set Young's modulus $E_g = 70$ GPa, Poison's ratio $\nu_g = 0.22$, and glass thickness = 2 mm. The thickness of the PVB interlayer varied from 0.38 to 3.04 mm during our parametric experimental study (a thickness of 0.76 mm is for common passenger cars).

The specimen was prepared using the same manufacturing process (with compression at 10 bar and 120 °C) as that used in automotive windshields. Glass is a typical linear elastic brittle material, and its facture strain is about 0.1%. Meanwhile, PVB is a rubber-like material that can sustain a much larger deformation during tension. The unique design allows for possible protection against impact with a small amount of energy and dissipates large impact energy through the large deformation of the interlayer.

The specimen was clamped within two metal cover sheets with thin layers of rubber pad inside to avoid possible scratches and stress concentration on the sample surface and to distribute the boundary force uniformly. The tightening torque for the screw bolt was set to 4 Nm, which properly mimics the clamping boundary condition. Six screws penetrating the metal cover sheets were fixed to ensure that the metal cover sheets are placed in the right horizontal position and to reduce the rebound under impact. A force direction conversion part (indicated in Fig. 4.1) was employed to convert the vertical impact force into a horizontal one, thus providing out-of-plane dynamic loading to the sample.

4.3.3 Experimental Results and Discussion

(1) Crack morphology

Two types of macroscopic crack patterns, namely, radial and circular, were observed, and all of the pieces of the glass plates after impact stayed connected to the PVB layer. Hoop stress was larger than radial stress, such that a radial crack appeared before a circular crack. Therefore, we had an opportunity to capture and study the evolution history of radial and circular cracks separately. Figure 4.12 shows a typical crack propagation process that contains a series of selected image sequence depicting radial and circular crack growth in the PVB laminated glass specimen. Time $t = 0\ \mu s$ was used to denote crack tip initiation for both radial and circular cracks. Crack propagation velocity v was determined as the crack length increase Δl over the time increase Δt. For radial cracks, at time t_i, multiple major crack tips conformed approximately a circle with radius r_i on the average, as shown in Fig. 4.12a. Therefore, radial crack propagation velocity $v_r = \Delta l/\Delta t = (r_{i+1} - r_i)/(t_{i+1} - t_i)$. For circular cracks, the total length of n main cracks at t_i was added and divided by the crack number as $\bar{l}_i$. Hence, circular crack propagation velocity $v_c = \Delta l/\Delta t = (\bar{l}_{i+1} - \bar{l}_i)/(t_{i+1} - t_i)$. In addition, the B-spline interpolation method was used to smooth the experimental points in this study.

(2) Crack propagation

A dynamic strain indicator was introduced to enhance the credibility of the experimental results. Locations near the force sensor and at the top of the impactor were fitted with strain gauges to monitor in situ strains (and stresses) transmitted during the impact process. Thus, we could estimate the delay of the impact force from the force sensor to the glass surface and depict the accurate force–time curve acting directly on the glass surface (dashed line in Fig. 4.13). Figure 4.13 shows typical crack propagation velocity time history curves for radial and circular crack propagation in response to 2 kg of weight from 500 mm of drop height (impact velocity: 3.13 m/s) to a specimen with a 0.76 mm PVB interlayer. Time $t = 0\ \mu s$ was employed to denote crack tip initiation for the radial crack. Therefore, the negative time corresponded to pre-initiation conditions (Fig. 4.13). The crack only existed on the glass sheets, such that the stress component normal to the crack front σ_{11} was the major driving force for crack initiation and propagation. Dynamic crack propagation velocity v may be expressed as

$$v^2 = \frac{\sigma_{11} + \Sigma_{11}}{\rho_g(1 + \varepsilon_{11})}, \tag{4.2}$$

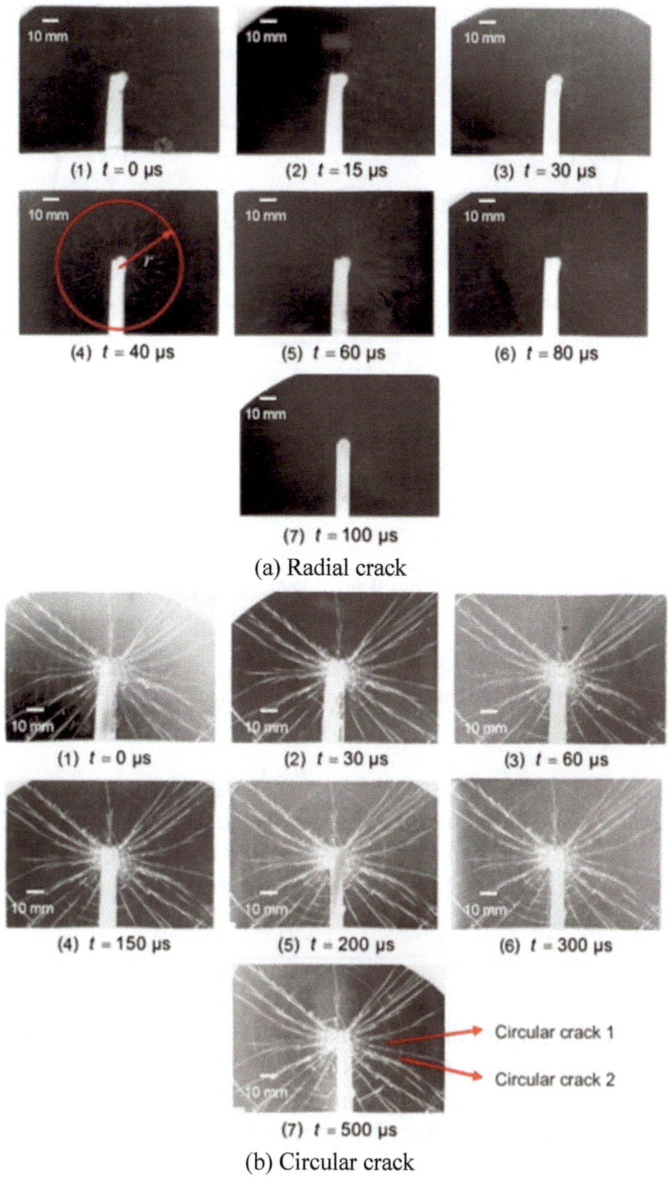

Fig. 4.12 Selected sequence of images depicting crack growth at impact velocity of 3.13 m/s with PVB thickness of 0.76 mm

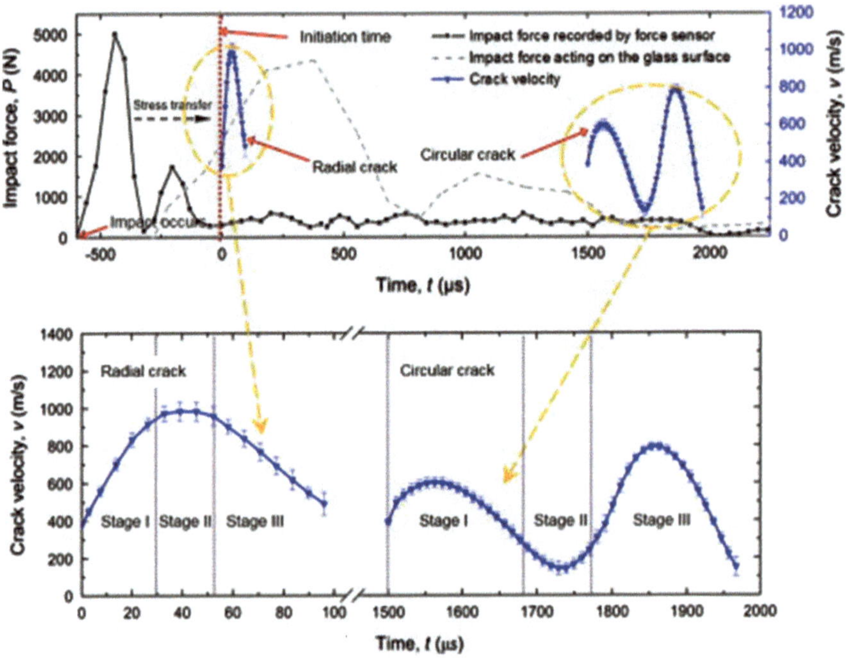

Fig. 4.13 Load–time curve and crack growth velocity history of the radial and circular cracks in response to 500 mm drop height (impact velocity: 3.13 m/s) and 0.76 mm PVB thickness

where the contact stress at the crack tip can be calculated based on an existing dynamic contact model by attaching a strain gauge near the impactor. Σ_{11} is the strain component. By supposing the proportionality coefficient D in σ_{11} within the framework of linear elasticity [24], the above equation may be rewritten as

$$v^2 = \frac{1 + D}{\rho_g(1/\sigma_{11} + 1/E_g)}. \tag{4.3}$$

In this impact loading condition, the maximum crack propagation velocity is 990 and 790 m/s for radial and circular cracks, respectively. Reference [10] indicated that σ_{11} is larger in hoop stress than in circumferential stress, so radial cracks propagate faster than circular cracks. Radial crack propagation velocity is lower than that in pure soda–lime glass [11], where no obvious circular crack occurs. The PVB interlayer acts as an energy-absorbing material to reduce the crack propagation velocity. The PVB interlayer also provides a connection medium for the cracked glass panel, such that a circular crack could continue to dissipate impact energy after the full growth of radial cracking.

Impact occurred on the outer surface of the glass plate at around −250 μs (Fig. 4.13). A compressive loading wave front traveled down at around 5300 m/s [25] in the glass layer. The PVB layer presented dual advantages of delaying the

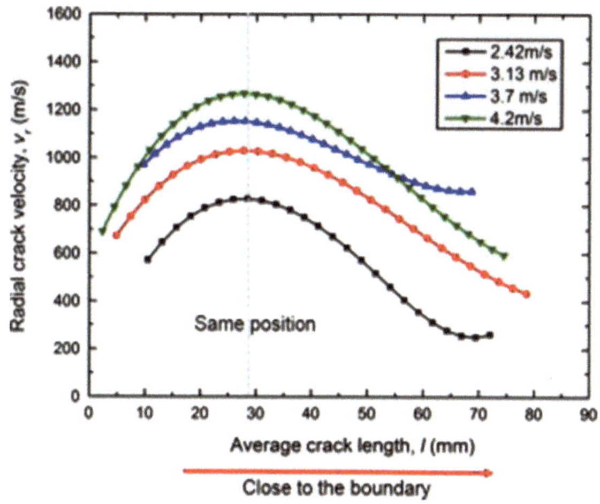

Fig. 4.14 Velocity–length curve of the radial crack at various impact velocities from 2.42 to 4.2 m/s with the PVB thickness of 0.76 mm

passage of the elastic wave into the backing glass layer and reducing the amplitude and wavelength of the wave that eventually passed through by internal friction caused by its viscoelasticity. These effects are beneficial in utilizing the energy absorbing capabilities of the PVB interlayer to the fullest.

The compressive waves arrived at the free surface of the backing layer and were reflected as tensile waves. Thus, the delay in crack initiation probably resulted from the transmission and reflection of the stress wave and the increasing process of SIF caused by increasing elastic energy stored near the crack tip. Similar phenomena were observed in Ref. [22]. More and more tensile waves continually arrived at the backing plate; thus, radial cracks initiated while reaching the limit of dynamic crack initiation toughness K_{IC}^{d}. For a rate-sensitive material, the PVB interlayer is rate dependent, whereas the fracture that occurs on the glass plate is rate independent. Considering that cracks grow in the glass layer, impact velocity exerts little effect on the K_{IC}^{d} value, which means the effect of rate sensitivity can be ignored. The propagation of the radial crack can be divided into three stages (Fig. 4.13). After initiation, the crack tip accelerated from about 400 to around 990 m/s within 30 μs. The impact force continued to increase after the initiation of the radial crack, which means the elastic energy stored near the crack tip kept increasing and caused acceleration sequences. Then, the crack velocity remained constant at about 990 m/s during the next 25 μs. However, it decelerated to less than 600 m/s in under 100 μs after initiation. The entry of the crack tip into the region of compressive stress as it approaches the boundaries and the arrival of less stress wave at the crack tip are two reasons for the deceleration phenomenon [22].

Further investigation showed that the reason for the deceleration and crack arrestment phenomenon is relevant to the boundaries. The history of radial crack velocity at various impact velocities from 2.42 to 4.2 m/s with the PVB thickness of 0.76 mm is plotted as a function of crack length in Fig. 4.14. Crack velocity started decreasing at the same position.

For the circular crack, crack initiation occurred 1750 μs after the impact on the glass surface (Fig. 4.13). Combined with the images depicting radial and circular crack growth, we believe that circular cracks initiated long after radial crack growth. Observation of the specimen at the end of the test showed that circular cracks merely existed on the loaded layer, according to which we can conclude that circular cracks are nuclear under the influence of surface acoustic waves (namely "Rayleigh waves"). The propagation of this type of elastic waves is limited to the loaded surface of the solids. Rayleigh waves spread at a much lower speed compared with longitudinal compression waves. Moreover, the lateral dimensions of the sample were much larger than those in the longitudinal direction. Therefore, the circular crack appeared later than the radial crack, which is confirmed by a previous finding [10]. The circular crack propagated under the tensile stress caused by the reflected tensile Rayleigh waves, which resulted in the situation in which circular cracks always emerged first close to the boundaries. The propagation of the circular cracks can be divided into three stages as well (Fig. 4.13). The first and third stages involve acceleration, maintenance, and deceleration, similar to that in radial cracks. Circular cracks have more space to grow than radial cracks, such that the total crack lengths are longer in circular cracks. Besides, circular cracks have the lower peak value. However, a special stage exists between the two stages, and this stage can be called "slowing-down stage." In this stage, the crack propagation velocity decelerates to almost 0. This is a direct consequence of the fracture at the glass interface caused by the radial crack. Circular crack propagation always delays at the relatively large fracture where the stress waves have to propagate through the interlayer, without which the circular crack cannot transfer effectively across the interface [26].

(3) Effect of impact velocity

Parametric experiments were conducted to investigate the effect of impact velocity v_d on radial and circular crack propagation characteristics. Figure 4.15 shows the crack propagation velocity time history of radial and circular cracks under various impact velocities from 2.42 to 4.2 m/s (drop height from 300 to 900 mm), and 2 kg of drop weight and 0.76 mm PVB thickness were employed. Three repeated experiments under each condition were carried out to ensure the accuracy of the results (error bars in Fig. 4.15). Time $t = 0$ μs was employed to denote crack tip initiation.

For the radial crack, in the case with high impact velocity, a short acceleration time is needed to reach the top propagation velocity, where the top value is also high. This result is consistent with the fact that at a high impact velocity with a large wave amplitude, the load applied to a specimen increases at a rapid rate; thus, crack acceleration can be large during the dynamic loading process.

Additional experiments also indicated that high impact velocity requires a short delay time to initiate (Fig. 4.16). However, the case with a high impact velocity would also decelerate rapidly and early due to the limited available cracking space on the panel plane. Similar to the situation in the radial crack, high impact velocity results in a short initiation time in the circular crack; the shorter the acceleration time is, the larger the peak value is.

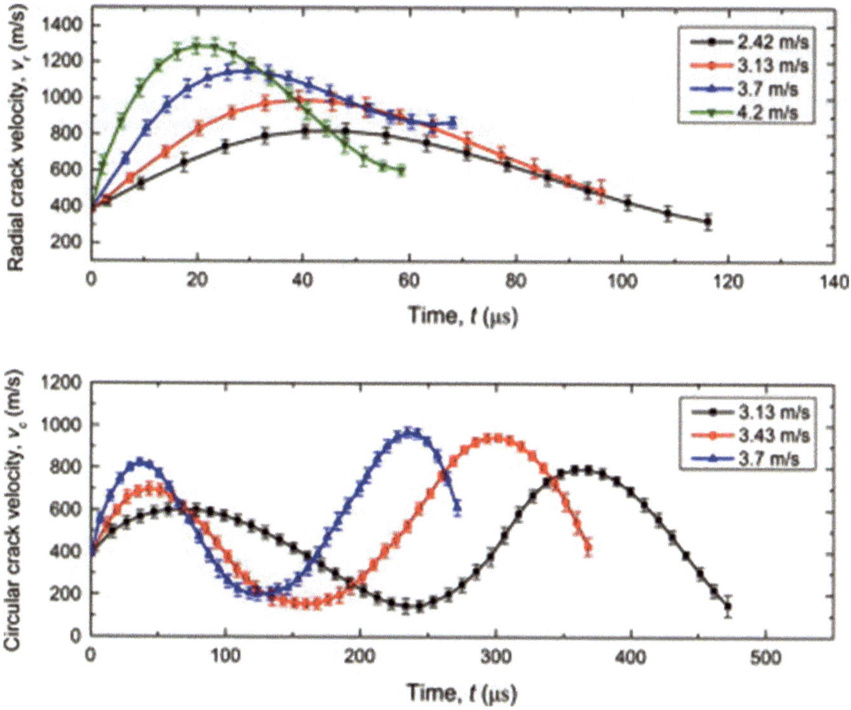

Fig. 4.15 Crack velocity history of radial and circular cracks at various impact velocities from 2.42 to 4.2 m/s with the PVB thickness of 0.76 mm

To establish the model that combines crack velocity with the impact conditions, we chose the variable "maximum velocity" to quantify our study; this variable depends on both the impact conditions and the material property. We can suppose that the maximum velocity $v_{\max}$ at the crack tip during the whole propagation process has an implicit relationship with the impact force, the duration of the loading time, and the dynamic initiation fracture toughness of the material in this investigation.

According to the principle of dimensionless unifying, the equation of $v_{\max}$ could have the following form:

$$v\max = \frac{AP_d^{2/3}}{(k_{IC}^d)^{2/3} t_d} \tag{4.4}$$

where p_d is the average contact force over the contact period, t_d is the duration of the loading, and A is a macroscopic crack pattern-dependent coefficient. According to the momentum theorem, we have

$$p_d t_d = m_d v_d \tag{4.5}$$

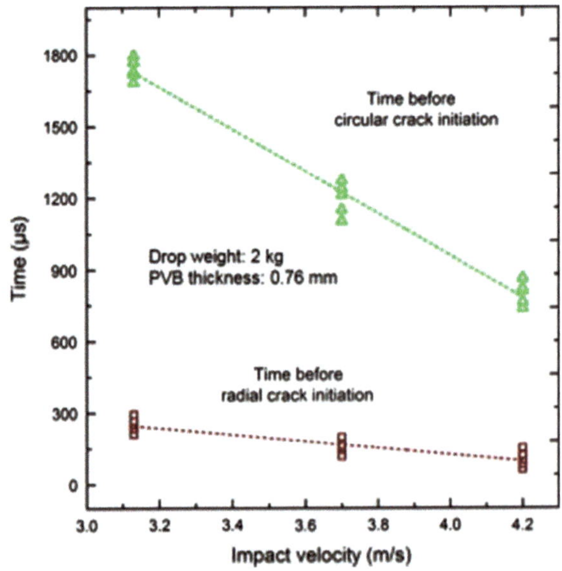

Fig. 4.16 Initiation time of radial and circular cracks at various impact velocities from 3.13 to 4.2 m/s with the PVB thickness of 0.76 mm

This result means that the maximum velocity at the crack tip can be expressed in terms of the impact velocity v_d and the drop weight m_d

$$v\text{max} = \frac{Am_d^{2/3} v_d^{2/3}}{(k_{IC}^d)^{2/3} t_d^{5/3}} \tag{4.6}$$

Thus, the maximum velocity contains two model parameters: the dynamic initiation fracture toughness k_{IC}^d, which can be geometry dependent, and the macroscopic crack pattern-dependent parameter $A > 0$. Both parameters are subject to experimental data through best fitting.

(4) Effect of PVB thickness

A series of parametric experiments is conducted to study the effect of PVB thickness t_p on the crack propagation characteristics. Three different PVB thicknesses, i.e., 0.38, 0.76, and 1.52 mm, are chosen for the parametric experiments. A drop weight of 2 kg is employed. The crack propagation velocity time history of radial and circular cracks under three PVB thicknesses are shown in Fig. 4.17. The impact velocity that is used to investigate the radial crack is 2.42 m/s (i.e., the drop height is set to 300 mm), while the impact velocity is 3.13 m/s (i.e., the drop height is set to 500 mm) for circular crack.

Martínez et al. [27] reported that the ability to transmit impact energy depends on the thickness and type of the adhesive used. Therefore, the variation in PVB thickness will have a great effect on delaying the passage of the elastic wave into the backing glass layer and on reducing the amplitude and the wavelength of wave

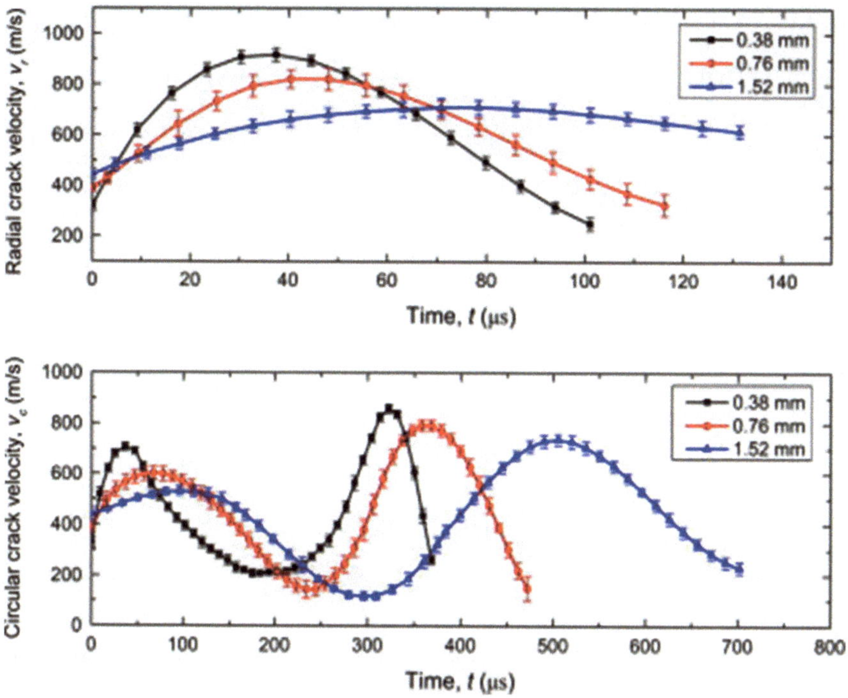

Fig. 4.17 Crack velocity history of the radial and circular crack with various PVB thicknesses from 0.38 to 1.52 mm at the impact velocity of 2.42 and 3.13 m/s, respectively

through internal friction [28], thereby reducing the maximum radial crack speed, as shown in Fig. 4.17.

Considering the wave speed in the through-thickness direction of around 5320 m/s in the glass layer and 300 m/s in the PVB interlayer (which means the PVB interlayer plays an important role in slowing down the stress wave), the transit time is around 3, 6, and 11 μs with a PVB thickness ranging from 0.76, 1.52, to 3.04 mm. Compared with the delay time in the order of 102 μs for radial crack initiation in Fig. 4.18, we can conclude that the delay of the radial crack initiation mainly results from the increasing process of the stress intensity factor. The high PVB thickness would reduce the amplitude of an elastic wave crossing the PVB interlayer by internal friction, thereby weakening the increasing process of the stress intensity factor. Therefore, in the case with a thick PVB layer, a long delay time is needed for the crack initiation. A similar phenomenon is observed for the circular crack caused by Rayleigh wave. Considering the Rayleigh surface wave speed in the in-plane direction of around 3370 m/s in the glass layer [29], the transit time is around 37 μs, which is much longer than the transit of the elastic wave in the medium. Similarly, the delay of the circular crack initiation mainly also results from the increasing process of the stress intensity factor. In addition, from the force–time curve, we can see that the

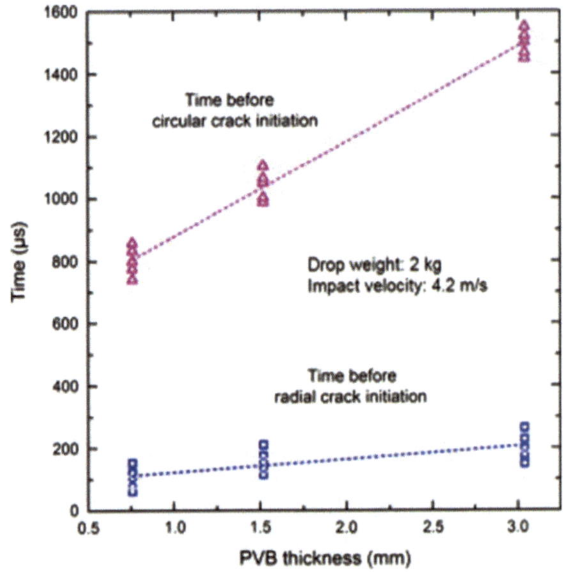

Fig. 4.18 Initiation time of the radial and circular crack with various PVB thicknesses from 0.76 to 3.04 mm at the impact velocity of 4.2 m/s

compressive waves stay around the loaded side for a relatively long time during the impact, which also affects the increasing process of the stress intensity factor.

The glass sheet with a thicker PVB layer starts at a higher initiation velocity v_0 for both crack patterns. The exact expression of v_0 is not clarified yet. Thus, we may implicitly represent the expression to the initiation velocity v_0 as $v_0(t_p)$. Given that $k_{IC}^d \propto \sigma_c \propto v_0$ (see Eq. 4.3), where σ_c is the fracture stress, we have the following expression for the initiation velocity [24]:

$$v_0^2(t_p) = c_R^2(1 + D)\left(1 - \left(1 + \frac{(k_{IC}^d)^2}{M}\right)^{-1}\right) \tag{4.7}$$

where M is a constant coefficient, and C_R is the Rayleigh wave velocity. In this case, we arrive at an expression for the dynamic initiation fracture toughness in Eq. (4.6) in the form of

$$k_{IC}^d = \sqrt{\frac{M v_0^2(t_p)}{C_R^2(1 + D) - v_0^2(t_p)}} \tag{4.8}$$

Therefore, we may obtain the expression for the maximum velocity at the crack tip as follows:

$$v\text{max} = \left[\frac{Bm_d^2 v_d^2 t_d^{-5}(c_R^2(1 + D) - v_0^2(t_p))}{v_0^2(t_p)}\right] \tag{4.9}$$

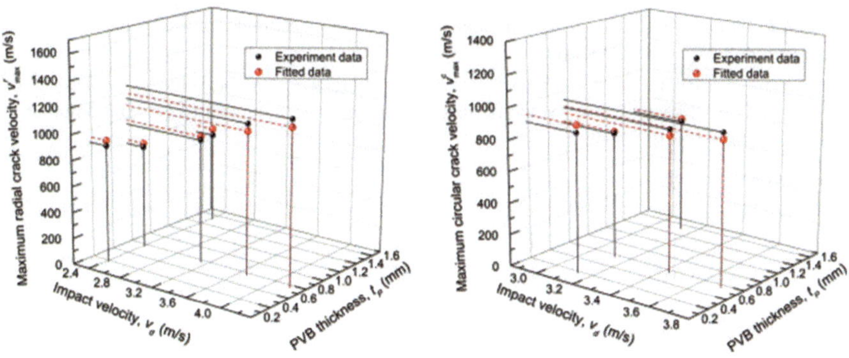

(a) radial crack (b) circular crack.

Fig. 4.19 Experiment data and fitted data of the maximum crack velocity from the semi-physical with various impact velocities from 2.42 to 4.2 m/s with PVB thickness from 0.38 to 1.52 mm

where $B > 0$ is another constant coefficient that depends on the macroscopic crack pattern. According to Eq. (4.9), a thick PVB layer corresponds to a low maximum velocity for both radial and circular crack, which is consistent with the experiment result (Fig. 4.17). This result is expected because the glass sheet with a thick PVB layer would absorb more impact energy via molecule vibration and friction by the interlayer as a polymer-like material, as verified by Ref. [28]. Given the Rayleigh wave velocity $C_R \approx 3370$ m/s [30], which is sufficiently high for the initiation velocity v_o, the expression may be simplified as

$$v\max = \left[\frac{Bm_d^2 v_d^2 t_d^{-5}(c_R^2(1+D))}{v_0^2(t_p)}\right]^{1/3} = c\left[\frac{m_d^2 v_d^2 t_d^{-5}}{v_0^2(t_p)}\right]^{1/3} \tag{4.10}$$

Agreement between theoretical and experimental results may be achieved with an appropriate choice of the values of the constant coefficient c (c_1 and c_2 for radial and circular crack, respectively). The fitted models for the maximum velocity of radial crack $v^r_{\max}$ and circular crack $v^c_{\max}$ are given in the following form:

$$\begin{cases} v^r_{\max} \approx c_1\left[\dfrac{m_d^2 v_d^2 t_d^{-5}}{v_0^2(t_p)}\right]^{1/3} = 0.0657\left[\dfrac{m_d^2 v_d^2 t_d^{-5}}{v_0^2(t_p)}\right]^{1/3} \\ v^c_{\max} \approx c_2\left[\dfrac{m_d^2 v_d^2 t_d^{-5}}{v_0^2(t_p)}\right]^{1/3} = 0.0545\left[\dfrac{m_d^2 v_d^2 t_d^{-5}}{v_0^2(t_p)}\right]^{1/3} \end{cases} \tag{4.11}$$

The fitted data and the experimental data are shown together in Fig. 4.19 (the experimental data in the figure are the average maximum crack velocity). The good agreement between the two types of data proves that the model is dependable.

Table 4.2 Parametric statistical experiments: linear regression estimates of the Weibull parameters under different experiment conditions

Crack type	PVB thickness (mm)	Impact velocity (m/s)	Mean crack number	Weibull char. number (N_0)	Weibull modulus (m)
Radial crack	0.38	2.42	87	96	4.4
	0.76	2.42	81	89	4.0
	1.52	2.42	70	79	3.1
	3.04	2.42	66	75	2.7
	0.76	3.13	90	99	4.4
	0.76	3.7	99	107	5.1
	0.76	4.2	118	128	6.0
Circular crack	0.38	2.42	2.5	2.7	2.8
	0.76	2.42	2	2.2	2.5
	1.52	2.42	1.7	2.0	2.2
	3.04	2.42	1.5	1.6	1.7
	0.76	3.13	2.7	3.1	2.7
	0.76	3.7	3.9	4.2	5.1
	0.76	4.2	7.0	7.4	9.3

(5) Statistical analysis of macroscopic cracking morphology

To confirm the prior crack propagation mechanism analysis and understand the crack propagation mechanism from another perspective, the relationship between the number of cracks and the impact velocity or the PVB thickness is established. All cracks are located in the surface of the glass sheet, and glass is a material with internal stochastic flaws that may serve as crack nuclei. Thus, a statistical model may be needed to describe the two macroscopic crack patterns. Table 4.2 shows the detailed combination of the statistical experiments. A total of 100 repeated experimental data for each specified condition were used for each experiment condition to obtain a statistical result. The two-parameter Weibull model is employed to study the distribution of both radial and circular crack number under each experiment condition. Weibull statistics are commonly used in the engineering community, especially in fracture statistics. In the specific case of Weibull statistics analysis in this study, the cumulative probability function is written as follows:

$$p_f = 1 - \exp\left(-\left(\frac{N}{N_0}\right)^m\right) \tag{4.12}$$

where p_f is the cumulative probability, N is the number of radial or circular crack (N_r or N_c), and m and N_0 are the distribution shape parameter and characteristic crack number, respectively. To improve the statistical accuracy, the number of circular cracks is counted with decimals during the statistical process (that is, the number of circular cracks with an incomplete circle is determined by the length ratio of an incomplete circle and a complete one).

For ease of information access, the double logarithm of the Weibull equation (4.14) is used in number analysis, thereby obtaining the Weibull parameters in a simple graphical representation of the data. Figure 4.20 shows the graphs with $X = \ln N$ plotted on the horizontal axis, and $Y = \ln\left[\ln\left(1/\left(1 - p_f\right)\right)\right]$ on the vertical axis. A simple linear regression analysis can be performed to obtain the Weibull parameters under different experiment conditions. The slope of the line is the Weibull modulus, m. The characteristic crack number, N_0 is the value of the crack number for which $\ln\left[\ln\left(1/\left(1 - p_f\right)\right)\right]$ is zero. In this study, linear regression analysis is used for the Weibull parameter estimates under each experiment condition (see Table 4.2).

$$\ln(1 - p_f) = -\left(\frac{N}{N_0}\right)^m \tag{4.13}$$

$$\ln\left[\ln\left(\frac{1}{1 - p_f}\right)\right] = m \ln N - m \ln N_0 \tag{4.14}$$

First, the effect of different impact velocities on the distribution of both radial and circular crack number is investigated. Figure 4.21a shows the probability density curves of radial and circular crack number under different impact velocities. The characteristic crack number indicates the distribution location along the X-axis and is expected to vary according to the impact velocity or the PVB thickness. The distributions for both radial and circular crack number are proved to move to the right (higher value) as impact velocity increases. This result is reasonable because a higher impact velocity produces more elastic energy. Thus, a large new surface needs to be created to release the excess energy. In addition, the probability density curves considering different PVB thicknesses to describe the radial and circular crack number are shown in Fig. 4.21b. Contrary to the effect of impact velocity, the distributions for both radial and circular crack number move to the left (lower value) as PVB thickness increases because a thicker PVB layer absorbs more elastic energy, as previously stated. This statistical method could also be applied to engineering applications and play an important role in laminated glass design.

(1) Relation between radial crack velocity and crack numbers on the backing glass layer

Here, the effect of the radial crack number on their own crack propagation velocity and the crack generation on the other glass layer is thoroughly investigated.

First, the radial crack velocity on the backing glass layer with different crack numbers is calculated. A theoretical model from the perspective of energy conversion is established to depict the cracking process of the backing glass layer and to elucidate the fundamental reason for the variation of the crack velocity.

The specimen in response to impact speed $V = 3.7$ m/s is chosen for 100 repeated experiments by considering the intrinsic stochastic flaws in glass laminated samples. A series of images that depict the crack growth on both glass plates is recorded at the set time intervals. A typical crack propagation process is shown in Fig. 4.22. The glass

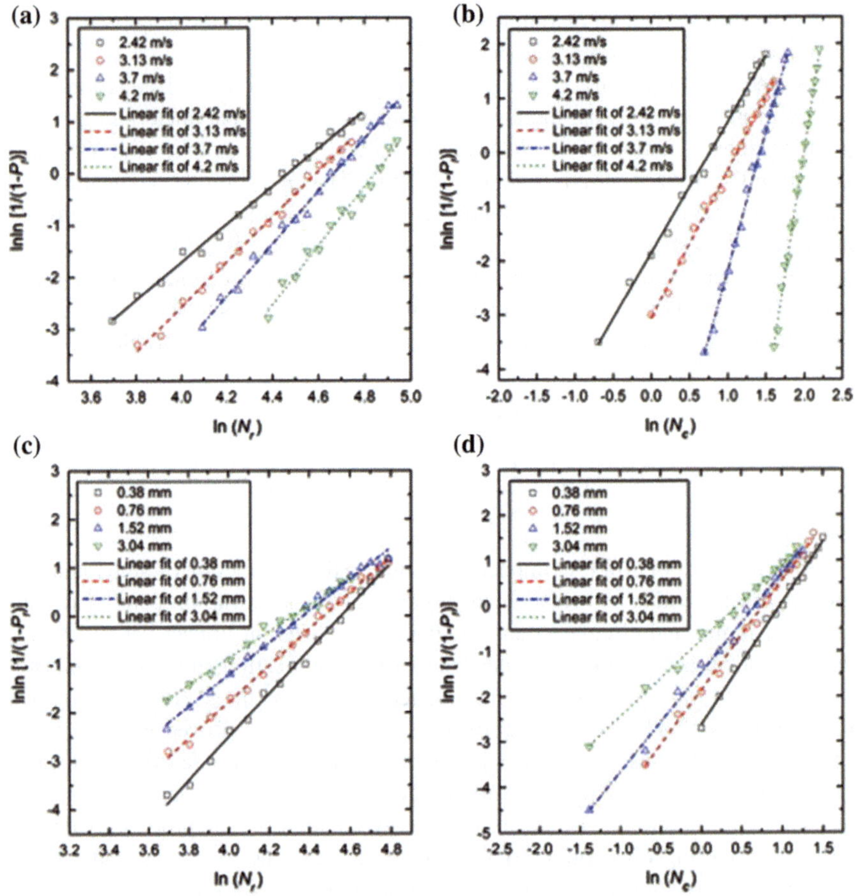

Fig. 4.20 Fitted Weibull model plots for macroscopic cracking morphology: **a** radial crack at various impact velocities from 2.42 to 4.2 m/s; **b** circular crack at various impact velocities from 2.42 to 4.2 m/s; **c** radial crack with different PVB thicknesses from 0.38 to 3.04 mm; and **d** circular crack with different PVB thicknesses from 0.38 to 3.04 mm

plate that directly contacts with the impactor is referred to as "impacted glass layer," while the glass plate on the other side is referred as "backing glass layer." $t = 0$ μs represents the time when impact is triggered. Figure 4.22a–d demonstrate the whole cracking process on the backing glass layer where cracks always initiate first. Only radial crack pattern is observed, while both radial and circular cracks appear on the impacted glass layer (under the effect of Rayleigh waves, whose propagation is limited to the loaded surface of the solids) long after the cracking of the backing glass layer (Fig. 4.22e–f). However, the radial crack number on the backing glass plate varies from 18 to 112 within 100 repeated experiments, which is a large deviation.

Furthermore, qualitative observation indicates that the number of radial cracks has a great effect on the later crack generation on the impacted glass layer and on

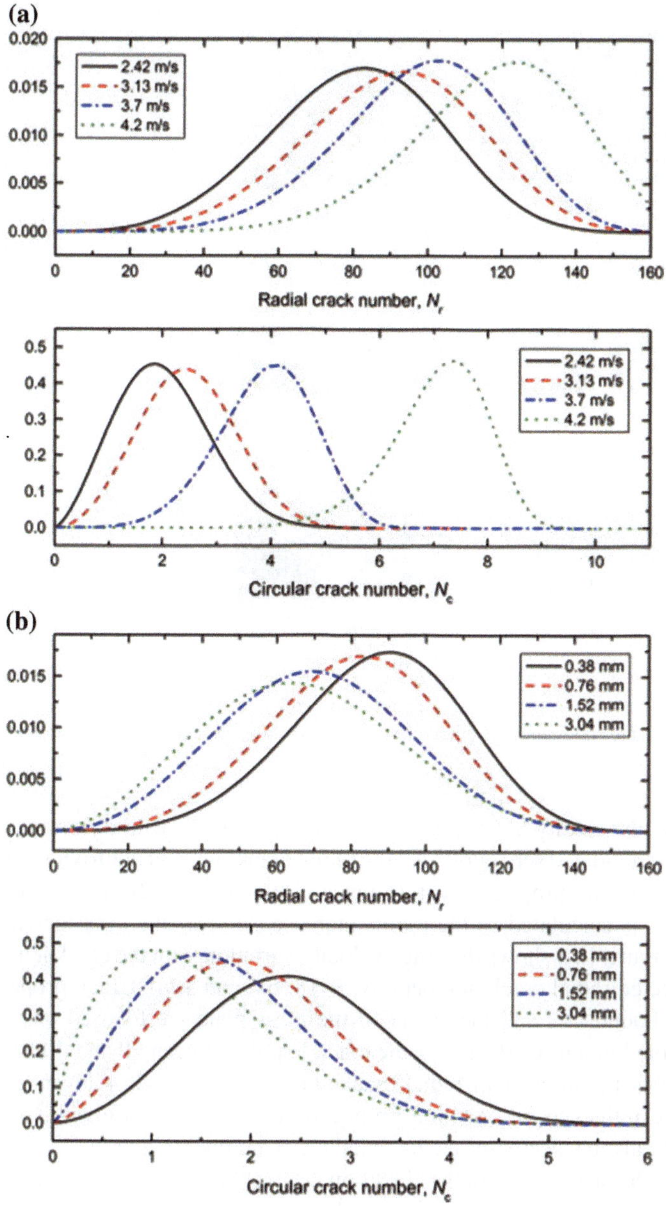

Fig. 4.21 Probability density curves of the radial and circular crack number under various boundary conditions

the propagation characteristics of radial crack on the backing glass layer itself. In the following section, the effect of the different radial crack numbers of backing glass layer will be discussed.

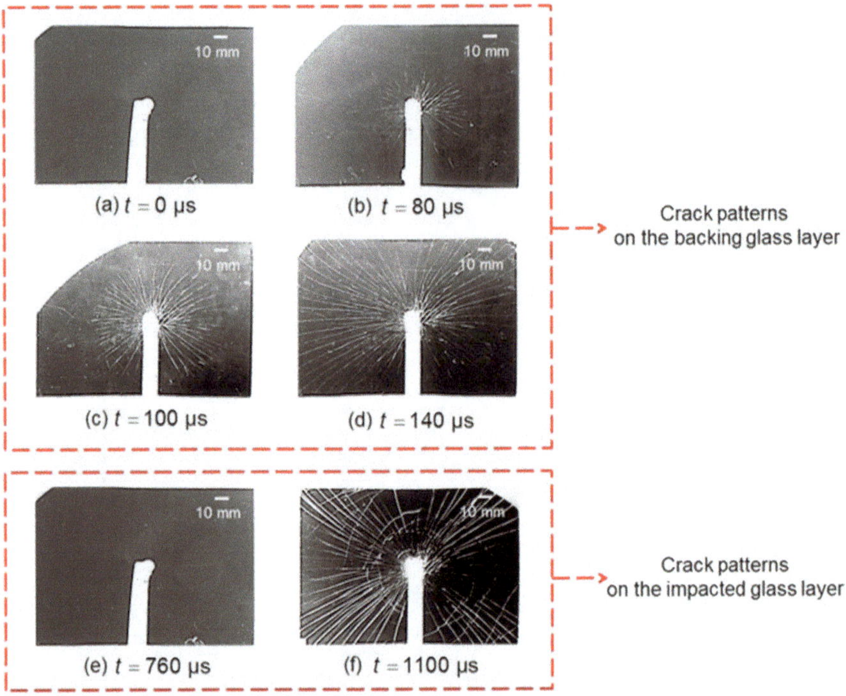

Fig. 4.22 Images depicting the crack growth on the two glass layers ($V = 3.7$ m/s)

The crack velocity history on the backing glass layer with different radial crack number is calculated separately. All the radial cracks present approximately the same length during the propagation, forming a circle (Fig. 4.22). Therefore, the crack velocity v_r is calculated as the radius increase Δr over the time increase Δt, i.e., $\Delta r/\Delta t$. Figure 4.23 shows the crack velocity variation with the crack length l_r under three selected radial crack numbers ($N_r = 18$, 64, and 108). Crack propagations far from the boundaries and the center of the glass plate are focused on to eliminate the possible boundary effect (i.e., the crack length between 20 and 40 mm). Results indicate that the number of radial cracks on the backing glass layer regularly has a negative influence on cracking propagation velocity themselves, i.e., the crack velocity v_r always decreases as the radial crack number increases.

During the propagation of the cracks on the backing glass layer, the impacted glass layer remains intact, as previously stated (Fig. 4.23). However, due to the adhesion by the PVB interlayer, the two glass layers have the same radius of deformed region r_f (Fig. 4.23). The bending energy of the impacted glass plate caused by the impact can be estimated by assuming that the backing glass layer is still intact, in which case the two glass plates can be regarded as "single one with $2h_g$ thickness." Therefore, the bending energy of the impacted glass layer is equal to half the energy in the single glass plate with $2h_g$ thickness. After the contact between the impactor and

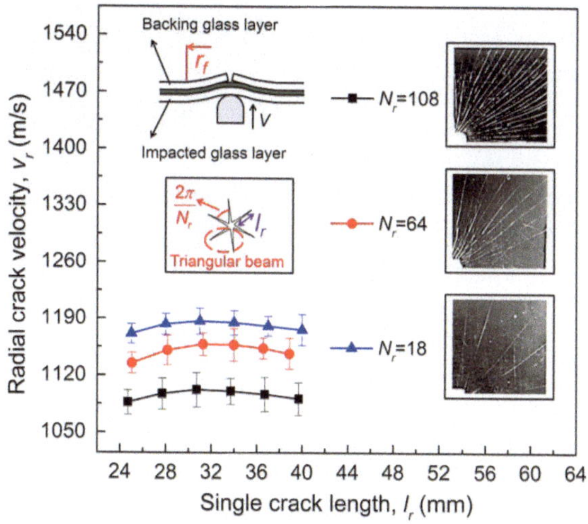

Fig. 4.23 Radial crack velocity versus crack length with different radial crack number ($V = 3.7$ m/s)

the glass plate, kinetic and bending energies balance $\rho_g \Omega V^2 \sim E_g (2h_g \kappa)^2 \Omega$ [31], where $\Omega = 2h_g \pi r_f^2$ is the volume of the deformed region and V is the impact speed, whose variation can be ignored during such a short period. Thus, the curvature $\kappa \sim w_0/r_f^2 \sim V/2h_g c_g$, yielding $r_f \sim (2h_g c_g t)^{1/2}$, where $w_0 = V_t$ is the indentation, and $c_g = \sqrt{E_g/\rho_g}$ is the sound speed in the glass material. The bending energy of impacted glass plate U_{b1} can be expressed as

$$U_{b1} = k_1 \pi E_g \left(\frac{V}{c_g} \right) r_f^2 h_g \tag{4.15}$$

where k_1 is a coefficient.

Considering the low Young's modulus of PVB material (i.e., $E_P = 0.1$ GPa), the bending energy of PVB U_{b3} is approximately 0.001 of that in the glass plate, which has been proved by FE simulation with a three-dimensional laminated plate FE model using a 1 mm × 1 mm surface size element. The boundary condition of the glass plate model is set to be clamped, which is consistent with the real experimental condition. Thus, U_{b3} is small enough to be ignored, unlike U_{b1}.

For the backing glass layer in the cracking process with crack length l_r, the number of radial cracks N_r on the backing layer is set and would not change as the cracks extend. Thus, the bending energy of the backing glass plate U_{b2} can be estimated by the energy of N_r triangular beams of length l_r without considering transverse bending [31]

$$U_{b2} = E_g h_g^3 w_0^2 N_r \tan\left(\frac{\pi}{N_r}\right)/3l_r^2$$
$$\approx k_2 \pi E_g h_g r_f^2 \left(\frac{V}{c_g}\right)^2 \left(\frac{1}{2} + \frac{\pi^2}{6N_r^2}\right) \tag{4.16}$$

where k_2 is another coefficient. Therefore, the bending energy of the whole laminated plates U_b can be estimated as

$$U_b = (U_{b1} + U_{b2})$$
$$\approx \pi E_g h_g r_f^2 \left(\frac{V}{c_g}\right)^2 \left[k_1 + k_2\left(\frac{1}{2} + \frac{\pi^2}{6N_r^2}\right)\right] \tag{4.17}$$

From the perspective of energy transfer, during the cracking process, the elastic energy of the laminated plates (i.e., the bending energy) caused by the impact is converted into glass surface energy through the fracture behavior. Here, we define the increasing rate of the plates elastic energy as "η_b" and the energy release rate as "η_c." Therefore, we have

$$\eta_b = \frac{dU_b}{dt} \approx \pi E_g h_g^2 \frac{V^2}{c_g} \left[k_1' + k_2'\left(\frac{1}{2} + \frac{\pi^2}{6N_r^2}\right)\right] \tag{4.18}$$

$$\eta_c = \frac{d(2N_r \Gamma h_g l_r)}{dt} = 2N_r \Gamma h_g v_r \tag{4.19}$$

where k_1' and k_2' are coefficients. Γ is the glass material fracture (surface) energy. A balance should exist between the input energy (i.e., the increasing elastic energy) and the output energy (i.e., the fracture energy). Suppose that the proportion between the two parameters η_b/η_c, which is called the energy conversion factor, indicates the stability of the system or the ability for further crack growth. Thus, a large proportion would obviously lead to increased instability for the whole system. Thus, for the system of PVB laminated plates in our study, the energy conversion factor η_b/η_c can be expressed as

$$\frac{\eta_b}{\eta_c} = \frac{\pi E_g h_g}{2\Gamma c_g} \cdot V^2 \cdot \frac{1}{v_r} \left\{\frac{1}{N_r}\left[k_1' + k_2'\left(\frac{1}{2} + \frac{\pi^2}{6N_r^2}\right)\right]\right\} \tag{4.20}$$

We can clearly see that the expression of the energy conversion factor contains three parts: (1) $\pi E_g h_g/2\Gamma c_g$ is determined by the material properties (which refers to the glass material in this instance); (2) V^2 refers to the impact condition; (3) the rest of the expression $f(v_r, N_r)$ refers to the impact responses of the laminates. Thus, the stability of the system is determined together by the material properties, impact condition, and the impact response of the material according to Eq. (4.20). Previous studies [32, 33] that investigated the effect of the impact speed on the crack velocity have demonstrated that, with a higher impact speed V that increases the system instability η_b/η_c, both crack velocity v_r and radial crack number N_r tend

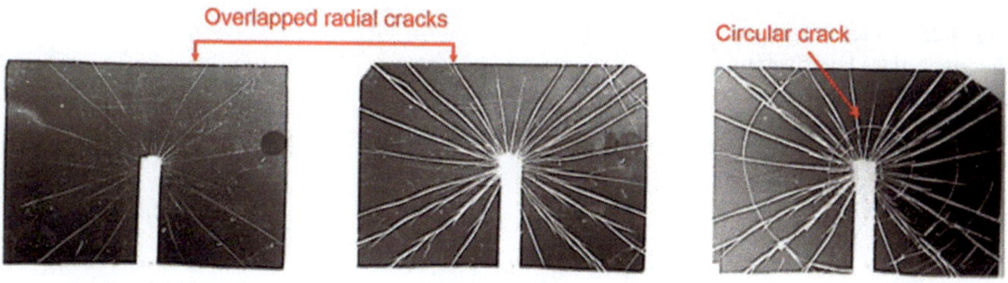

(a) $t = 200$ μs

Final crack pattern on the backing glass layer

(b) $t = 800$ μs

Middle crack pattern on the impacted glass layer

(c) $t = 1200$ μs

Final crack pattern on the impacted glass layer

Fig. 4.24 Selected sequence of images depicting the crack patterns on both glass layers at the loading speed $V_d = 3.7$ m/s, with PVB thickness $h = 0.76$ mm

to increase as well, thereby lowering the value of η_b/η_c to maintain the balanced energy input and output to a large extent. Therefore, because the crack number N_r on the backing glass layer decreases under the same impact condition, the value of η_b/η_c would correspondingly increase due to the result of Eq. (4.20), which means the instability of the system is rising. Thus, the system tends to response in a way to reduce this instability (i.e., the value of η_b/η_c) by propagating the cracks at a higher speed, thereby improving the value of v_r in Eq. (4.20), which is consistent with our experiment results (Fig. 4.23).

(2) Crack morphology on the impacted glass layer

The total crack length for the two crack patterns on the impacted glass layer is calculated to study the energy absorption ability of laminated glass influenced by the radial crack number on the backing glass layer.

Figure 4.24 shows the crack morphology on the impacted glass layer, where both crack patterns (i.e., radial and circular cracks) appear long after the cracking of the backing glass plate. The radial cracks on the backing layer overlap completely with the radial cracks on the impacted layer that are indirectly caused by the stress concentration brought by each generated crack on the backing glass layer as an initial flaw [34] (which is also proved by our experimental investigation) (Fig. 4.24). Thus, the morphology and the number of radial cracks in the impacted glass layer depend completely on those of the backing layer. The radial cracks on the impacted glass initiate from those on the backing glass plate through the depth direction [34].

A specimen in response to lower impact speed $V = 2.42$ m/s is chosen for 100 repeated experiments. As expected, the radial crack number on the backing glass plate varies from 15 to 65 in these experiments. The total crack length for both two patterns on the impacted glass layer is calculated with different crack numbers on the backing glass layer (Fig. 4.25). L_r and L_c are defined as the total radial and circular crack length on the impacted glass layer, respectively. Thus, $(L_r + L_c)$ refers to the total crack length on the impacted glass plate. The radial crack length linearly

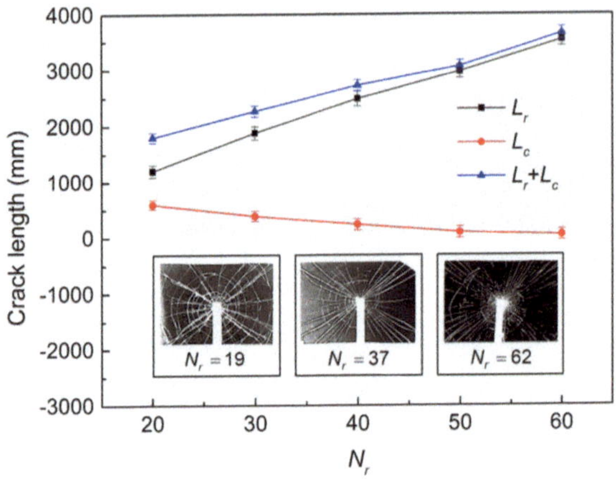

Fig. 4.25 Crack length on the inner glass layer versus radial crack number ($V = 2.42$ m/s)

increases with crack number on the backing glass layer due to the uniform growth of all cracks while the variation trend of circular crack length decreases (Fig. 4.25).

Circular cracks always initiate long after the propagation of radial cracks on the impacted glass has finished [10] (Fig. 4.24). Thus, before the initiation of circular cracks, the laminated plates can be regarded as N_r triangular beams with thickness $2h_g$ and length l_0 (which can be estimated as the side length of the specimen) because the radial cracks on the two glass plates are completely overlapped (Fig. 4.24). The bending energy U_{b0} of the plates is

$$U_{b0} = 4\pi E_g h_g^3 w_0^2 \left(1 + \pi^2/3N_r^2\right)/\left(3l_0^2\right) \tag{4.21}$$

In the case with a lower radial crack number N_r, the bending energy would increase, thereby strengthening the stress concentration in the radial direction, which facilitates the later generation of circular cracks. In addition, taking the law of energy conservation into account, the total fracture energy on the whole laminated plates can be estimated as $2\Gamma L h_g$, where L is the total crack length on the two glass plates. Thus the fracture energy released on the two glass plates can be estimated by calculating the crack length on each plate. For the plates with a shorter radial crack length, leaving more residual energy will also benefit the generation of circular cracks.

Although the increase of the circular crack number contributes positively to the energy absorption of the laminated plates, the effect is much less than that of the radial cracks (Fig. 4.25). The radial crack morphology on the impacted glass is directly determined by that on the backing glass layer. Thus, we can safely conclude that the backing glass plate is the key that decides the fracture morphology and also the energy absorption ability of the whole laminated structure such that the material selection and structure design for the backing layer are critical for impact protection.

4.4 Summary

In this chapter, we discuss the crack initiation and propagation in laminated glass subjected to foreign impact through experimental analysis.

First, by using high-speed photography, we conduct preliminary impact fracture experiments to investigate the crack propagation behavior of PVB laminated glass when it is subjected to lightweight impact. The time histories of the averaged radial crack tip position, propagation velocity, and acceleration are obtained. Our findings indicate that the steady-state cracking speed of PVB laminated glass is lower than that of pure glass and also increases with higher impactor speed and mass. Moreover, the supported glass layer would always initiate before the loaded layer, and the final morphologies of radial cracks on both sides completely overlap even if they propagate at different times. Two different mechanisms of crack propagation on different glass layers are explained. The cracks on a supported glass layer are predominantly motivated by in-plane crack-tip stress such that a relatively longer time is needed for its propagation. For the cracks that appear much later on a loaded layer and are caused by stress concentrated in the depth direction for each generated crack on the supported layer, the required propagation time is largely reduced.

Another experiment is conducted to thoroughly investigate the radial and circular crack propagation of PVB laminated glass. A parametric study on two dominant factors, i.e., impact velocity and PVB thickness, is conducted, with focus on the crack propagation of PVB laminated glass. Results show that the cracking speeds of radial and circular cracks increase with higher impact velocity. Radial crack appears earlier than the circular one, and it propagates at higher speeds. On the basis of the mechanism of crack initiation and propagation, a semi-physical model that describes the relationship between the maximum cracking velocity and influential factors is established. In addition, the Weibull statistical model is suggested, considering various factors to describe the macroscopic crack pattern, thereby providing theoretical evidence for engineering practice.

Finally, an investigation is conducted into the relation between radial crack velocity and crack numbers on the backing glass layer and the relation between the crack length and the energy absorption ability on the impacted glass layer. Results show that the radial crack velocity on the backing glass layer decreases with the crack number under the same impact conditions during numerous repeated experiments. Thus, the energy conversion factor is suggested to elucidate the physical relation between the cracking number and the crack propagation speed. However, the fracture energy of the whole laminated plates is mainly determined by the radial crack length, which means that the radial crack number generated first on the backing glass plate is the key that decides the energy absorption ability of the whole laminated structure.

References

1. J. Xu, Y. Li, B. Liu, M. Zhu, D. Ge, Experimental study on mechanical behavior of PVB laminated glass under quasi-static and dynamic loadings. Compos. B Eng. **42**, 302–308 (2011)
2. P. Forquin, F. Hild, A probabilistic damage model of the dynamic fragmentation process in brittle materials. Adv. Appl. Mech. **44**, 1–72 (2010)
3. S.M. Zhao, L.R. Dharani, L. Chai, S.D. Barbat, Analysis of damage in laminated automotive glazing subjected to simulated head impact. Eng. Fail. Anal. **13**, 582–597 (2006)
4. X. Sun, M.A. Khaleel, X. Sun, M.A. Khaleel, Effects of different design parameters on stone-impact resistance of automotive windshields. Proc. Inst. Mech. Eng. Part D J. Automobile Eng. **219**, 1059–1067 (2005)
5. S. Zhao, L.R. Dharani, L. Chai, S.D. Barbat, Dynamic response of laminated automotive glazing impacted by spherical featureless headform. Int. J. Crashworthiness **11**, 105–114 (2006)
6. S. Zhao, L.R. Dharani, X. Liang, L. Chai, S.D. Barbat, Crack initiation in laminated automotive glazing subjected to simulated head impact. Int. J. Crashworthiness **10**, 229–236 (2005)
7. P.A. Du Bois, S. Kolling, W. Fassnacht, Modelling of safety glass for crash simulation. Comput. Mater. Sci. **28**, 675–683 (2003)
8. M. Timmel, S. Kolling, P. Osterrieder, P.A. Du Bois, A finite element model for impact simulation with laminated glass. Int. J. Impact Eng. **34**, 1465–1478 (2007)
9. X. Sun, W. Liu, W. Chen, D. Templeton, Modeling and characterization of dynamic failure of borosilicate glass under compression/shear loading ☆. Int. J. Impact Eng **36**, 226–234 (2009)
10. J. Xu, Y. Li, X. Chen, Y. Yan, D. Ge, M. Zhu, B. Liu, Characteristics of windshield cracking upon low-speed impact: numerical simulation based on the extended finite element method. Comput. Mater. Sci. **48**, 582–588 (2010)
11. J.H. Nielsen, J.F. Olesen, H. Stang, The fracture process of tempered soda-lime-silica glass. Exp. Mech. **49**, 855 (2009)
12. E. Sharon, J. Fineberg, Confirming the continuum theory of dynamic brittle fracture for fast cracks. Nature **397**, 333–335 (1999)
13. J. Xu, Y. Li, Model of vehicle velocity calculation in vehicle-pedestrian accident based on deflection of windshield. J. Mech. Eng. **45**, 210–215 (2009)
14. J.H. Song, H. Wang, T. Belytschko, A comparative study on finite element methods for dynamic fracture. Comput. Mech. **42**, 239–250 (2008)
15. D. Grégoire, H. Maigre, A. Combescure, New experimental and numerical techniques to study the arrest and the restart of a crack under impact in transparent materials. Int. J. Solids Struct. **46**, 3480–3491 (2009)
16. N.P. Daphalapurkar, H. Lu, D. Coker, R. Komanduri, Simulation of dynamic crack growth using the generalized interpolation material point (GIMP) method. Int. J. Fract. **143**, 79–102 (2007)
17. S.K. Dwivedi, H.D. Espinosa, Modeling dynamic crack propagation in fiber reinforced composites including frictional effects ☆. Mech. Mater. **35**, 481–509 (2003)
18. K. Ravi-Chandar, W.G. Knauss, An experimental investigation into dynamic fracture: IV. On the interaction of stress waves with propagating cracks. Int. J. Fract. **26**, 189–200 (1984)
19. R.P. Singh, V. Parameswaran, An experimental investigation of dynamic crack propagation in a brittle material reinforced with a ductile layer. Opt. Lasers Eng. **40**, 289–306 (2003)
20. B.E. Clements, J.N. Johnson, R.S. Hixson, Stress waves in composite materials. Phys. Rev. E: Stat. Phys. Plasmas Fluids **54**, 6876–6888 (1996)
21. A. Thom, Experimental and modeling studies of stress wave propagation in multilayer composite materials: low modulus interlayer effects. J. Compos. Mater. **39**, 981–1005 (2005)
22. J. Lambros, A.J. Rosakis, Dynamic crack initiation and growth in thick unidirectional graphite/epoxy plates. Compos. Sci. Technol. **57**, 55–65 (1997)
23. X.X. Zhang, R.C. Yu, G. Ruiz, M. Tarifa, M.A. Camara, Effect of loading rate on crack velocities in HSC. Int. J. Impact Eng. **37**, 359–370 (2010)
24. A. Berezovski, G.A. Maugin, On the propagation velocity of a straight brittle crack. Int. J. Fract. **143**, 135–142 (2007)

25. S. Zhao, L.R. Dharani, L. Chai, S.D. Barbat, Dynamic response of laminated automotive glazing impacted by spherical featureless headform. Int. J. Crashworthiness **11**, 105–114 (2006b)
26. H. Park, W. Chen, Experimental investigation on dynamic crack propagation through interface in glass (2011)
27. M.A. Martínez, I.S. Chocron, J. Rodríguez, V.S. Gálvez, L.A. Sastre, Confined compression of elastic adhesives at high rates of strain. Int. J. Adhes. Adhes. **18**, 375–383 (1998)
28. J. Xu, Y.B. Li, X. Chen, D.Y. Ge, B.H. Liu, M.Y. Zhu, T.H. Park, Automotive windshield—-pedestrian head impact: energy absorption capability of interlayer material. Int. J. Automot. Technol. **12**, 687–695 (2011)
29. E. Sharon, G. Cohen, J. Fineberg, Crack front waves and the dynamics of a rapidly moving crack. Phys. Rev. Lett. **88**, 47–103 (2002)
30. J.A. Hauch, M.P. Marder, Energy balance in dynamic fracture. Investigated by a potential drop technique. Int. J. Fract. **90**, 133–151 (2010)
31. N. Vandenberghe, R. Vermorel, E. Villermaux, Star-shaped crack pattern of broken windows. Phys. Rev. Lett. **110**, 285–291 (2013)
32. J. Xu, Y. Sun, B. Liu, M. Zhu, X. Yao, Y. Yan, Y. Li, X. Chen, Experimental and macroscopic investigation of dynamic crack patterns in PVB laminated glass sheets subject to light-weight impact. Eng. Fail. Anal. **18**, 1605–1612 (2011)
33. J. Chen, J. Xu, X. Yao, B. Liu, X. Xu, Y. Zhang, Y. Li, Experimental investigation on the radial and circular crack propagation of PVB laminated glass subject to dynamic out-of-plane loading. Eng. Fract. Mech. **s 112–113**, 26–40 (2013)
34. J.W. Lee, I.K. Lloyd, H. Chai, Y.G. Jung, B.R. Lawn, Arrest, deflection, penetration and reinitiation of cracks in brittle layers across adhesive interlayers. Acta Mater. **55**, 5859–5866 (2007)

Chapter 5
Theoretical Analysis of Laminated Glass

Experimental investigations and theoretical analysis of laminated glass in different physical models were presented in the previous section. In this section, we introduce the different mathematical models of laminated glass beam or plate in detail based on numerous investigation results to ensure further understanding.

5.1 Theoretical Research on Laminated Glass Beam Model

5.1.1 Laminated Glass Beam

Laminated glass has a nearly plate-like geometry. Thus, most researchers focused on the mechanical model of laminated glass plate. However, the mechanical model of a beam is relatively simpler than that of a plate. As early as the 1970s, researchers have investigated the PVB laminated glass beam [1, 2]. In these studies, the boundary conditions of laminated glass beam were simply supported, and the displacement and load of the beam had a linear relationship.

According to Behr et al. [3], given that the magnitude of the shear modulus of PVB membrane is lower than that of glass, the PVB laminated glass beam can be regarded as a layered beam without the shear effect, and the thickness of PVB can be neglected.

Then, Behr et al. [4] investigated the structural behavior of architectural laminated glass beam experimentally and theoretically. Their findings indicate that architectural laminated glass behaves in a manner similar to that of monolithic glass under lateral pressures at room temperature or below. Similarly, it behaves like layered glass at a temperature higher than the transition temperature of PVB interlayer. Experimental results on the thickness of PVB and shear effect influence the layered beam under room temperature at least. Thus, Vallabhan [5] derived a mechanical model of bend-

J. Xu and Y. Li, *Impact Behavior and Pedestrian Protection of Automotive Laminated Windshield*, https://doi.org/10.1007/978-981-13-2441-3_5

ing load–displacement through the variation of the rotation of shear strain in x,y plane and in-plane displacement, which is based on minimum potential energy theory and also takes the membrane and bending stress of PVB into account.

Norville et al. [6] proposed a mathematical model of laminated glass. This model is a simple multilayer beam model for laminated glass and is established by assuming that the PVB membrane transfers the shear force in the two layers and maintains the space between that in which an identified coefficient governs the shear force. The model was modified by calibrating the effective material coefficient, thereby obtaining model calculation results that were close to the experimental results.

Calculating the strength of the laminated glass is difficult, and modeling laminated glass is also complicated because of the existence of the PVB interlayer. Therefore, the mathematical model will also be complicated and will need to be solved by using a numerical method with numerous solution times [7]. Thus, Ivanov et al. [7] derived a simple and reasonable analytical model laminated glass beam for structure optimization of laminated glass. This model shows that laminated glass could be superior to monolithic glass.

Moreover, Aşık and Tezcan [8] developed a mathematical model of laminated beams, which also employed the minimum total potential energy principle with the assumption of a large deflection for a laminated glass beam. These models were validated through experiments and FE models, thereby showing that the behavior of laminated glass beam varied linearly or nonlinearly through boundary conditions or constraints.

Recently, given the extensive use of laminated curved glass beam, Aşık et al. [9] derived three nonlinear coupled partial differential equations to investigate the behavior of laminated curved glass beam by using large deflection theory to predict the true behavior of laminated curved glass.

Heterogeneity causes laminated glass units to exhibit a complex mechanical response [10]. Thus, Zemanová et al. [10] proposed an FE approach to the laminated glass beam to study its immediate response with the use of the Newton method with consistent linearization. This method is more accurate than two-dimensional large-strain FE simulation.

5.1.2 Laminated Glass Plate

In essence, due to the different structures and assumptions of beam and plate, differences exist between the theoretical solutions of stress and displacement of beam and plate. However, few studies have conducted theoretical analysis of PVB laminated glass plate.

Behr et al. [3] investigated the behavior of laminated glass units (called "laminated glass plates") under lateral pressure and compared it with that of the monolithic glass plate and the layered glass unit (only two glass plates without the interlayer). Their experimental results indicate that the laminated glass unit behaves similar to the monolithic glass plate at room temperature. However, it behaves similar to the layered glass unit under high temperature, thereby indicating that the behavior of laminated glass unit lies between the two cases.

Vallabhan [5, 11] established the force and displacement model of PVB layered plate based on classical plate-shell theory. The convergence solution was solved by using the finite difference solution and cyclic method. Aşik and Vallabhan [12] discussed the solution method of von Karman equations by using the lateral displacement and Airy stress function.

Mohareb et al. [13] conducted nonlinear analysis of laminated glass plates subjected to lateral pressure by using the minimum potential energy theorem to acquire the field equation (five differential equations) and boundary conditions, in which the plane section is assumed to remain a plane for each individual layer for the laminated units. Their five equations for governing the mathematical model of laminated glass use the displacement approach, and the model solutions are compared with the experimental results. Their findings indicate that the shear modulus of the interlayer is nonlinear and affects the behavior of the laminated glass units; for example, the maximum deflection decreases as the interlayer shear modulus increases under a known load.

Similarly, Vallabhan et al. [14] developed five differential equations using the minimization of the total potential energy to model the behavior of laminated glass units (or laminated glass plates). Their work employed nonlinear theory because conventional linear theory will lead to erroneous results due to the larger displacements of the laminated glass plates. These equations are solved numerically with iteration, which agree well with the experimental results. Moreover, the maximum displacement of laminated glass units is between that of the layered and monolithic glass plates.

Aşık [15] proposed a variational method to derive the laminated glass plate model under larger deformation, in which the points on the plate are discretized and then the corresponding dynamic differential equation of each point is derived. They developed a nonlinear behavior mathematical model of laminated glass plates. This model is represented by five coupled nonlinear partial differential equations for obtaining the lateral and in-plane displacements. Model simulation results indicate that the location of maximum stress varied in the plates; it began at the center and then moved diagonally across the unit following the two perpendicular symmetry axes until it reached the corner.

The abovementioned mathematical models of laminated glass beams or plates will be introduced in the next section.

5.2 Mathematical Model of Laminated Glass Beam

To theoretically study the deflection of laminated glass on impact, the mathematical model for the behavior of laminated glass is discussed as follows, with the laminated glass beam adopted for simplicity.

The shapes of laminated glass for practical use in architecture, the automotive industry, and aerospace are varying and include flat and curved shapes, which are mainly designed to meet various acquirements. Therefore, the following discussion is divided into two parts: the laminated glass plane beam and the laminated curved glass beam.

5.2.1 Mathematical Model of Laminated Glass Plane Beam

Nomenclature [8]

A_1, A_2	cross-sectional area of top and bottom glass ply
b	width of beam
E	modulus of elasticity of glass
G	shear modulus of interlayer
h_1, h_2	thickness of top and bottom glass plies
I	cross-sectional moment of inertia of the laminated glass section
L	half length of beam
M	bending moment of beam
N_1, N_2	cross-sectional force in the top and bottom glass ply
P	point load applied at the middle of beam
q	uniformly distributed load applied over the length of beam
t	thickness of the interlayer
u_1, u_2	in-plane displacement for the top and bottom ply in the x direction
w	the lateral displacement of the glass beam
x, z	rectangular coordinates
α, β, λ	parameters related to section geometry and material properties
$\sigma_1^{top}, \sigma_1^{bot}$	bending stresses on the top and bottom surfaces of the top ply
$\sigma_2^{top}, \sigma_2^{bot}$	bending stresses on the top and bottom surfaces of the bottom ply

Aşık [8] used the minimum total potential energy principle to derive the mathematical model on the basis of large deformation theory because the laminated glass is thin and can undergo large deformation [16]. The model is validated by experimental data and FE models for simply and fixed supported beams, respectively. Experiment and mathematical model results show that the behavior of laminated glass beam could be linear or nonlinear because of the boundary condition or constraints.

The mathematical model derived by Aşık [8] presents the behavior of laminated glass beam under large deflection, as shown in Figs. 5.1 and 5.2, by using three coupled nonlinear differential equations, which express the behavior of the monolithic and layered glass beams and take the following forms:

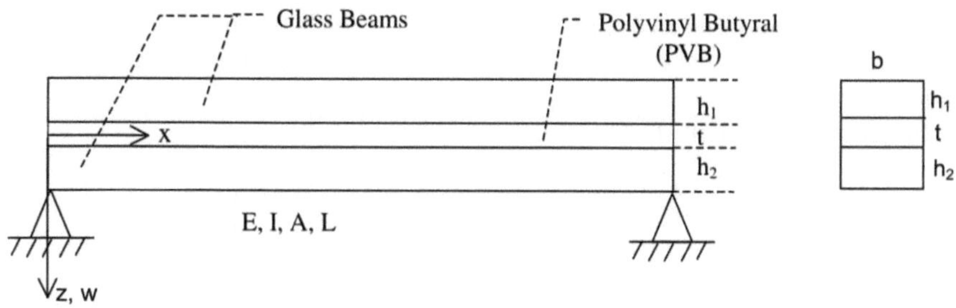

Fig. 5.1 Laminated glass beam [8]

Fig. 5.2 Undeformed and deformed sections of laminated glass beam [8]

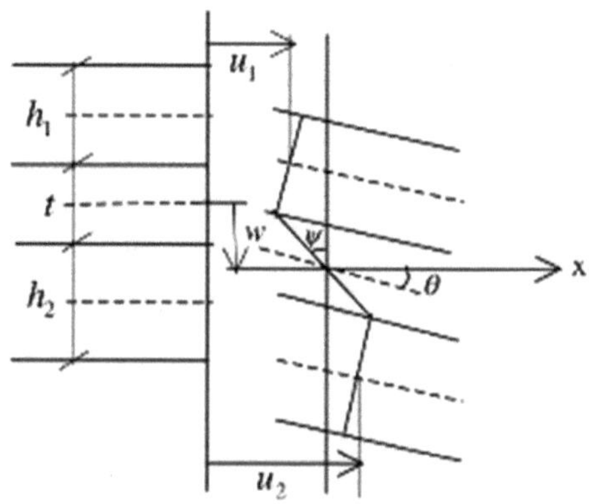

$$\left[\frac{d^2}{dx^2}\left(EI\frac{d^2}{dx^2}\right) - \frac{Gb}{t}\left(\frac{h_1}{2} + \frac{h_2}{2} + t\right)^2 \frac{d^2}{dx^2}\right] w$$

$$= q + (N_1 + N_2)\frac{d^2 w}{dx^2} - \frac{Gb}{t}\left(\frac{h_1}{2} + \frac{h_2}{2} + t\right)\left(\frac{du_1}{dx} - \frac{du_2}{dx}\right) \quad (5.1)$$

$$\frac{dN_1}{dx} = \frac{Gb}{t}\left[u_1 - u_2 - \left(\frac{h_1}{2} + \frac{h_2}{2} + t\right)\frac{dw}{dx}\right] \quad (5.2)$$

$$\frac{dN_2}{dx} = -\frac{Gb}{t}\left[u_1 - u_2 - \left(\frac{h_1}{2} + \frac{h_2}{2} + t\right)\frac{dw}{dx}\right] \quad (5.3)$$

N_1 and N_2 are the axial forces in the top and bottom layered glass, respectively, which are given as

$$N_1 = EA_1\left[\frac{du_1}{dx} + \frac{1}{2}\left(\frac{dw}{dx}\right)^2\right] \quad (5.4)$$

$$N_2 = EA_2\left[\frac{du_2}{dx} + \frac{1}{2}\left(\frac{dw}{dx}\right)^2\right] \quad (5.5)$$

Aşık et al. [8] used an analytic solution to analyze the laminated glass for a simply supported beam subjected to a uniformly distributed load q and a point load *P* at the center.

Finally, the lateral displacement $w(x)$ is obtained as follows:

$$w(x) = \frac{1}{h_1}\frac{A_1 + A_2}{EA_1A_2}\left\{\frac{\beta q}{2\lambda^4}\left(x^2 - 2Lx\right) - \frac{\beta}{6\lambda^2}\left(qL + \frac{P}{2}\right)\left(x^3 - 3L^2x\right) + \frac{\beta}{\lambda^2}\frac{q}{24}\left(x^4 - 4L^3x\right) - \frac{\beta q}{\lambda^6}\left[\frac{\cosh\lambda(x-L)}{\cosh\lambda L} - 1\right] + \frac{\beta}{\lambda^4}\frac{P}{2}\left[\frac{1}{\lambda}\frac{\sinh\lambda x}{\cosh\lambda L} - x\right]\right\} - \frac{1}{h_1}\frac{t}{Gb}N_1(x) \tag{5.6}$$

The stresses at the surfaces of the plies are given

$$\begin{aligned}
\sigma_1^{top} &= -\frac{M}{I}\frac{h_1}{2} + \frac{N_1}{A_1}\\
\sigma_1^{bot} &= \frac{M}{I}\frac{h_1}{2} + \frac{N_1}{A_1}\\
\sigma_2^{top} &= -\frac{M}{I}\frac{h_2}{2} + \frac{N_2}{A_2}\\
\sigma_2^{bot} &= \frac{M}{I}\frac{h_2}{2} + \frac{N_2}{A_2}
\end{aligned} \tag{5.7}$$

By solving the nonlinear equation, Aşık et al. found that the laminated glass beam behaves linearly with a simply supported laminated glass beam. They used this finding to predict the behavior of laminated glass beam with fixed support. Results show that this behavior is different from that of the simply supported laminated glass beam. The nonlinear behavior was dominant when the laminated glass beam had fixed support under large deformation.

Moreover, Aşık and Dural [16] investigated the effect of support condition on the vibration of laminated composite beams through a mathematical model for the dynamic behavior of laminated glass beam, which is verified under static loads. Similarly, the simply supported and fixed-support laminated glass beams were considered. Results show that the fundamental natural frequency of the laminated glass beam with fixed support is twice that of the simply supported laminated glass beam.

5.2.2 *Simple Mathematical Model of Laminated Glass Plane Beam*

Nomenclature [7]

- E Young's modulus of the glass
- g acceleration gravity
- h_0 thickness of interlayer
- h_1 thickness of top glass
- h_2 thickness of bottom glass
- l length of laminated glass beam
- q uniform transverse load
- w deflection of the glass beam
- x rectangular coordinates
- σ normal stress

A simple but reasonable mathematical model of laminated glass beam of the deflection and extremal stress derived for structure optimization by Ivanov [7] is shown as follows, in which the simply supported laminated glass beam is subjected to uniform transverse load:

The deflection is given as

$$w(x) = \frac{q}{ED}\left[\frac{g}{k^4}\frac{\cosh(kx)}{\cosh\left(k\frac{l}{2}\right)} - \frac{g-1}{24}x^4 + \frac{(g-1)l^2k^2 - 8g}{16k^2}x^2 + \frac{-5(g-1)l^4k^4 + 48gl^2k^2 - 384g}{384k^4}\right] \tag{5.8}$$

where

$$k = \sqrt{\frac{G_0}{Eh_0}\left(\frac{h_1+h_2}{h_1h_2} + \frac{h^2}{4D}\right)} \tag{5.9}$$

$$D = \frac{h_1^3 + h_2^3}{12} \tag{5.10}$$

$$h = 2h_0 + h_1 + h_2 \tag{5.11}$$

In the above equation, the maximum deflection can be described as

$$\max w = \frac{q}{384EDk^4}\left[g\left(-5l^4k^4 + 48l^2k^2 - 384\right) + \frac{384g}{\cosh\left(k\frac{l}{2}\right)} + 5l^4k^4\right] \tag{5.12}$$

The extremal stress is derived as follows:

$$\sigma_{\substack{\min\\\max^i}} = \frac{q}{2hh_iD}\left[A_{\substack{\min\\\max^i}}\frac{\cosh(kx)}{k^2\cosh\left(k\frac{l}{2}\right)} - \left(A_{\substack{\min\\\max^i}} \mp hh_i^2\right)\frac{x^2}{2} + \frac{A_{\substack{\min\\\max^i}}\left(l^2k^2-8\right) \mp hh_i^2l^2k^2}{8k^2}\right] \tag{5.13}$$

where

$$A_{\substack{\min\\\max^i}} = g\left[(-1)^i 4D \pm hh_i^2\right] \tag{5.14}$$

The mathematical model was verified by using MATLAB. The results show good agreement with other solutions that used the model in the optimization process to provide a lightweight structure of laminated glass. Finally, they found that the inner layer of laminated glass should be thinner than the external glass layer to meet light weight requirements.

5.2.3 Mathematical Model of Laminated Curved Glass Beam

Nomenclature [9]

A_1, A_2	cross-sectional area of top and bottom glass plies
b	width of beam
E	modulus of elasticity of glass
G	shear modulus of interlayer
h_1, h_2	thickness of top and bottom glass plies
r_1, r_2	radius of top and bottom glass plies
r_I	radius of interlayer
I_1, I_2	moment of inertia of top and bottom glass arches
N_1, N_2	cross-sectional force at the top and bottom glass plies
P	point load applied at the middle of beam
q	uniformly distributed load applied over the length of beam
s	arc length measured on the centroidal axis
t	thickness of interlayer
U_m^i	membrane strain energy for the top and bottom plies
U_b^i	bending strain energy for the top and bottom plies
U_I	shear strain energy for the interlayer
u_1, u_2	displacements for the top and bottom plies in the θ direction
V	volume of glass or interlayer

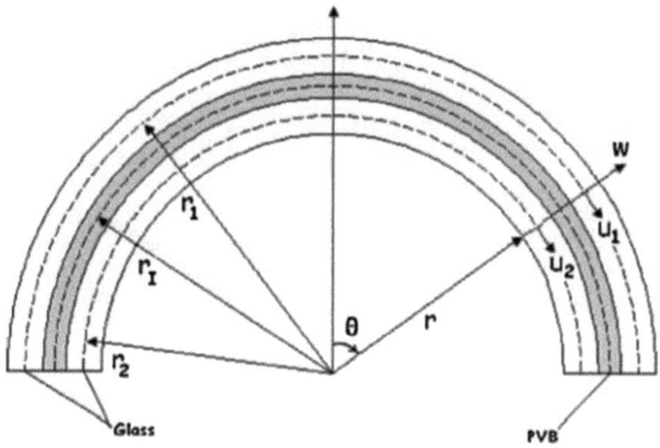

Fig. 5.3 Laminated circular glass beam with an interlayered PVB [9]

Ω	potential energy of applied loads
w	lateral displacement of glass beam
θ, r	polar coordinates
γ_1	shear strain in the interlayer
α	under-relaxation parameter for convergence
Π	total potential energy of the system

Given the extensive use of laminated curved glass, Aşık et al. [9] discussed the mathematical model of laminated curved glass beam by using three nonlinear coupled partial differential equations derived in polar coordinates by applying variational and energy principles.

Aşık et al. [9] considered that the total potential energy of the unit, as shown in Fig. 5.3, involves the sum of the membrane, bending, shear, and force potential energies, thereby obtaining the differential equations according to the displacements, which take the following form:

$$\Pi = U_m^1 + U_b^1 + U_m^2 + U_b^2 + U_1 + \Omega \tag{5.15}$$

where U_m^1 and U_b^1 are the membrane and the bending strain energies for the outer glass arch, respectively; U_m^2 and U_b^2 are the membrane and the bending strain of the inner glass arch, respectively.

Then, the equation in terms of displacements can be taken as

$$\Pi = \int_{V_1} \frac{1}{2}E\left(\frac{1}{r_1}\left(\frac{du_1}{d\theta}+w\right)+\frac{1}{2}\left(\frac{1}{r_1}\frac{dw}{d\theta}\right)^2\right)^2 dV$$
$$+\int_{V_1}\frac{1}{2}E\left(-z\frac{1}{r_1^2}\frac{d^2w}{d\theta^2}\right)^2 dV+\int_{V_2}\frac{1}{2}E\left(\frac{1}{r_2}\left(\frac{du_2}{d\theta}+w\right)\right.$$
$$\left.+\frac{1}{2}\left(\frac{1}{r_2}\frac{dw}{d\theta}\right)^2\right)^2 dV+\int_{V_2}\frac{1}{2}E\left(-z\frac{1}{r_2^2}\frac{d^2w}{d\theta^2}\right)^2 dV$$
$$+\int_{V_1}\frac{1}{2}G\left(\frac{u_1-u_2-\left(\frac{1}{r_1}\frac{h_1}{2}+\frac{1}{r_2}\frac{h_2}{2}+\frac{t}{r_1}\right)\frac{dw}{d\theta}}{t}\right)^2 dV$$
$$-\int_0^s qwds - P_a w(a) \tag{5.16}$$

where P_a is the concentrated radial load applied on the top of the outer glass at $s = a$.

Aşık et al. [9] derived three nonlinear coupled differential equations and boundary conditions for a simply supported laminated glass arch according to w, u_1 and u_2 by first varying the total potential energy.

$$\frac{d}{d\theta^2}\left(EI\frac{d^2w}{d\theta^2}\right)-\frac{d}{d\theta}\left(\left(\frac{N_1}{r_1}+\frac{N_2}{r_2}\right)\frac{dw}{d\theta}\right)+(N_1+N_2)$$
$$+Gbr_1\left(\frac{1}{r_1}\frac{h_1}{2}+\frac{1}{r_2}\frac{h_2}{2}+\frac{t}{r_1}\right)\frac{d\gamma_1}{d\theta}-qr-P_a\delta(a-s)=0 \tag{5.17}$$
$$\frac{dN_1}{d\theta}-Gbr_1\gamma_1=0 \tag{5.18}$$
$$\frac{dN_2}{d\theta}-Gbr_1\gamma_1=0 \tag{5.19}$$

where

$$N_1 = EA_1\left(\frac{1}{r_1}\left(\frac{du_1}{d\theta}+w\right)+\frac{1}{2}\left(\frac{1}{r_1}\frac{dw}{d\theta}\right)^2\right) \tag{5.20}$$
$$N_2 = EA_2\left(\frac{1}{r_2}\left(\frac{du_2}{d\theta}+w\right)+\frac{1}{2}\left(\frac{1}{r_2}\frac{dw}{d\theta}\right)^2\right) \tag{5.21}$$

where $I = \frac{I_1}{r_1^3}+\frac{I_2}{r_2^3}$.

The above model can predict the linear or nonlinear behavior of the laminated curved glass beam. Findings indicate that the behavior of the laminated curved glass beam ranges from the behavior of curved monolithic and layered glass beams.

5.3 Mathematical Model of Laminated Glass Plate

Nomenclature [15]

$D_1 = \frac{Eh_1^3}{12(1-v^2)}$	flexural rigidity of the top plate
$D_2 = \frac{Eh_2^3}{12(1-v^2)}$	flexural rigidity of the bottom plate
E	modulus of elasticity of the glass plate
e_{1x}, e_{1y}, e_{1xy}	membrane strains for the top plate
e_{2x}, e_{2y}, e_{2xy}	membrane strains for the bottom plate
G	shear modulus for glass
G_I	shear modulus for interlayer
h_1	thickness of the top plate
h_2	thickness of the bottom plate
q	uniformly distributed applied pressure
t	thickness of interlayer
u_1	in-plane displacement of the middle surface of the top plate
u_2	in-plane displacement of the middle surface of the bottom plate
v_1	in-plane displacement of the middle surface of the top plate
v_2	in-plane displacement of the middle surface of the bottom plate
w	lateral displacement of a generic point in the middle surface of the plate
x, y, z	rectangular coordinates
v	Poisson's ratio of the glass plate

References [13–15] investigated the behavior of laminated glass beam, which is nonlinear, and expressed this behavior by using five coupled nonlinear partial differential equations, which are the equilibrium equations in continuous and are valid at every specific point in the domain [15]

$$
\begin{aligned}
&\left[(D_1 + D_2)\nabla^4 - \frac{G_I}{t}\left(\frac{h_1}{2} + \frac{h_2}{2} + t\right)^2 \nabla^2\right] w \\
&= q + \frac{Eh_1}{1-v^2}[(e_{1x} + ve_{1y})w_{,xx} + (e_{1y} + ve_{1x})w_{,yy} \\
&\quad + 2(1-v)e_{1xy}w_{,xy}] + \frac{Eh_2}{1-v^2}[(e_{2x} + ve_{2y})w_{,xx} \\
&\quad + (e_{2y} + ve_{2x})w_{,yy} + 2(1-v)e_{2xy}w_{,xy}] \\
&\quad - \frac{G_I}{t}\left(\frac{h_1}{2} + \frac{h_2}{2} + t\right)\left(\frac{\partial u_1}{\partial x} - \frac{\partial u_2}{\partial x} + \frac{\partial v_1}{\partial y} - \frac{\partial v_2}{\partial y}\right),
\end{aligned}
\tag{5.22}
$$

$$\left[\frac{\partial^2}{\partial x^2}+\frac{1-v}{2}\frac{\partial^2}{\partial y^2}-\frac{G_{\rm I}(1-v)}{2Gh_1t}\right]u_1$$
$$=-\frac{1+v}{2}\frac{\partial^2 v_1}{\partial x\partial y}-\frac{G_{\rm I}(1-v)}{2Gh_1t}u_2$$
$$-\frac{\partial w}{\partial x}\left[\frac{\partial^2 w}{\partial x^2}+\frac{1-v}{2}\frac{\partial^2 w}{\partial y^2}\right]-\frac{1+v}{2}\frac{\partial^2 w}{\partial x\partial y}\frac{\partial w}{\partial y}$$
$$-\frac{G_{\rm I}(1-v)}{2Gh_1t}\left(\frac{h_1}{2}+\frac{h_2}{2}+t\right)\frac{\partial w}{\partial x}, \tag{5.23}$$

$$\left[\frac{1-v}{2}\frac{\partial^2}{\partial x^2}+\frac{\partial^2}{\partial y^2}-\frac{G_{\rm I}(1-v)}{2Gh_1t}\right]v_1$$
$$=-\frac{1+v}{2}\frac{\partial^2 u_1}{\partial x\partial y}-\frac{G_{\rm I}(1-v)}{2Gh_1t}v_2$$
$$-\frac{\partial w}{\partial x}\left[\frac{\partial^2 w}{\partial y^2}+\frac{1-v}{2}\frac{\partial^2 w}{\partial x^2}\right]-\frac{1+v}{2}\frac{\partial^2 w}{\partial x\partial y}\frac{\partial w}{\partial x}$$
$$-\frac{G_{\rm I}(1-v)}{2Gh_1t}\left(\frac{h_1}{2}+\frac{h_2}{2}+t\right)\frac{\partial w}{\partial y}, \tag{5.24}$$

$$\left[\frac{\partial^2}{\partial x^2}+\frac{1-v}{2}\frac{\partial^2}{\partial y^2}-\frac{G_{\rm I}(1-v)}{2Gh_2t}\right]u_2$$
$$=-\frac{1+v}{2}\frac{\partial^2 v_2}{\partial x\partial y}-\frac{G_{\rm I}(1-v)}{2Gh_2t}u_1$$
$$-\frac{\partial w}{\partial x}\left[\frac{\partial^2 w}{\partial x^2}+\frac{1-v}{2}\frac{\partial^2 w}{\partial y^2}\right]-\frac{1+v}{2}\frac{\partial^2 w}{\partial x\partial y}\frac{\partial w}{\partial y}$$
$$+\frac{G_{\rm I}(1-v)}{2Gh_2t}\left(\frac{h_1}{2}+\frac{h_2}{2}+t\right)\frac{\partial w}{\partial x}, \tag{5.25}$$

$$\left[\frac{1-v}{2}\frac{\partial^2}{\partial x^2}+\frac{\partial^2}{\partial y^2}-\frac{G_{\rm I}(1-v)}{2Gh_2t}\right]v_2$$
$$=-\frac{1+v}{2}\frac{\partial^2 u_2}{\partial x\partial y}-\frac{G_{\rm I}(1-v)}{2Gh_2t}v_1$$
$$-\frac{\partial w}{\partial y}\left[\frac{\partial^2 w}{\partial y^2}+\frac{1-v}{2}\frac{\partial^2 w}{\partial x^2}\right]-\frac{1+v}{2}\frac{\partial^2 w}{\partial x\partial y}\frac{\partial w}{\partial x}$$
$$+\frac{G_{\rm I}(1-v)}{2Gh_2t}\left(\frac{h_1}{2}+\frac{h_2}{2}+t\right)\frac{\partial w}{\partial y}, \tag{5.26}$$

In the above equation [13]

$$\nabla^4=\frac{\partial^4}{\partial x^4}+2\frac{\partial^4}{\partial x^2\partial y^2}+\frac{\partial^4}{\partial y^4} \tag{5.27}$$

and

$$\nabla^2 = \frac{\partial^2}{\partial x^2} + \frac{\partial^2}{\partial y^2} \tag{5.28}$$

The relationship of strain and displacements can be described as

$$e_{1y} = \frac{\partial v_1}{\partial y} + \frac{1}{2}\left(\frac{\partial w}{\partial y}\right)^2 \tag{5.29}$$

$$e_{2y} = \frac{\partial v_2}{\partial y} + \frac{1}{2}\left(\frac{\partial w}{\partial y}\right)^2 \tag{5.30}$$

$$e_{1xy} = \frac{\partial v_1}{\partial y} + \frac{\partial u_1}{\partial y} + \frac{\partial w}{\partial x}\frac{\partial w}{\partial y} \tag{5.31}$$

$$e_{2xy} = \frac{\partial v_2}{\partial x} + \frac{\partial u_2}{\partial y} + \frac{\partial w}{\partial x}\frac{\partial w}{\partial y} \tag{5.32}$$

5.4 Numerical Analysis of Laminated Glass

5.4.1 Stress–Strain Relationship

Some researchers have attempted to establish the stress–strain relationship of PVB laminated glass on the basis of continuum damage mechanics (CDM). Sun and Kahaleel [17] proposed the stress–strain constitutive model of standard single-layered glass.

The damage results from normal stress can be described as

$$\begin{cases} 0 & \sigma_i \le \sigma_{\text{threshold}} \\ D_{ij} = \frac{\sigma_i - \sigma_{\text{threshold}}}{\sigma_{\text{crit}} - \sigma_{\text{threshold}}} & \sigma_{\text{threshold}} < \sigma_i < \sigma_{\text{threshold crit}} \\ 1 & \sigma_i > \sigma_{\text{crit}} \end{cases} \tag{5.33}$$

The damage results from shear stress can be described as

$$\begin{cases} 0, & \sigma_i \le \sigma_{\text{threshold}}, \text{ or } \max(\sigma_i) > 0 \\ D_{ij} = \frac{\sigma_i - \sigma_{\text{threshold}}}{\sigma_{\text{crit}} - \sigma_{\text{threshold}}} & \\ \text{when } \sigma_{\text{threshold}} < \sigma_i < \sigma_{\text{threshold crit}}, \text{ and } \max(\sigma_i) < 0 & \\ 1, & \sigma_i \ge \sigma_{\text{threshold}}, \text{ or } \max(\sigma_i) < 0 \end{cases} \tag{5.34}$$

where D_{ij} is the damage variable, which ranges from 0 to 1; σ_i is the normal stress; σ_{ij} is the shear stress; $\sigma_{\text{threshold}}$ is the threshold value of stress; and σ_{crit} is the critical stress. Moreover, $i = 1, 2, 3$, $j = 1, 2, 3$ and $i \neq j$.

In the model, when stress exceeds its threshold value, material damage occurs; when stress exceeds the critical stress, the material fails completely. Calculation results show that the damage area and the damage degree on the glass results from the sphere agrees well with the experimental results, thereby showing the accuracy and feasibility of the theoretical analysis. Sun and Khaleel [18] developed a CDM-based embeddable limited program to predict the damage of a windshield that is impacted by stone. Similarly, Zhao [19, 20] calculated the normal and shear stresses of PVB laminated windshield subjected to headform and calibrated the damage criterion of the model.

FEM, which is a numerical simulation method, is widely used for simulation studies. Bois and Timmel [21, 22] used LS-DYNA to simulate PVB laminated glass; the element types of glass and PVB were shell element and membrane, respectively. The connection of layers adopted the principle of common node sharing. The overall equivalent thickness and density are calculated as follows:

$$t_E = \sqrt[3]{t_G^3 + 3t_G(t_G + t_{\text{PVB}}) + \frac{E_{\text{PVB}}}{2E_G} t_{\text{PVB}}^3} \tag{5.35}$$

where t and E are the thickness and the elastic modulus, respectively. The subscripts PVB, G, and E denote the PVB, glass, and equivalent overall layered plates, respectively.

$$\rho_E = \left(\rho_G t_G + \frac{1}{2}\rho_{\text{PVB}} t_{\text{PVB}}\right) / t_E \tag{5.36}$$

where ρ is the density. The glass layers adopt a linear elastic material, and the material of PVB membrane adopts the material similar to that in the experimental results in LS-DYNA material library and fits its material parameters.

Simulation results show that when the stress is less than 100%, the Blatz-Ko and Mooney–Rivlin material models are relatively close to the material behavior of PV. However, the six-order Ogden material model can better describe the behavior of PVB under impact. Zhao [19, 23, 24] embedded the constitutive model mentioned above into Abaqus to investigate the size and distribution of the damage area and the stress under different damage patterns. The effect of the thickness of glass and PVB and the size of the windshield on the damage area were also investigated. Lawn and Wilshaw [25] pointed out that a plastic area exists at the point where the head and windshield meet; in this area, the shear and compression stress were maximum, except the Hertzian cone crack and aperture. Knight et al. [26] found that when the shear stress exceeds the field stress, the plastic stress area will exist around the impact point. The PVB laminate glass is a brittle high-polymer composite material. Thus, simulating the damage pattern and the crack propagation by using the traditional numerical simulation method is difficult for this material, and the effect is not satisfactory.

Zang et al. [27–29] designed a numerical simulation program based on 3D discrete element to simulate the damage of PVB laminated windshield subjected to headform. The simulation results were satisfactory because of the discrete element method, which can simulate non-continuum fracture and damage and is better than the traditional FEM.

5.4.2 Numerical Simulation Algorithms

(1) Numerical algorithms

Chen et al. [30] conducted numerical analysis of laminated glass undergoing impact failure with six numerical algorithms, namely, element deletion method, CDM, discrete element method, combined discrete and FE methods, XFEM, and cohesive zone models.

The failure of the element in the element deletion method is achieved by setting the stress of the element to zero or removing the mass of the element from the global mass matrix. This method is widely used in commercial software solvers such as LS-DYNA [30].

The fundamental idea of CDM is to assume the damage effect of the material through its stress–strain behavior [30, 31]. Sun et al. [18, 32] proposed a constitutive law considered the contribution of damage that can be described as

$$\sigma_{ij} = \left\{ \left[K^{e}_{ijkl} \right] + \left[K^{d}_{ijkl} \right] \right\} \cdot \{ \varepsilon_{kl} \} \tag{5.37}$$

where the undamaged and damage parts, K^{e}_{ijkl} and K^{d}_{ijkl}, are

$$K^{e}_{ijkl} = \lambda \sigma_{ij} \delta_{kl} + \mu \left(\delta_{ik} \delta_{jl} + \delta_{il} \delta_{kj} \right) \tag{5.38}$$

$$K^{d}_{ijkl} = C_1 \left(\delta_{ij} D_{kl} + \delta_{kl} D_{ij} \right) + C_2 \left(\delta_{jk} D_{il} + \delta_{il} D_{jk} \right) \tag{5.39}$$

where λ and μ are Lame's constants; C_1 and C_2 are damage parameters; and D_{ij} denotes damage components.

Discrete element method describes the movements of discrete elements by using Newton's law, where the interactions of discrete elements are addressed by using contact algorithms.

Combined discrete and FE methods used in laminated glass are based on the penalty method proposed by Lei et al. [33], where the PVB is modeled by FEs, the two glasses are modeled by DEs, and the combination patterns of DE and FE involve combining DE with one FE node.

XFEM is used to simulate cracking without remeshing the crack growth [30, 34].

Cohesive zone models are divided into intrinsic and extrinsic models [35, 36] and are governed by phenomenological traction–separation law [30].

(2) Laminated glass models

The numerical model structures of laminated glass are normally divided into three structures: one-layered, two-layered, and three-layered models.

The glasses and PVB of the one-layered model structure are a whole structure modeled by shell elements [37]. The two-layered model of laminated glass is presented by a glass modeled on layered shell elements and PVB modeled on membrane elements [37, 38]. The three-layered model of laminated glass is adopted extensively and is modeled according to the real structure of laminated glass in practical application, where the glasses are modeled by shell or solid elements and PVB is modeled by membrane, shell, or solid [21, 22, 37, 39–42].

5.5 Summary

This chapter presented a theoretical analysis of laminated glass based on different physical models to ensure further understanding of the dynamic response of laminated glass.

First, the laminated glass beam is described based on the minimum total potential energy principle because the mechanical model of the beam is relatively simpler than plate and boundary conditions of the simply supported laminated glass beam. Moreover, the displacement and the load of the beam have a linear relationship. A simple and reasonable mathematical model of laminated glass beam with deflection and extremal stress is also described, where the simply supported laminated glass beam is under uniform transverse load. In addition, we introduce a mathematical model of laminated curved glass beam with three nonlinear and coupled partial differential equations based on variational and energy principles.

Given the different structures and assumptions of beam and plate, differences exist between the theoretical solutions of stress and displacement of the beam and the plate. The plate model is then described by using five coupled nonlinear partial differential equations.

Then, the stress–strain relationship of laminated glass based on CDM is described. In the model, when stress exceeds its threshold value, material damage occurs; when stress exceeds the critical stress, the material fails completely. The calculation results of other studies show that the damage area and the damage degree on the glass as a result of the sphere agree well with the experimental results, thereby showing the accuracy and feasibility of the theoretical analysis.

Lastly, six numerical algorithms for numerical analysis of laminated glass undergoing impact failure are briefly introduced.

References

1. S. Hallström, J.L. Grenestedt, Failure analysis of laminated timber beams reinforced with glass fibre composites. Wood Sci. Technol. **31**, 17–34 (1997)
2. J.A. Hooper, On the bending of architectural laminated glass. Int. J. Mech. Sci. **15**, 309–323 (1973)

3. R.A. Behr, J.E. Minor, M.P. Linden, C.V.G. Vallabhan, Laminated glass units under uniform lateral pressure. J. Struct. Eng.-ASCE **111**, 1037–1050 (1985)
4. R.A. Behr, J.E. Minor, H.S. Norville, Structural behavior of architectural laminated glass. J. Struct. Eng. **119**, 202–222 (1993)
5. C.V.G. Vallabhan, Iterative analysis of nonlinear glass plates. J. Struct. Eng. **109**, 489–502 (1983)
6. H.S. Norville, K.W. King, J.L. Swofford, Behavior and strength of laminated glass. J. Eng. Mech. **124**, 46–53 (1998)
7. I.V. Ivanov, Analysis, modelling, and optimization of laminated glasses as plane beam. Int. J. Solids Struct. **43**, 6887–6907 (2006)
8. M.Z. Aşık, S. Tezcan, A mathematical model for the behavior of laminated glass beams. Comput. Struct. **83**, 1742–1753 (2005)
9. M.Z. Aşık, E. Dural, M. Yetmez, T. Uzhan, A mathematical model for the behavior of laminated uniformly curved glass beams. Compos. B Eng. **58**, 593–604 (2014)
10. A. Zemanová, J. Zeman, M. Šejnoha, Numerical model of elastic laminated glass beams under finite strain. Arch. Civ. Mech. Eng. **14**, 734–744 (2014)
11. C.V.G. Vallabhan, J.E. Minor, S.R. Nagalla, Stresses in layered glass units and monolithic glass plates. J. Struct. Eng. **113**, 36–43 (1987)
12. M.Z. Aşik, C.V.G. Vallabhan, On the convergence of nonlinear plate solutions. Comput. Struct. **65**, 225–229 (1997)
13. M.E. Mohareb, A mathematical model for the nonlinear analysis of laminated glass plates subjected to lateral press. Texas Tech University (1990)
14. C.V.G. Vallabhan, Y.C. Das, M. Magdi, M. Asik, J.R. Bailey, Analysis of laminated glass units. J. Struct. Eng. **119**, 1572–1585 (1993)
15. M.Z. Aşık, Laminated glass plates: revealing of nonlinear behavior. Comput. Struct. **81**, 2659–2671 (2003)
16. M.Z. Asik, E. Dural, Effect of support conditions on the vibration of laminated composite beams. J. Vib. Control **13**, 1361–1368 (2007)
17. X. Sun, M.A. Khaleel, Effects of different design parameters on stone-impact resistance of automotive windshields. Proc. Inst. Mech. Eng. Part D: J. Automob. Eng. **219**, 1059–1067 (2005)
18. X. Sun, M.A. Khaleel, R.W. Davies, Modeling of stone-impact resistance of monolithic glass ply using continuum damage mechanics. Int. J. Damage Mech. **14**, 165–178 (2005)
19. S. Zhao, L.R. Dharani, L. Chai, S.D. Barbat, Analysis of damage in laminated automotive glazing subjected to simulated head impact. Eng. Fail. Anal. **13**, 582–597 (2006)
20. S. Zhao, L.R. Dharani, X. Liang, Analysis of damage in laminated architectural glazing subjected to blast loading. Adv. Struct. Eng. **11**, 129–134 (2008)
21. M. Timmel, S. Kolling, P. Osterrieder, P.A.D. Bois, A finite element model for impact simulation with laminated glass. Int. J. Impact Eng **34**, 1465–1478 (2007)
22. P.A.D. Bois, S. Kolling, W. Fassnacht, Modelling of safety glass for crash simulation. Comput. Mater. Sci. **28**, 675–683 (2003)
23. S. Zhao, L.R. Dharani, X. Liang, L. Chai, S.D. Barbat, Crack initiation in laminated automotive glazing subjected to simulated head impact. Int. J. Crashworthiness **10**, 229–236 (2005)
24. S. Zhao, L.R. Dharani, L. Chai, S.D. Barbat, Dynamic response of laminated automotive glazing impacted by spherical featureless headform. Int. J. Crashworthiness **11**, 105–114 (2006)
25. B. Lawn, R. Wilshaw, Review: indentation fracture: principles and applications. J. Mater. Sci. **10**, 1049–1081 (1975)
26. C.G. Knight, M.V. Swain, M.M. Chaudhri, Impact of small steel spheres on glass surfaces. J. Mater. Sci. **12**, 1573–1586 (1977)
27. M.Y. Zang, L.I. Jun, Z. Lei, Study on impact fracture behavior of Bi-layer laminated glass based on DEM. Sci. Technol. Eng. (2009)
28. M. Zang, Static characteristic and impact fracture behavior of automobile glass. Chin. J. Mech. Eng. **45**, 268–272 (2009)

29. M.Y. Zang, Z. Lei, S.F. Wang, Investigation of impact fracture behavior of automobile laminated glass by 3D discrete element method. Comput. Mech. **41**, 73–83 (2007)
30. S. Chen, M. Zang, D. Wang, S. Yoshimura, T. Yamada, Numerical analysis of impact failure of automotive laminated glass: a review. Compos. B Eng. **122**, 47–60 (2017)
31. L.M. Kachanov, *Introduction to Continuum Damage Mechanics* (M. Nijhoff, 1986)
32. X. Sun, M.A. Khaleel, Modeling of glass fracture damage using continuum damage mechanics-static spherical indentation. Int. J. Damage Mech. **13**, 263–285 (2004)
33. Z. Lei, M. Zang, An approach to combining 3D discrete and finite element methods based on penalty function method. Comput. Mech. **46**, 609–619 (2010)
지광습, 송정훈, 이상호A new method for growing multiple cracks without remeshing and its application to fatigue crack growth. Model. Simul. Mater. Sci. Eng. **12**, 901–915 (2004)
35. X.P. Xu, A. Needleman, Numerical simulations of fast crack growth in brittle solids. J. Mech. Phys. Solids **42**, 1397–1434 (1994)
36. G.T. Camacho, M. Ortiz, Computational modelling of impact damage in brittle materials. Int. J. Solids Struct. **33**, 2899–2938 (1996)
37. Y. Peng, J. Yang, C. Deck, R. Willinger, Finite element modeling of crash test behavior for windshield laminated glass. Int. J. Impact Eng **57**, 27–35 (2013)
38. M.Y. Zang, Research on finite element model of PVB laminated glass
39. L. Qi, J. Liu, M. Qiang, D. Wang, X. Tang, Simulation and test validation of windscreen subject to pedestrian head impact. J. Automot. Saf. Energy (2011)
40. X. Xu, J. Xu, J. Chen, P. Li, B. Liu, Y. Li, Investigation of dynamic multi-cracking behavior in PVB laminated glass plates. Int. J. Impact Eng. **100**, 62–74 (2017)
41. M.Y. Zang, Z.L. Song, Z.G. Yang, Finite element model of windscreen for pedestrian protection. J. South China Univ. Technol. **42**, 143–148 (2014)
42. D. Wang, Y. Wang, Y. Shi, W.U. Shenrong, Z. Zhao, W.U. Te, C.A. Co, Numerical simulation and test validation of automotive laminated windshield glass. J. Automot. Saf. Energy **5**, 263–269 (2014)

Chapter 6
Numerical Computation Based on Finite Element Method

In accordance with the previously obtained experimental results, two windshield FE models, namely, the single-layered and triple-layered models, are established to investigate the mechanical behavior of laminated windshield subject to headform impact. The modeling steps are described in detail, and the influencing factors of simulation for the two models are discussed. The simulation results of the two FE models are compared, and the better results are applied on a vehicle with different laminated windshield thicknesses to verify the accuracy of the windshield FE model.

6.1 Headform and Cutoff Vehicle FE Models

To mimic a real headform impact test scenario and save computational resources, numerical models for the headform and the cutoff vehicle are established according to the headform impactor and the cutoff vehicle body used in the aforementioned tests.

As shown in Fig. 6.1, this cutoff vehicle FE model includes the components involved in the test, e.g., the entire laminated windshield, a pillar, adhesive joints, and laminated windshield frameworks. This model has been validated in previous full-scale vehicle crashworthiness simulations. The detailed boundary conditions for this FE model are illustrated in Fig. 6.1a. The front, end, and bottom boundaries of this cutoff vehicle model are clamped completely. The same five impact points on the laminated windshield as those in the tests in Fig. 6.1b are selected. Figure 6.1c shows that the size of the mesh of the entire model is 10 mm. Moreover, as indicated in Fig. 6.1d, the windshield and the vehicle body are connected using adhesive elements attached on a pillar and the windshield frame. The "Contact_Automatic_Surface_to_Surface" option is selected to simulate the contact between the head and the windshield

J. Xu and Y. Li, *Impact Behavior and Pedestrian Protection of Automotive Laminated Windshield*, https://doi.org/10.1007/978-981-13-2441-3_6

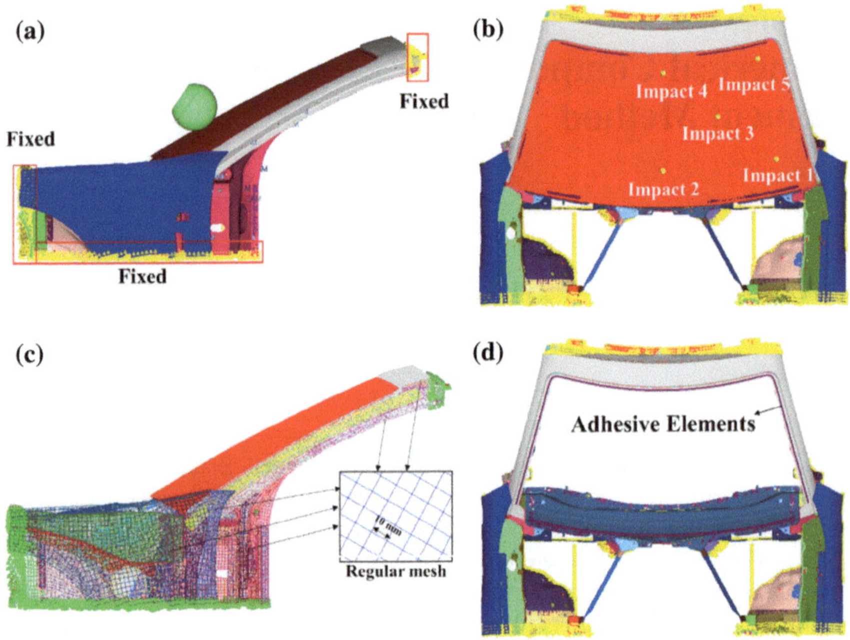

Fig. 6.1 Headform and cutoff vehicle FE model

shell layer in the single-layered model and also to simulate the contact between the head and the glass/PVB layer in the triple-layered model. A friction coefficient of 0.3 is defined according to empirical experience from previous literature [1, 2]. Figure 6.2 presents the single-layered and triple-layered models established in this study. Figure 6.2a shows that the single-layered windshield FE model consists of only one layer of shell elements. Figure 6.2b shows that the triple-layered windshield FE model consists of two glass layers of shell elements and one PVB layer of solid elements, which are connected by sharing nodes between the shell and solid element layers. As Fig. 6.2c shows, the bottom surface of the upper three-dimensional shell element layer is selected as the reference surface, and the upper surface of the lower three-dimensional shell element layer is selected as the reference surface, which avoids the scenario of a half-thickness offset with the shell mid-plane to connect the nodes with the solid-element layer. As shown in Fig. 6.3, a regular mesh with an element size of 5 mm is adopted for both windshield FE models.

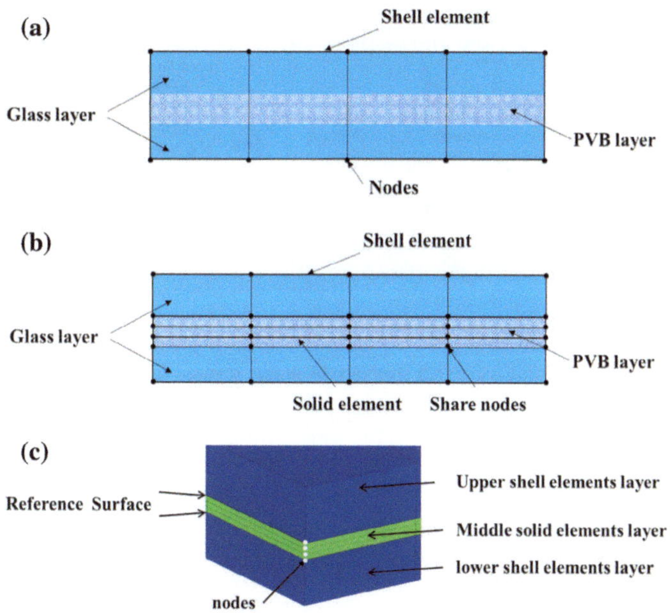

Fig. 6.2 **a** Single-layered windshield FE model for numerical simulation. **b** Triple-layered windshield FE model for numerical simulation. **c** Relative location of shell and solid layers

Fig. 6.3 Mesh type and element size of the windshield

6.2 Process of Numerical Modeling of Laminated Glass

6.2.1 Modeling of Single-Layered Laminated Glass

The standard PVB laminated glass consists of a PVB interlayer sandwiched by two mono soda-limes, which is a typical sandwich structure. To improve the calculation efficiency of the models, single-layered modeling of laminated glass is proposed. Two problems should be considered: (1) how to simulate the sandwich structure by using only the single-layered shell element and (2) how to present the different

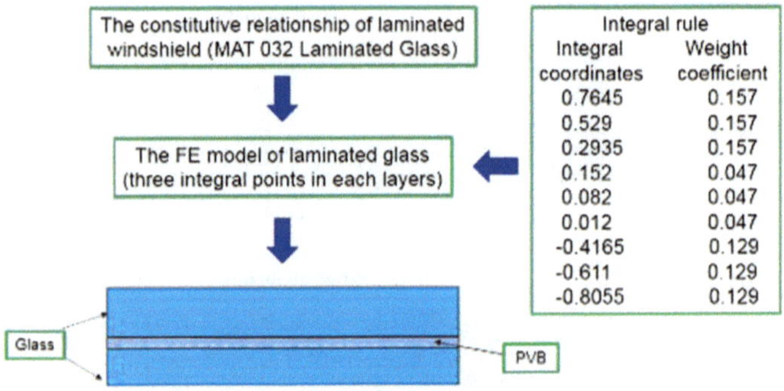

Fig. 6.4 Diagram of defining sandwich structure by single-layered shell element

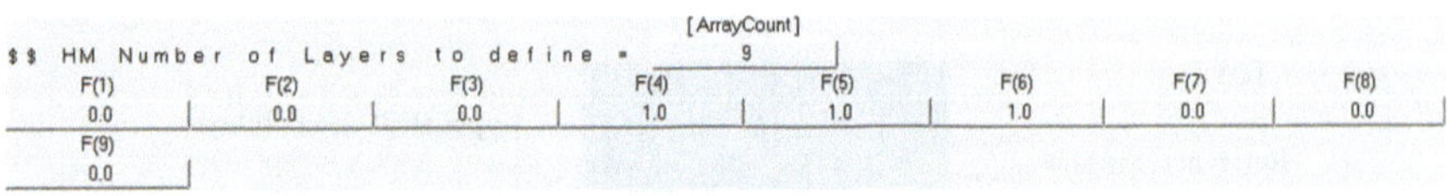

Fig. 6.5 MAT 032 ("MAT_LAMINATED_GLASS") to define the layer number of shell elements in the material card

material properties on the single-layered shell element. Thus, the proper material card needs to be selected in LS-DYNA.

6.2.1.1 FE Model Structure of Single-Layered Laminated Glass

Figure 6.2a shows the single-layered windshield FE model for numerical simulation, which combines the single-layered shell element into a three-layered shell structure to mimic the real structure of laminated glass.

Figure 6.4 shows the flowchart of defining the multilayered structure of laminated glass by MAT_032 and the integral rule in LS-DYNA ("Integration_shell" card), with single-layered shell elements combined into a three-layered shell element.

To define the sandwich structure of the shell element, the first step is to define the number of integral layers by using MAT_032 ("MAT_LAMINATED_GLASS") material card. The shell element is defined as nine piles of integral, as shown in Fig. 6.5. When the $F_n = 0$, the layer of integral presents for the glass layers and PVB layer when $F_n = 1$.

$$F_n = 0\text{:Glass} \tag{6.1}$$

$$F_n = 1\text{:Polymer} \tag{6.2}$$

Otherwise, the structure property (the coordinates and weight coefficient of each integral layer) of the shell element in laminated glass needs to be defined by using the integral rule "Integration_shell card" in LS-DYNA according to the thickness

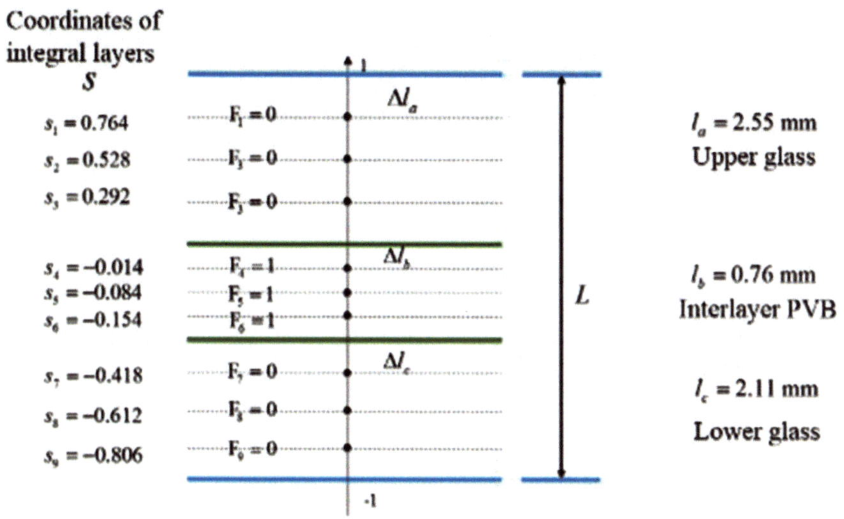

$$s_1 = 1 - \frac{\Delta l_a}{l} \times 2 \quad s_2 = s_1 - \frac{\Delta l_a}{l} \times 2 \quad s_3 = s_2 - \frac{\Delta l_a}{l} \times 2 \quad s_4 = s_3 - \frac{\Delta l_a}{l} \times 2 - \frac{\Delta l_b}{l} \times 2 \quad \ldots\ldots$$

Fig. 6.6 Coordinate value between −1 and 1 of each layer in single-layered shell element of windshield

of each glass layer of laminated glass, expect for the number of integral layers in material card MAT_032. In this example, the thickness of the windshield of the vehicle body, the upper glass, the interlayer PVB, and the lower glass is 5.41, 2.55, 0.76, and 2.11 mm, respectively.

Figure 6.6 shows the coordinates S of each integral layer of laminated glass of vehicle body in this example according to the real thickness of laminated glass and the calculation equation of integral rule in LS-DYNA (Eqs. 6.1 and 6.2).

The calculation equation of coordinates S of integral points is

$$\Delta l_a = l_a/4, \quad \Delta l_b = l_b/4, \quad \Delta l_c = l_c/4 \tag{6.3}$$

where $l_a = 2.55$ mm, $l_b = 0.76$ mm, $l_c = 2.11$ mm denote the upper glass, interlayer, and lower glass, respectively.

$$\begin{aligned} s_1 &= 1 - \frac{\Delta l_a}{L}, \quad s_2 = s_1 - \frac{\Delta l_a}{L} \times 2, \\ s_3 &= s_2 - \frac{\Delta l_a}{L} \times 2, \quad s_4 = s_3 - \frac{\Delta l_a}{L} \times 2 - \frac{\Delta l_b}{L} \times 2 \ldots \end{aligned} \tag{6-4}$$

Figure 6.7 shows a schematic of the method of solving the weight coefficient of each integral layer by using the integral rule in LS-DYNA. The weight coefficient is calculated as

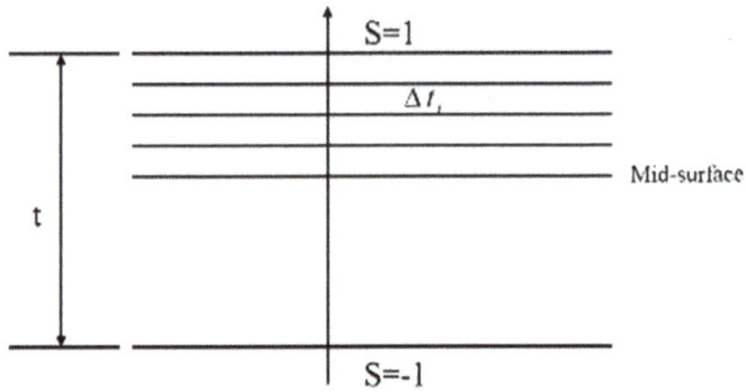

Fig. 6.7 Calculation method of defining the weight coefficient of each integral layers in "Integration_shell card" [3]

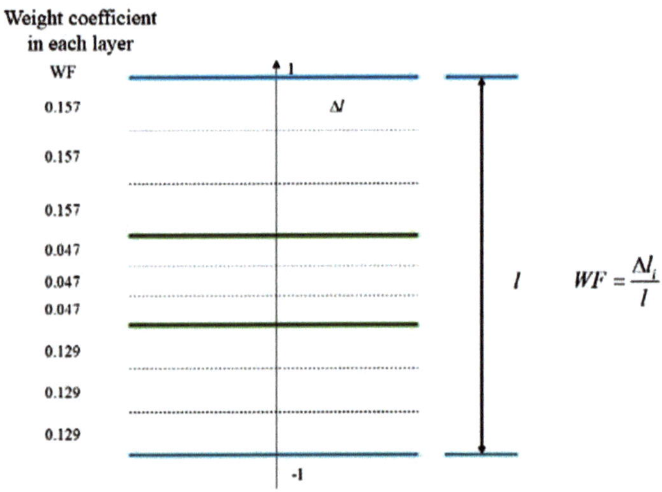

Fig. 6.8 Calculation results of the weight coefficient of each integral layer in the single-layered shell element

$$WF = \Delta t_i / t \tag{6.5}$$

WF is the weight coefficient of the integral layer calculated by the integral rule equations according to the real thickness of each glass layer in laminated glass.

The weight coefficient of shell element is described as

$$WF = \frac{\Delta l_i}{L}, \quad i = 1, 2, 3 \tag{6.6}$$

The results are shown in Fig. 6.8.

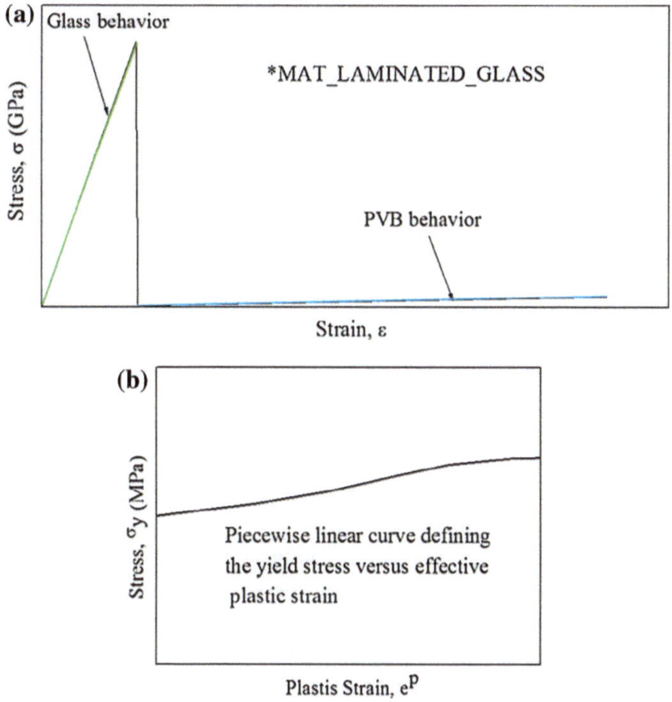

Fig. 6.9 **a** LS-DYNA multilayered shell for laminated glass behavior [5]. **b** LS-DYNA plastic stage stress–strain curve to describe the plastic deformation of PVB

6.2.1.2 Constitutive Description of Materials for Single-Layered Laminated Glass

The MAT_032 ("MAT_LAMINATED_GLASS") material model in LS-DYNA may be able to describe the multilayered structure with only one shell element [4]. The multilayered formulation of MAT_032 can describe the mechanical behavior of glass and PVB individually and systematically through the property card (integration_shell card) with various properties associated with different layers (with their own integration points).

The shell layer initially exhibits glass behavior that bears the main loading until it achieves the predefined "glass plastic failure strain (EFG)" in the MAT_032 material model. Then, the mechanical behavior of the shell layer is mainly dominated by PVB, as shown in Fig. 6.9a [5]. The material model permits the failure of glass. However, the single-layered shell elements in this material model cannot be deleted. Otherwise, the load capacity of PVB and the remaining glass layer will be lost completely. As such, it is expected to exhibit an unreasonable response to extremely large deformations for PVB, which is not discussed in this paper.

Table 6.1 Mechanical properties of the single-layered windshield FE model

	Single-layered windshield FE model					
Glass	*Rho* (kg/m^3)	*EG* (GPa)	*PRG*	*SYG* (MPa)	*ETG* (GPa)	*EFG*
	2500	100	0.23	110	50	0.0004
PVB	–	*EP* (MPa)	*PRP*	*SYP* (MPa)	*ETP* (MPa)	–
	–	280	0.495	10	2	–

Thus, the aforementioned material model MAT_032 is adopted in this study to characterize the laminated windshield glass modeled with a single layer of shell elements. The laminated glass specimen with a total thickness of 5.41 mm is split into nine layers, i.e., three layers each for a 2.55 mm-thick upper glass, a 0.76 mm-thick PVB interlayer, and a 2.1 mm-thick lower glass. Accordingly, nine integration points are set in each layer.

As shown in Figs. 6.10a–d, the mechanical properties of glass and PVB in this study are determined qualitatively by referring to a series of test curves obtained from previous studies and the correlation condition between simulation and test [6, 7]. These properties are also expressed by the defined curve with corresponding referenced curves. The consistent trend between the test and the defined curves is good, thereby proving that the mechanical property is reasonable.

With only one sheet of shell elements for the single-layered model, the mechanical properties of glass are determined qualitatively by referring to dynamic compression test curves at different strain rates and correlation condition [6], as shown in Fig. 6.10a. The plastic failure strain of glass in the single-layered model cannot be evaluated accurately based on compressive test curves because this material includes not only a glass failure phase but also a compaction phase where failed glass shatters into fine fragments [6]; as a result, the real failure strain is not evident initially. Previous tests and simulation studies show that the windshield glass plastic failure strain is in the range of 0.01–0.1% [4–6, 8–10]. Given the correlation between the simulations and tests, the plastic failure strain of 0.04% is appropriate and is therefore adopted.

In addition, PVB exhibits dynamic tensile deformation mainly when laminated glass is impacted; thus, the mechanical properties of PVB in the single-layered model are determined qualitatively by a series of test curves with different strain rates obtained from dynamic tensile tests and correlation condition [7], as shown in Fig. 6.10b.

The mechanical properties of glass and PVB in the single-layered model are expressed by inputting the nominal Young's modulus, yield stress, tangent modulus, and plastic failure strain into the single-layered model, as shown in Table 6.1 and by the defined curves in Fig. 6.10a, b.

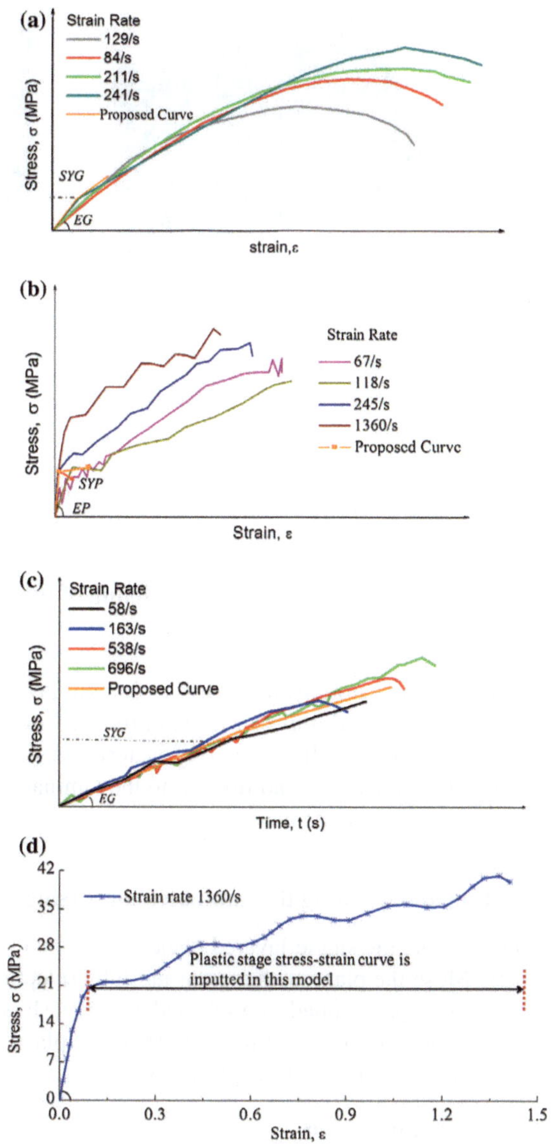

Fig. 6.10 Mechanical properties of the glass and PVB in the FE model. **a** Stress–strain curves obtained from the glass dynamic compression test with a simplified curve defined to express the mechanical properties of glass in the single-layered model. **b** Stress–strain curve obtained from the PVB dynamic tensile test with a simplified curve defined to express the mechanical properties of PVB in the single-layered model. **c** Stress–strain curves obtained from the glass dynamic tensile test with a simplified curve defined to express the mechanical properties of the lower glass in the triple-layered model. **d** Stress–strain curve obtained from the PVB dynamic tensile test to express the mechanical properties of PVB in the triple-layered model

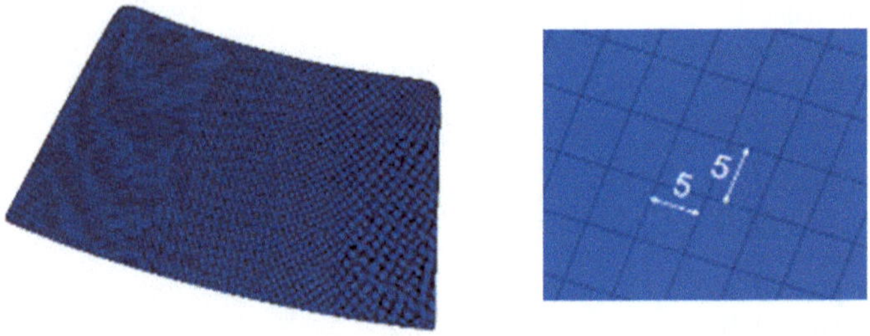

Fig. 6.11 FE model of laminated glass

Table 6.2 Dimensions in HyperMesh

	Mass	Length	Time	Force	Density	Young's modulus	Velocity	Gravity
Eg	kg	m	s	N	1E+03	2E+011	10.5	9.81
Unit	ton	mm	s	N	1E+03	2E+5	1.05E+04	9.81E+03

6.2.1.3 Tutorial on Numerical Modeling of Single-Layered Laminated Glass

In this series of tutorials, the process of single-layered numerical modeling of laminated glass on head impact will be introduced step by step. The main contents of the tutorials are as follows: editing material, defining the element properties, and defining the boundary condition e.t. to the laminated glass and the contact between the laminated glass and head.

Creating and meshing the components
Creating and meshing the laminated glass (single-layered model)

(1) Create the single-layered plane.
(2) Mesh the plane with regular mesh by selecting **2D** → **automesh**. In the 2D mesh panel, quads are selected as the mesh type, and the element size is 5 × 5 mm, as shown in Fig. 6.11. Then, the plane is meshed by choosing "**surfs**," and the surface of the plane is selected.

1. **Defining materials**

Defining the material of laminated glass
MAT_032 is chosen as the material of laminated glass and will be assigned to this component. The dimensions in HyperMesh are shown in Table 6.2.

(1) Click the  button, and the dialog box of materials will appear. Input the material's name "windshield." The "**type**" is defined as "**ALL**," then edit the "**card image**" and choose the MAT_032 ("MAT_LAMINATED_GLASS") as the "**card image**" selection, as shown in Fig. 6.12. MAT_032, which stands for

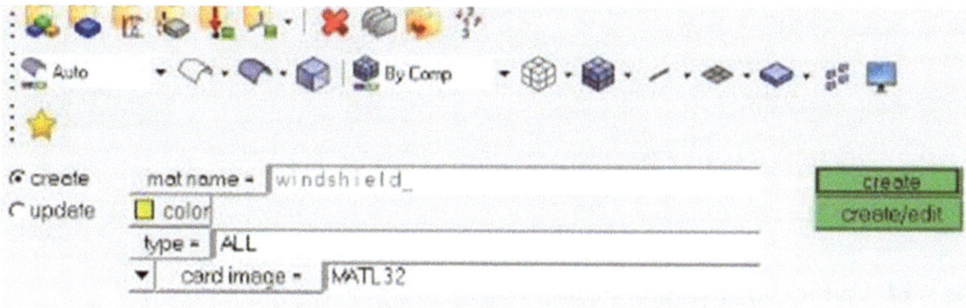

Fig. 6.12 Creating the material of laminated glass

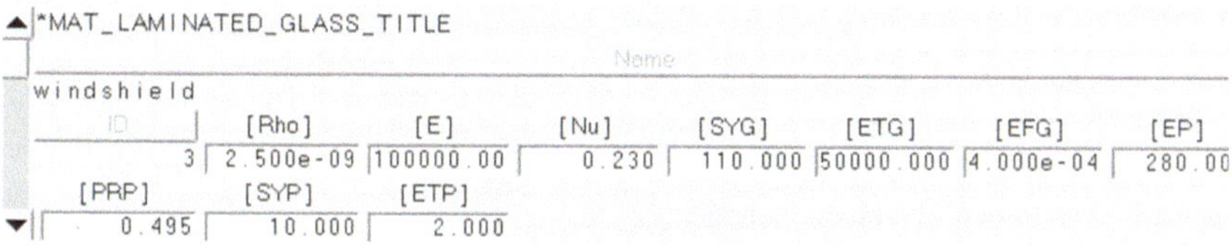

(a) Data form of MATL32

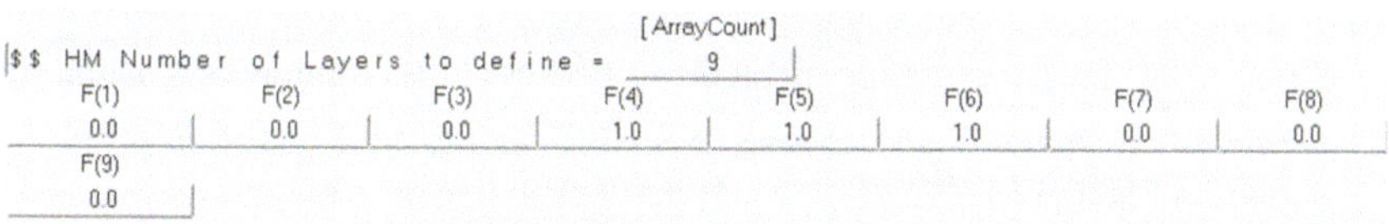

(b) Defining the integral layer

Fig. 6.13 Material editor of MAT_032

the laminated glass material model in LS-DYNA, may be able to describe the multilayered structure with only one shell element.

(2) Click the "**create/edit**" button, and enter the corresponding values for the parameters of materials, as shown in Fig. 6.13a. The definition of the integral layer is shown in Fig. 6.13b. The following important parameters are described:
For glass, "**Rho**" is the mass density, "**E**" is the Young's modulus, "**Nu**" is the Poisson's ratio, "**SYG**" is the yield stress, and "**ETG**" is the tangent modulus.
For PVB, "**EP**" is the Young's modulus, "**PRP**" is the Poisson's ratio, "**SYP**" is the yield stress, and "**ETP**" is the tangent modulus.
The definition of the multilayered structure of laminated glass by using MAT_032 and integral rule in LS-DYNA ("Integration_shell" card) for combining single-layered shell elements into a three-layered shell element is described in detail in Sect. 6.2.1.1. As shown in Fig. 6.13b, when the $F_n = 0$, the layer of integral denotes the glass layers and PVB layer when $F_n = 1$.

(3) Click "**return**" to exit the material editor.

Defining the section properties
Defining the section properties of laminated glass

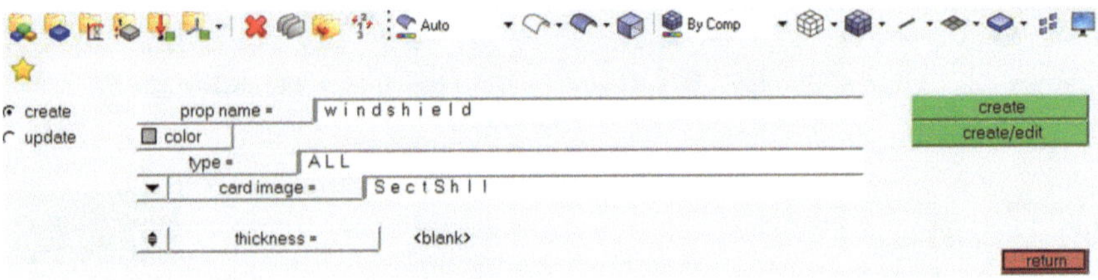

Fig. 6.14 Creating the property of laminated glass

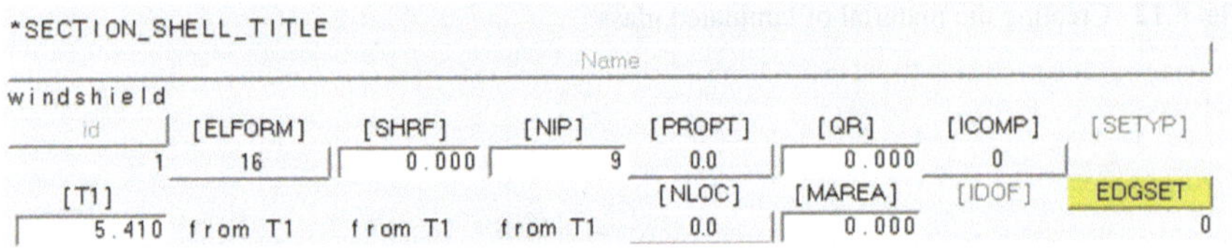

Fig. 6.15 Data form of shell element property

(1) Click  button to create element properties. In the dialog box, name the element "windshield" and choose "**Sectshll**" as the card image selection, as shown in Fig. 6.14.
(2) Click the "**create/edit**" button. The properties editor will then appear. Enter the value of each parameter, as shown in Fig. 6.15. Some important parameters are described as follows:
"**ELFORM**" is the element formulation options, and "16" stands for fully integrated shell element (very fast).
"**NIP**" is the number of through-thickness integration points.
"**IRID**" is the quadrature rule or integration rule ID.
"**T1**" is the shell thickness at node n1, unless the thickness is defined on the *ELEMENT_SHELL_OPTION card.
"**NLOC**" stands for the location of the reference surface for three-dimensional shell elements. "**1.0**," "**0.0**," and "**−1.0**" denote the element offset direction from the reference surface of the top surface, the mid-surface, and the bottom surface of the shell element, respectively.
(3) Click "**return**" to exit the properties editor.
Adding materials and properties to components
Adding material and properties to laminated glass

(1) In the model tree, click "**windshield**," and the editor box will appear.
(2) Add the "**Materials**" and "**Properties**" to "**windshield**." Click "**Property**," and the "**Select Property**" box will appear. Choose "**windshield**." Then, click "**Material**" and choose "**windshield**," as shown in Fig. 6.16.
(3) Click "**OK**" to exit the select editor.

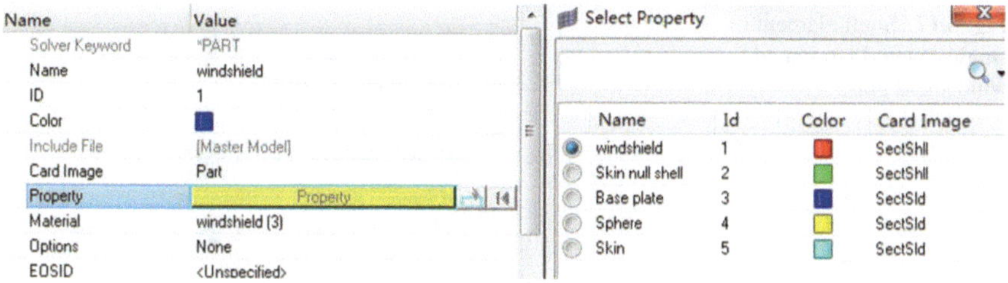

(a) Adding property to "windshield"

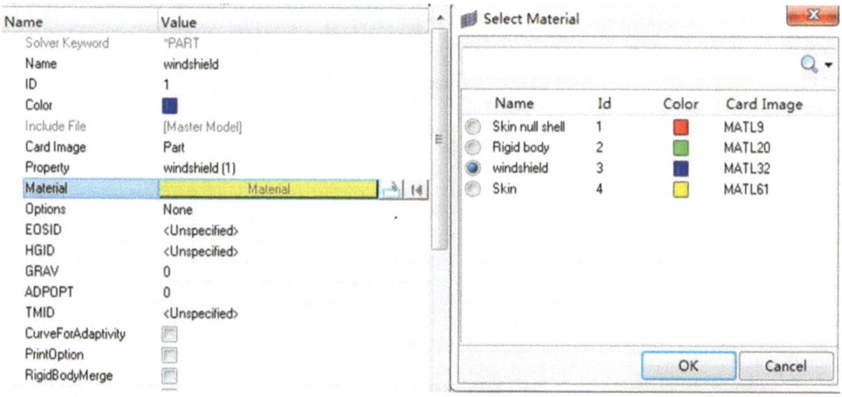

(b) Adding material to "windshield"

Fig. 6.16 Adding materials and properties to the windshield

The shell element with a thickness of 5.41 mm stands for laminated glass, as shown in Fig. 6.17.

2. **Importing the headform model**

Click , and the **import** dialog box will appear. For **File type**, choose LS-DYNA (keyword) and **Headform .k** files, as shown in Fig. 6.18 (without the CAD process in detail here).

Then, the center line of the headform model is translocated by 35° above horizontal, as shown in Fig. 6.18a. The head model is an assembly with four individual components, as shown in Fig. 6.18b.

The processes of meshing and applying materials and properties to headform model are the same as steps 2, 3, and 4. The material and properties of the four individual components are shown in Fig. 6.19.

We mainly define the constrained, contact, and initial conditions on headform and laminated windshield in detail.

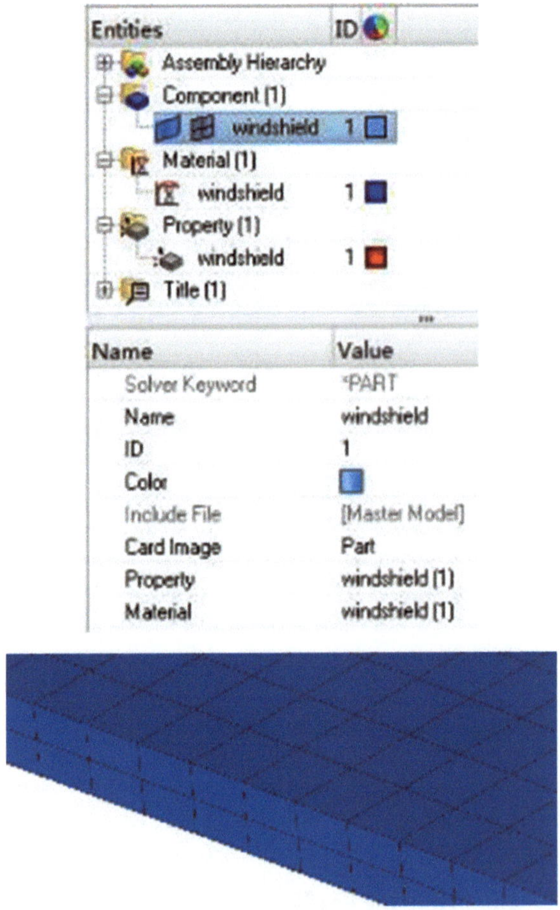

Fig. 6.17 Shell element of single-layered laminated windshield glass

Defining sets

The set of part on the headform model for defining the contact condition is defined next.

(1) Click **Tools** → **Create cards** → ***SET** → ***SET_PART** → ***SET_PART_LIST** on the toolbars at the top to define the set of the base plate and the sphere.
(2) In the "Create set" dialog box, name the set "**Contact part set(head)**" and choose "**Base plate**" and "**Sphere**," as shown in Fig. 6.20.

The node set on the headform model is then defined to define the contact condition.

(1) Click **Tools** → **Create cards** → ***SET** → ***SET_NODE** → ***SET_NODE_LIST** on the toolbars at the top.
(2) In the "Create set" dialog box, name the set "**Extra rigid**" and choose the node of contact area of "**Skin**" and "**Sphere**" on "**Skin**," as shown in Fig. 6.21.

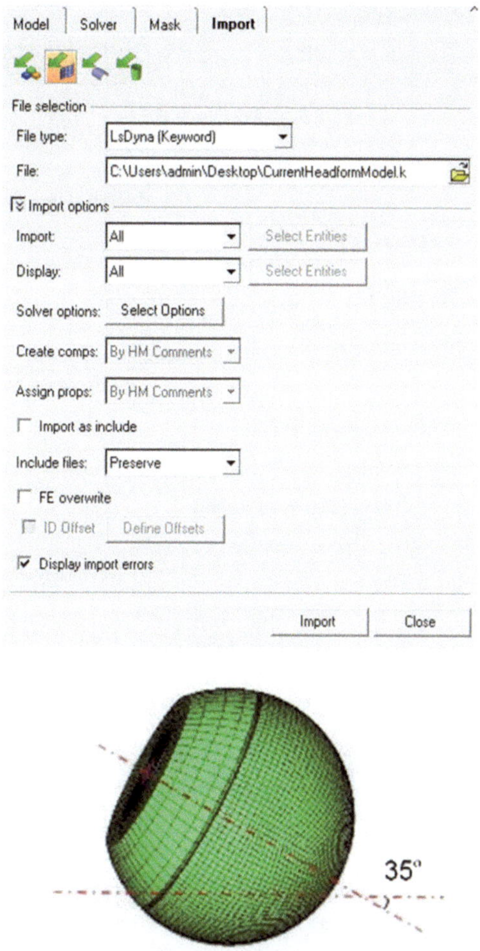

(a) Importing the headform model

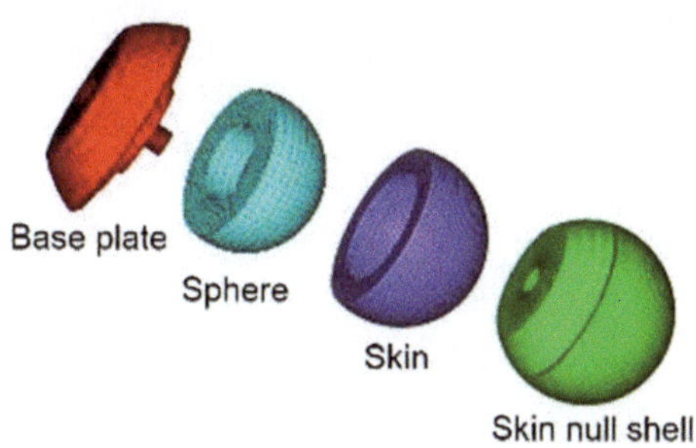

(b) Four individual components of headform

Fig. 6.18 Importing the headform model with four individual components

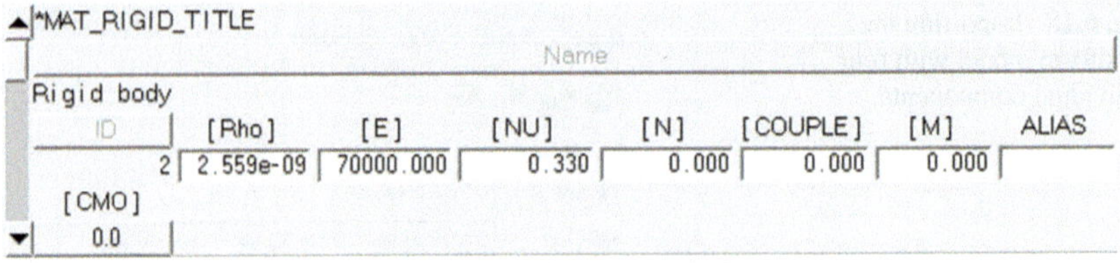

(a) Materials of base plate and sphere

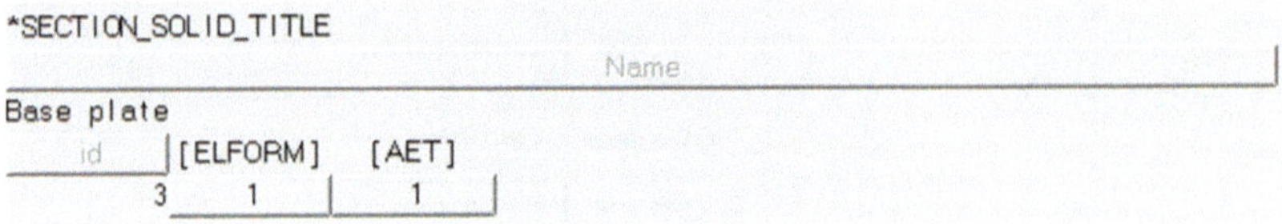

(b) Property of base plate

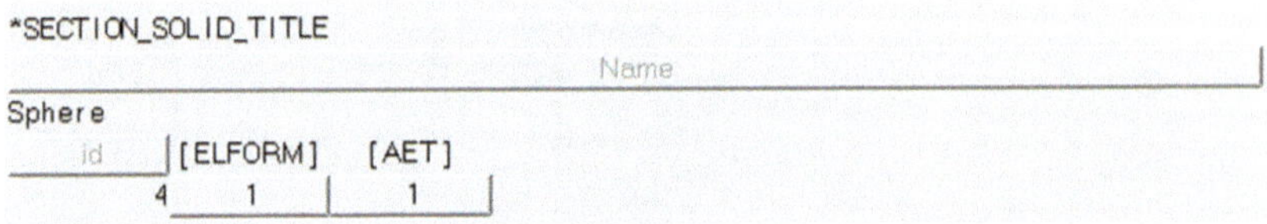

(c) Property of sphere

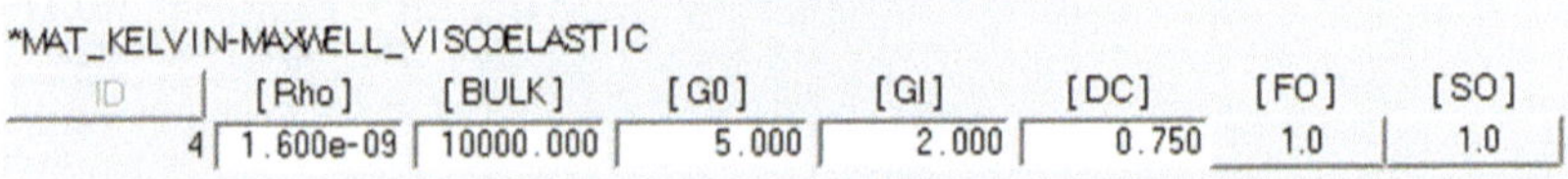

(d) Material of skin

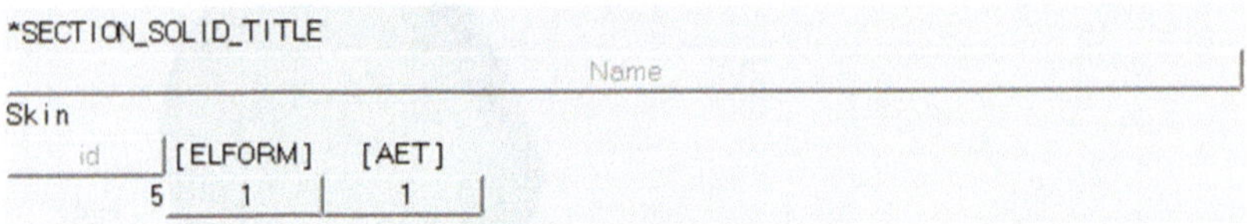

(e) Property of skin

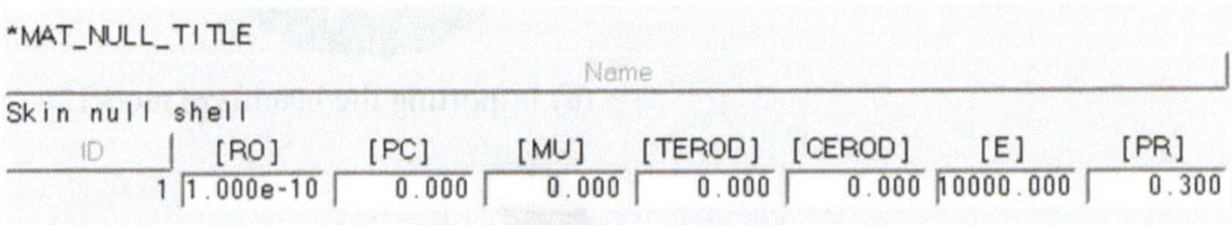

(f) Material of skin null shell

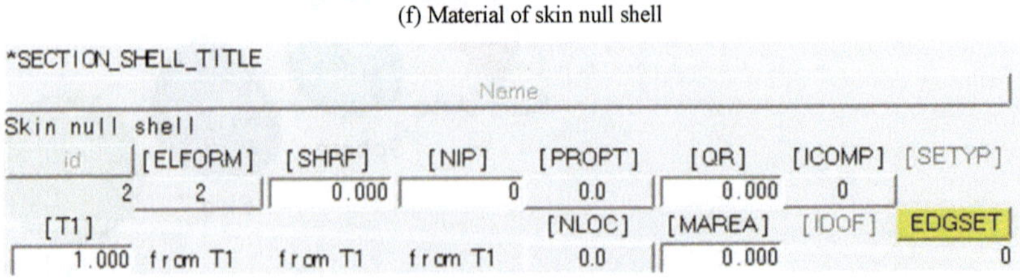

(g) Property of skin null shell

Fig. 6.19 Materials and properties of four individual components of the headform model

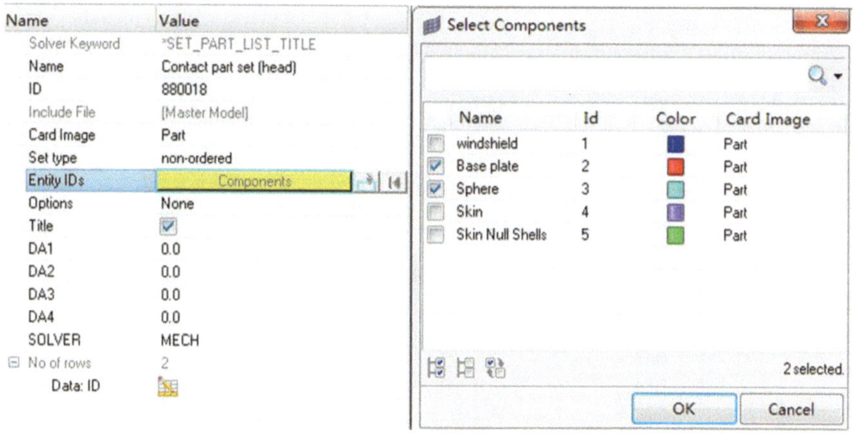

Fig. 6.20 Defining the part set: choosing base plate and sphere

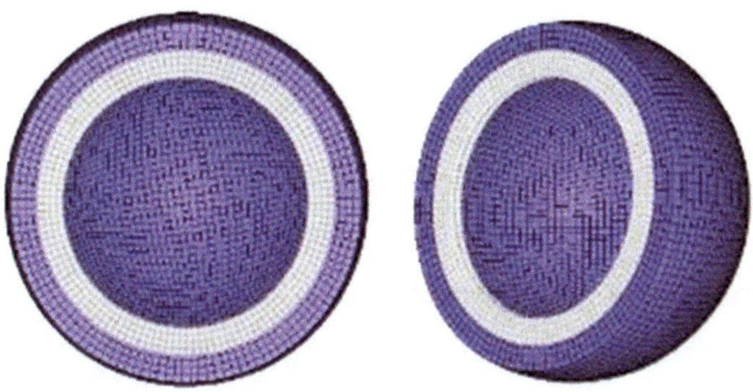

Fig. 6.21 Defining the extra rigid node set of sphere

The node set on the headform model is then defined to define the initial condition.

(1) Click **Tools** → **Create cards** → ***SET** → ***SET_NODE** → ***SET_NODE_LIST** on the toolbars at the top.
(2) In the "Create set" dialog box, name the set "**Head**" and choose the whole nodes of the headform, as shown in Fig. 6.22.

Defining the constrained conditions to components
Applying boundary condition to the laminated glass

(1) Click **Tools** → **Create cards** → ***BOUNDARY** → ***BOUNDARY_SPC_NODE** on the toolbars at the top to apply constraints to the laminated glass.
(2) In the "Create boundary" dialog box, choose the nodes of the four edges of laminated glass, and then the laminated glass is fixed, as shown in Fig. 6.23.

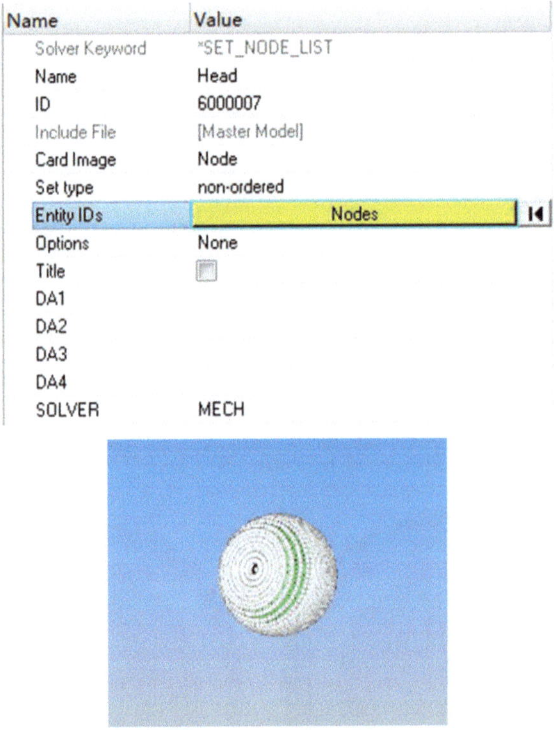

Fig. 6.22 Defining the set of the whole headform model

Fig. 6.23 Creating the boundary of windshield

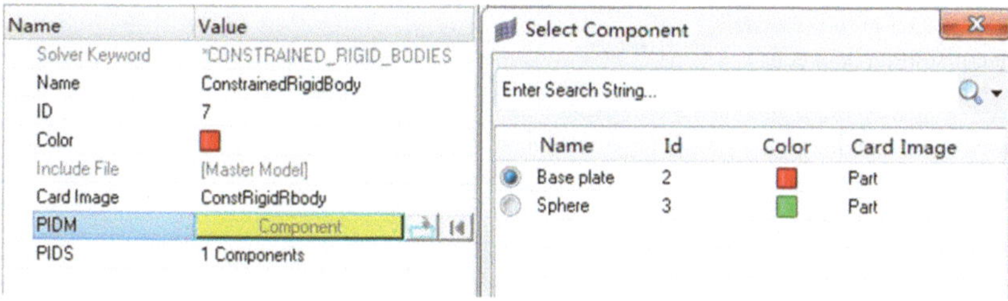

(a) "**PIDM**" choosing the component "**Base plate**"

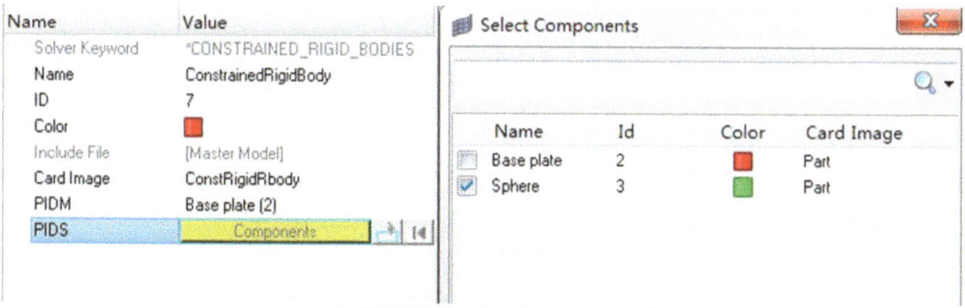

(b) "**PIDS**" choosing the component "**Sphere**"

Fig. 6.24 Defining the rigid-body constraints between base plate and sphere

Defining constraints on the headform

(1) Click **Tools** → **Create cards** → ***CONSTRAINED** → * **CONSTRAINED_RIGID_BODIES** on the toolbars at the top to define the rigid-body constraints of "**Base plate**" and "**Sphere**."
(2) In the "Create boundary" dialog box, "**PIDM**" chooses the "**Base plate**," and "**PIDS**" chooses the "**Sphere**," as shown in Fig. 6.24.
(3) Then, click **Tools** → **Create cards** → ***CONSTRAINED** → * **CONSTRAINED_EXTRA_NODES_SET** on the toolbars at the top to define the extra node constraints between "**Sphere**" and "**Skin**."
(4) In the "Create boundary" dialog box, "**PID**" chooses the component "**Sphere**," and "NSID" chooses the set of "**Extra rigid**," as shown in Fig. 6.25.

Defining the contact conditions on the components
First, the contact on the headform is defined.

(1) Choose the "***CONTACT_AUTOMATIC_SURFACE_TO_SURFACE**" card to create contact on the headform named "**sef_contact_head**."
(2) In the contact dialog box, "**SSID**" chooses "**Contact part set (head)**" and "**MISD**" chooses "**skin**." Input the friction value of **FS**, **FD**, as shown in Fig. 6.26.

"**SSID**" stands for slave segment, and "MISD" stands for master segment.

"**FS**" is the static coefficient of friction, and "FD" is the dynamic coefficient of friction.

Then, the single contact of "**skin null shell**" on the headform is defined.

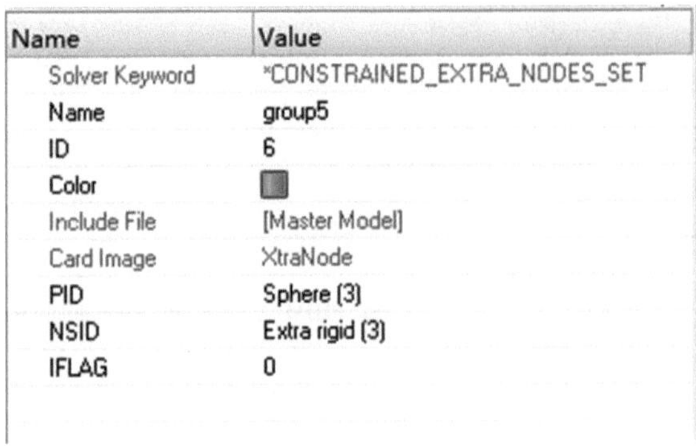

Name	Value
Solver Keyword	*CONSTRAINED_EXTRA_NODES_SET
Name	group5
ID	6
Color	
Include File	[Master Model]
Card Image	XtraNode
PID	Sphere (3)
NSID	Extra rigid (3)
IFLAG	0

Fig. 6.25 Defining the extra node constraints between sphere and skin

(1) Choose the "*CONTACT_AUTOMATIC_SINGLE_SURFACE" card to create contact of "**skin null shell**" and name it "**skin null shell**."
(2) In the contact dialog box, "SSID" chooses "**skin null shell**," and input the friction value of FS, FD, as shown in Fig. 6.27.

The contact type between laminated glass and head is surface to surface.

(1) Choose the "*CONTACT_AUTOMATIC_SURFACE_TO_SURFACE" card to create contact between head and windshield. Name this contact "head_windshield."
(2) In the contact dialog box, "SSID" chooses "windshield," and "MISD" chooses "skin null shell." Input the friction value of FS, FD, as shown in Fig. 6.28.

Applying velocity to the head

(1) Click **Tools** → **Create cards** → ***INITIAL** → ***INITIAL_VELOCITY_GENERATION** on the toolbars at the top.
(2) In the "Create load" dialog box, the **Load Collector** editor appears so that the load initial velocity can be named. "**NSID**" chooses the set of "**Head**." Then, input the velocity values of X, Y, Z in the blanks, as shown in Fig. 6.29. The initial velocity is 9.4 m/s.

Setting the output and control card
Defining the accelerometer on the headform for obtaining the acceleration data

(1) Click **Tools** → **Create cards** → ***ELEMENT_SEATBELT_ACCELEROMETER** → * on the toolbars at the top.
(2) In the "Create load" dialog, click the "**node**" button and choose three nodes on the "**Base plate**" component to create the shape of an isosceles right triangle element to represent the accelerometer, as shown in Fig. 6.30a, b. The three nodes of the element must be rigid nodes on the rigid bodies.
(3) Click  and choose the accelerometer element, as shown in Fig. 6.30c. Then, check the keyword menu of the accelerometer and input the values of "**IGRAV**" and "**INTOPT**," as shown in Fig. 6.30d.

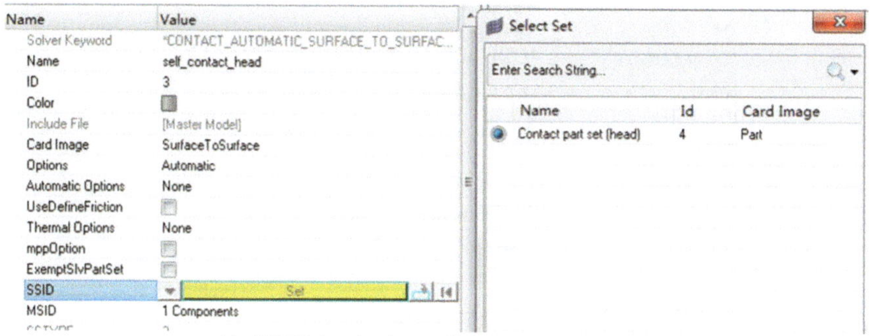

(a) "**SSID**" choosing the set of "**Contact part set(head)**"

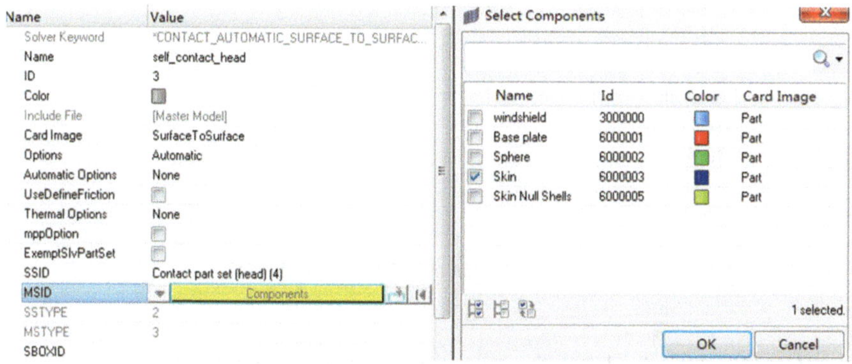

(b) "**MSID**" choosing the component of "skin"

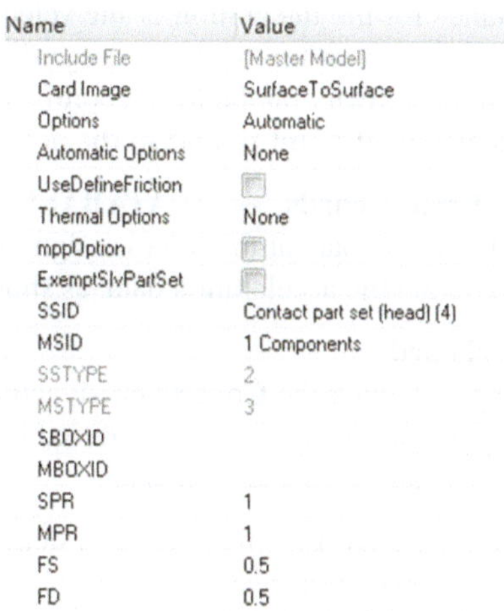

(c) Inputting the friction coefficient values

Fig. 6.26 Defining the contact of head

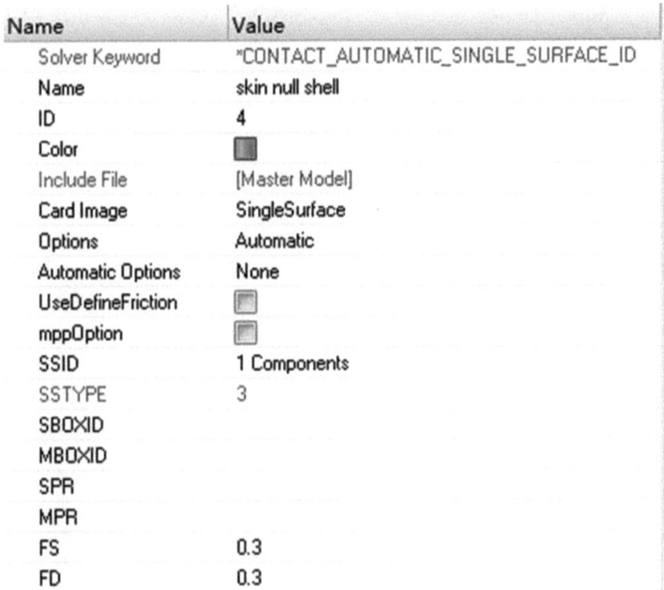

Name	Value
Solver Keyword	*CONTACT_AUTOMATIC_SINGLE_SURFACE_ID
Name	skin null shell
ID	4
Color	
Include File	[Master Model]
Card Image	SingleSurface
Options	Automatic
Automatic Options	None
UseDefineFriction	
mppOption	
SSID	1 Components
SSTYPE	3
SBOXID	
MBOXID	
SPR	
MPR	
FS	0.3
FD	0.3

Fig. 6.27 Defining the single contact of "skin null shell" on headform

"**IGRAV**": "0" stands for the output data including gravity; "1" stands for the output data without gravity.

"**INTOPT**": "0" stands for the integration of the velocity on the overall coordination system followed by conversion to the local coordination system; "1" stands for direct integration of the velocity on the local coordination system.

Assigning the measurement point to obtain the acceleration data

(1) Click **Tools** → **Create cards** → ***DATABASE** → ***DATABASE_HISTORY_NODE** on the toolbars at the top to assign the node "**N1**" as the measurement point to obtain the acceleration data, as shown in Fig. 6.31.

Defining the control card

Enter the **Analysis** panel, click the **Control card**, and then set the keywords, as shown in Fig. 6.32. Each control card [3] is introduced below. More detailed steps are given in the LS-DYNA keyword users' manual.

(a) ***CONTROL_ACCCURACY**: This is used to define the control parameters that can improve the calculation accuracy.

"OSU" is the global flag for second-order objective stress updates.
"INN" is the invariant node numbering for shell and solid elements.
"PIDOSU" is the part set ID for objective stress updates.

(b) ***CONTROL_BULK_VISCOSITY**: This is used to reset the default values of the bulk viscosity coefficients globally.

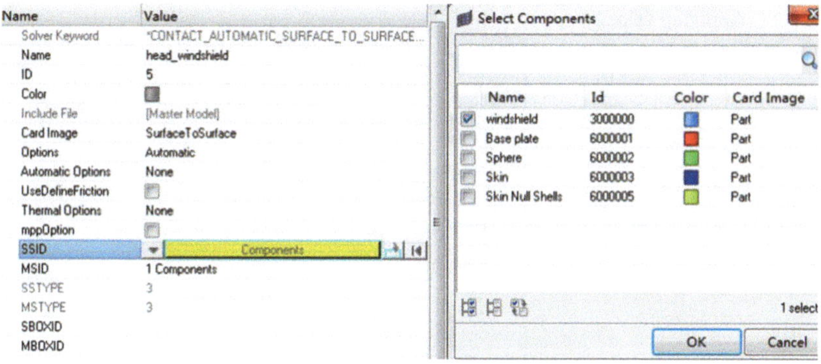

(a) "**SSID**" choosing the component "**windshield**"

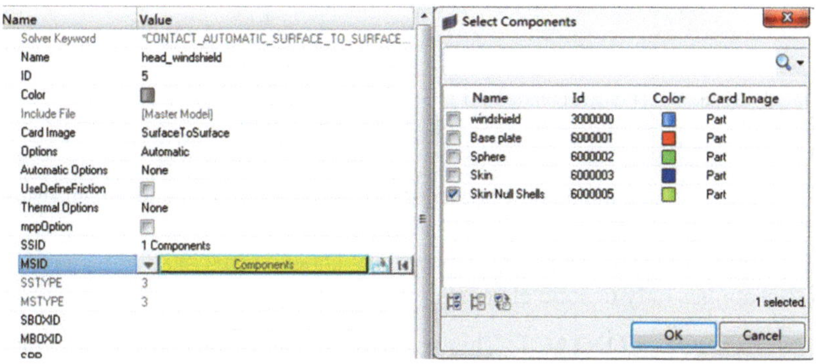

(b) "**MSID**" choosing the component "**skin null shells**"

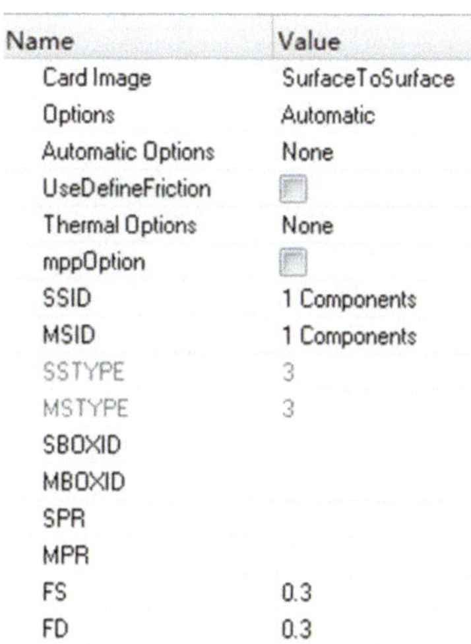

(c) Inputting the friction coefficient values

Fig. 6.28 Defining the contact between head and windshield

Fig. 6.29 Applying velocity to the headform

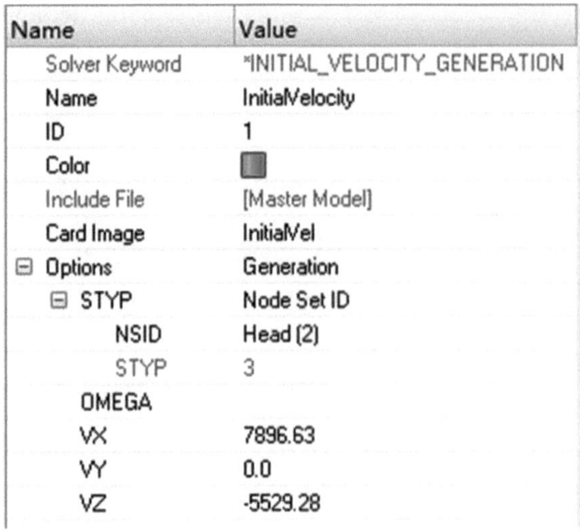

Name	Value
Solver Keyword	*INITIAL_VELOCITY_GENERATION
Name	InitialVelocity
ID	1
Color	
Include File	[Master Model]
Card Image	InitialVel
⊟ Options	Generation
⊟ STYP	Node Set ID
NSID	Head (2)
STYP	3
OMEGA	
VX	7896.63
VY	0.0
VZ	-5529.28

"Q1" and "Q2" are the default quadratic viscosity coefficient and linear viscosity coefficient, respectively.

The bulk viscosity is necessary to propagate shock waves in solid materials and therefore applies only to solid elements.

(c) ***CONTROL_CONTACT**: This is used to change defaults for computation with contact surfaces.

"SLSFAC" is the scale factor for sliding interface penalties.

"RWPNAL" is the scale factor for rigid wall penalties.

"ISLCHK" is the initial penetration check in contact surfaces with indication of initial penetration in output files.

The meanings of other keywords can be viewed in the LS-DYNA keyword users' manual.

(d) ***CONTROL_ENERGY**: This provides controls for energy dissipation options. It aims to maintain stable energy during simulation and ensure the accuracy of the simulation results.

"HGEN" is the hourglass energy calculation option.

"RWEN" is the rigid wall energy dissipation option.

"SLNTEN" is the sliding interface energy dissipation option.

"RYLEN" is the Rayleigh energy dissipation option.

(e) ***CONTROL_OUTPUT**: This is used to set the miscellaneous output parameters. It aims to control the measurement points and output data in simulations.

"NPOPT" is the print suppression during the input phase flag for the "d3hsp" file.

"NREFUP" is the flag to update the reference node coordinated for beam formulations.

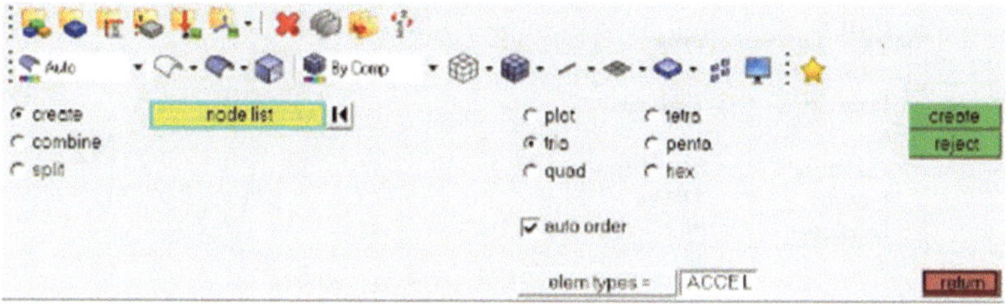

(a) Interface for creating accelerometer

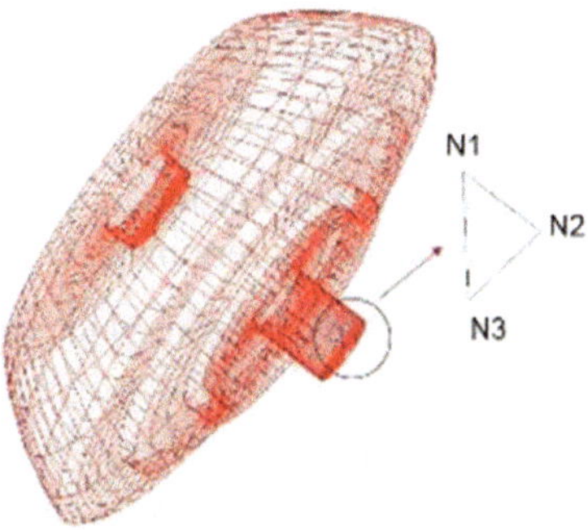

(b) Choosing three nodes on the "**Base plate**" component

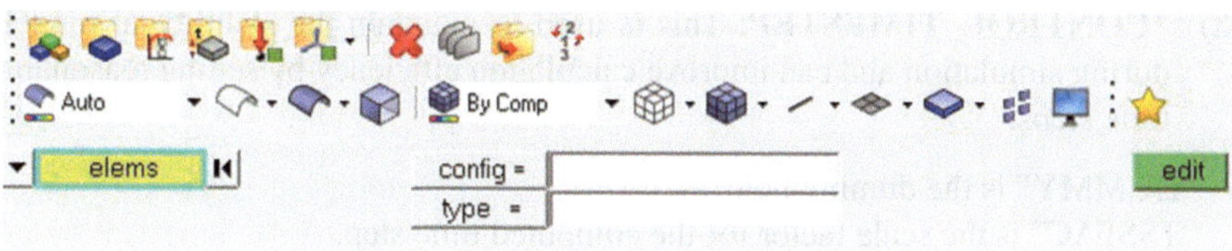

(c) Interface for choosing the accelerometer element

(d) Keyword menu of accelerometer

Fig. 6.30 Creating the accelerometer

The meanings of other keywords can be viewed in the LS-DYNA keyword users' manual.

(f) ***CONTROL_SHELL**: This is used to control the shell element. It aims to change the failure parameters NFAIL1 and NFAIL2 if necessary.

"NFAIL1" is the flag to check for highly distorted under-integrated shell elements, print a message, and delete the element or terminate.

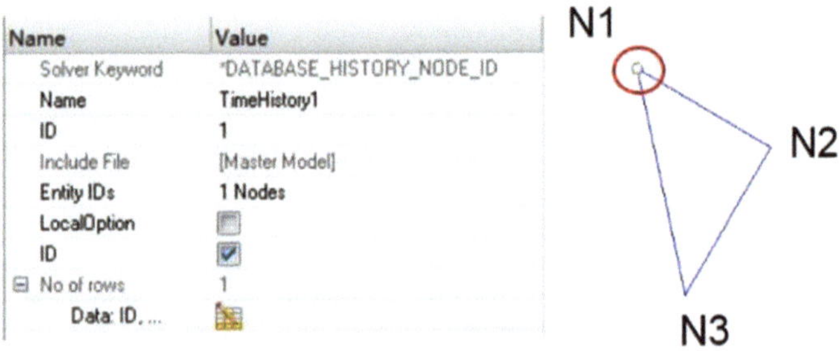

Fig. 6.31 Setting the output measurement point

"NFAIL2" is the flag to check for highly distorted, fully integrated shell elements, print a message, and delete the element or terminate.

(g) ***CONTROL_TERMINATION**: This is used to calculate the simulation time and stop the job.

"ENDTIM" is the termination time.

"ENDCYC" is the termination cycle. The termination cycle is optional and will be used if the specified cycle is reached before the termination time.

(h) ***CONTROL_TIMESTEP**: This is used to maintain the stability of models during simulation and can improve calculation efficiency by setting reasonable time steps.

"DUMMY" is the dummy field.

"TSSFAC" is the scale factor for the computed time step.

"DT2MS" is the new time step for mass-scaled calculation. The mass scaling must be active in the time zero analysis.

(i) ***DATABASE_BINARY_D3PLOT**: This is used to output the d3plot files.

"DT" is the time step of output animation.

(j) ***DATABASE_BINARY_D3THDT**: This is used to store the history data of elements.

"DT" is the time step of output history data.

(k) ***DATABASE_EXTENT_BINARY**: This is used to control the content of binary output databases d3plot, d3thdt, and d3part to some extent.

The meanings of keywords can be viewed in the LS-DYNA keyword users' manual.

Exporting the LS-DYNA Key file

(1) Click , and then click **Export** to export the file as a key file, as shown in Fig. 6.33.

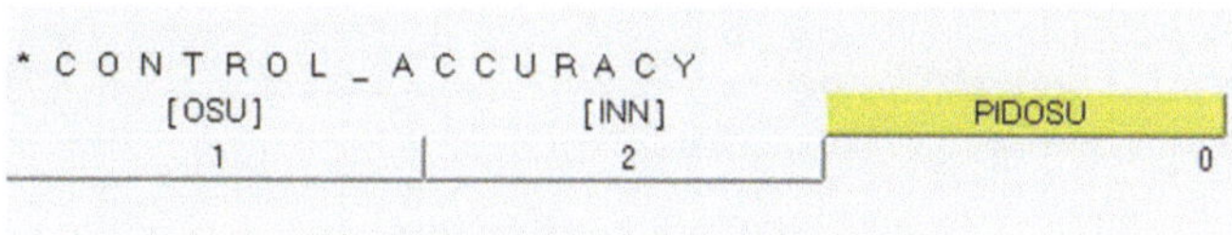

(a) CONTROL_ACCCURACY

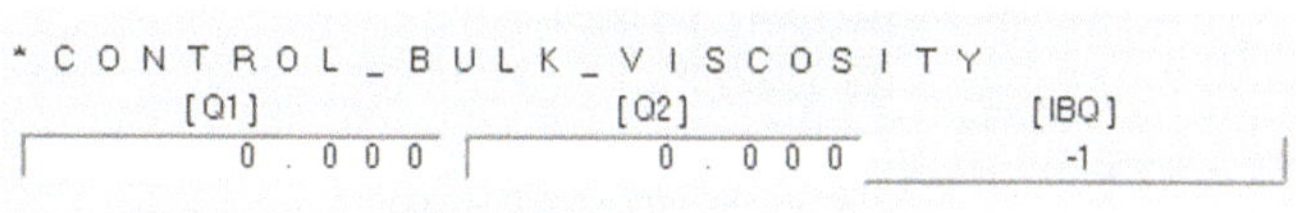

(b) CONTROL_BULK_VISCOSITY

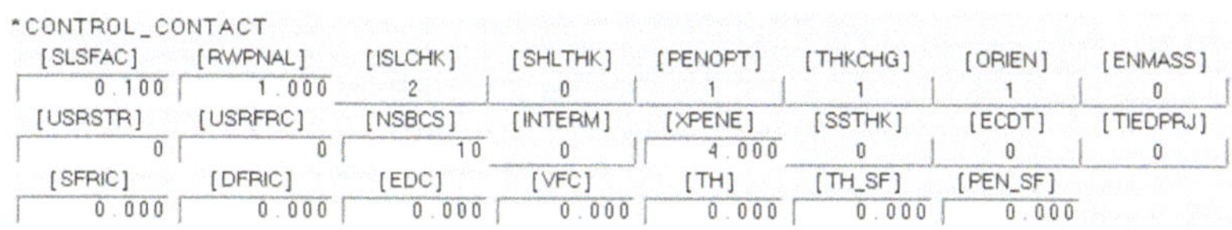

(c) CONTROL_CONTACT

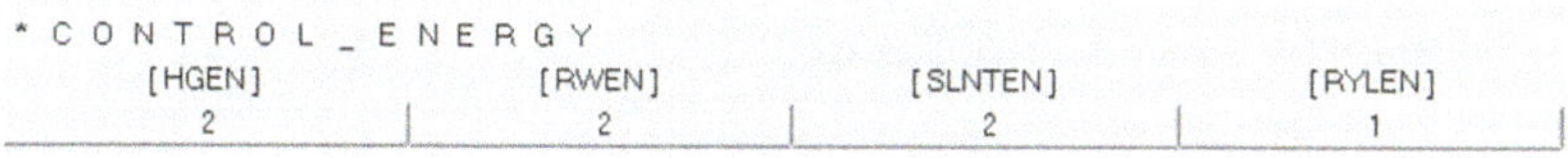

(d) CONTROL_ENERGY

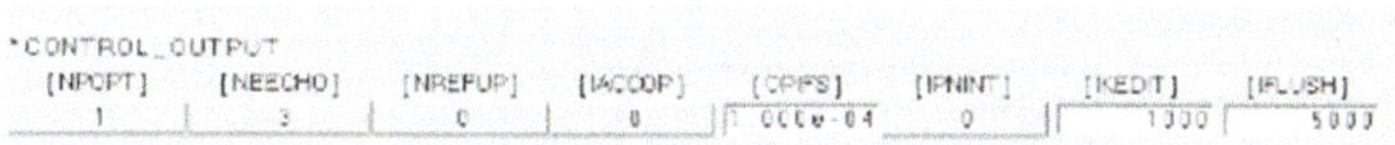

(e) CONTROL_OUTPUT

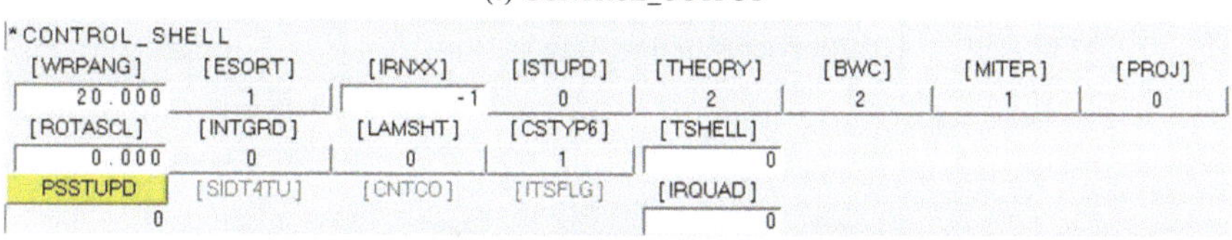

(f) CONTROL_SHELL

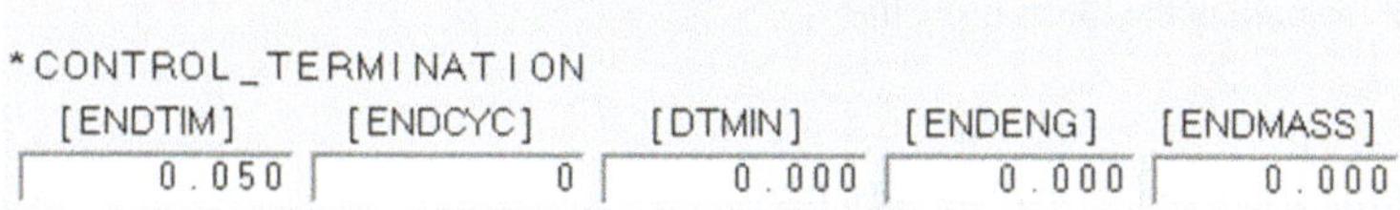

(g) CONTROL_TERMINATION

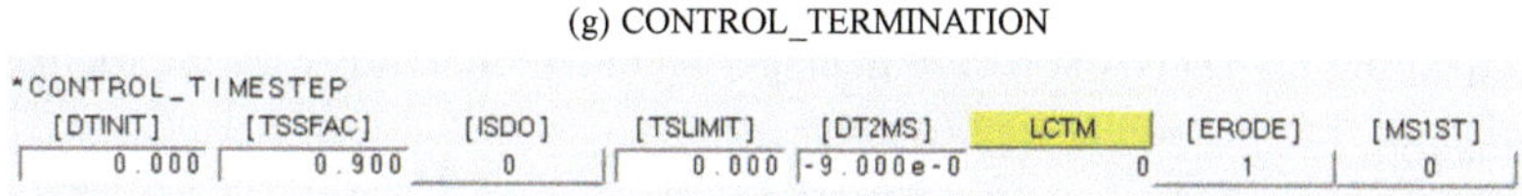

(h) CONTROL_TIMESTEP

Fig. 6.32 Setting the keywords

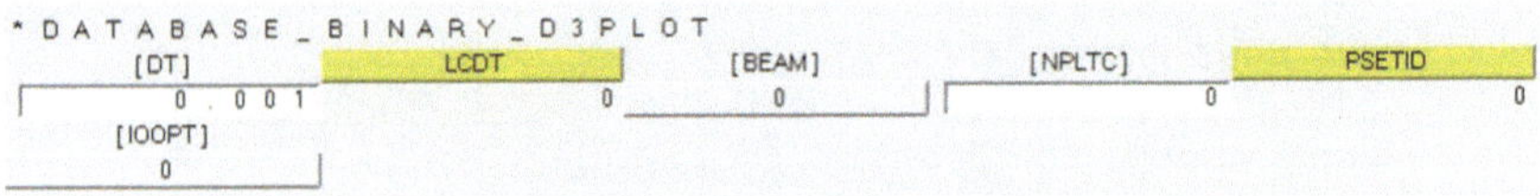

(i) DATABASE_BINARY_D3PLOT

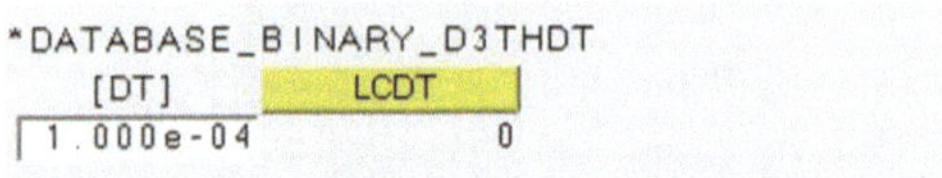

(j) DATABASE_BINARY_D3THDT

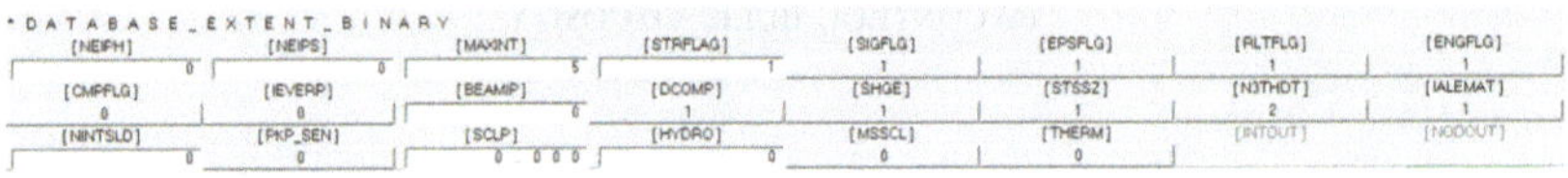

(k) DATABASE_EXTENT_BINARY

Fig. 6.32 (continued)

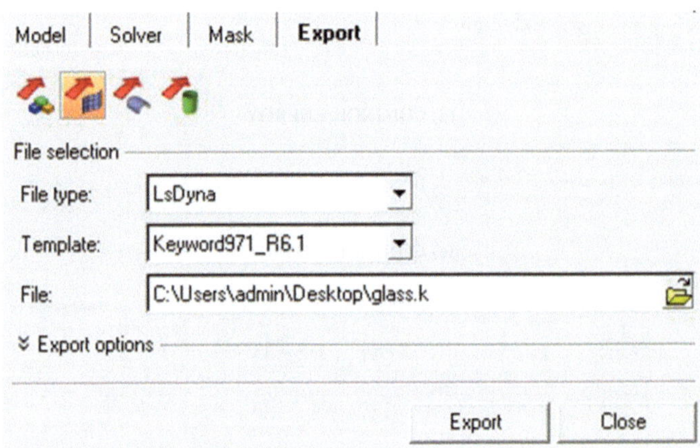

Fig. 6.33 Exporting the file as a key file

Submitting the LS-DYNA key file

(1) Open the LS-DYNA software manager and then click on the toolbars at the top.
(2) Submit the key file and click "RUN," as shown in Fig. 6.34.

Dealing with the simulation results in LS-DYNA Prepost

(1) Open the LS-DYNA Program Manager and click on the toolbars at the top.
(2) Click **File → Open → LS-DYNA Binary Plot**, as shown in Fig. 6.35, to input the d3plot file.

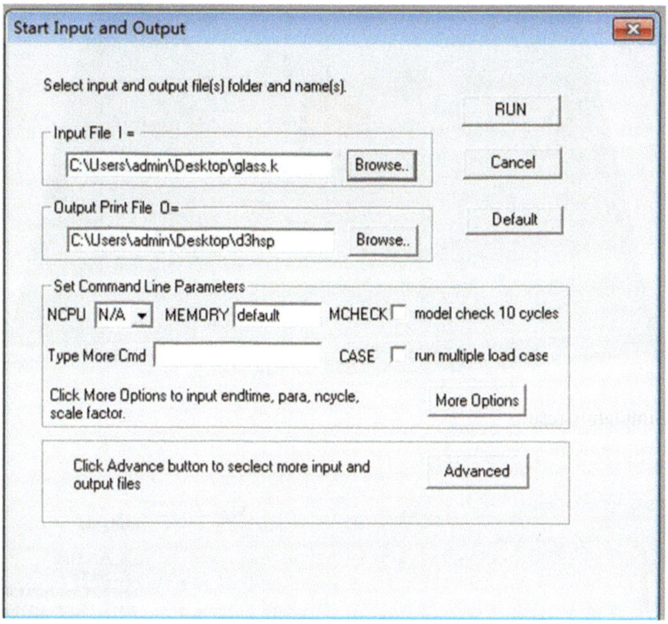

Fig. 6.34 Submitting the key file in LS-DYNA

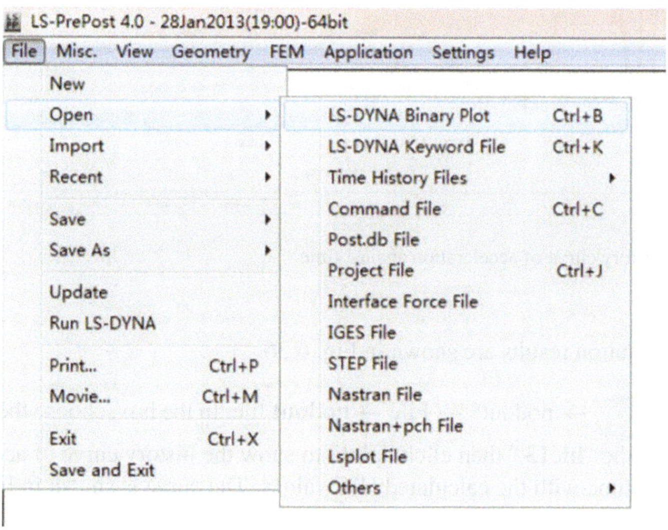

Fig. 6.35 Inputting the plot file

Fig. 6.36 Simulation results

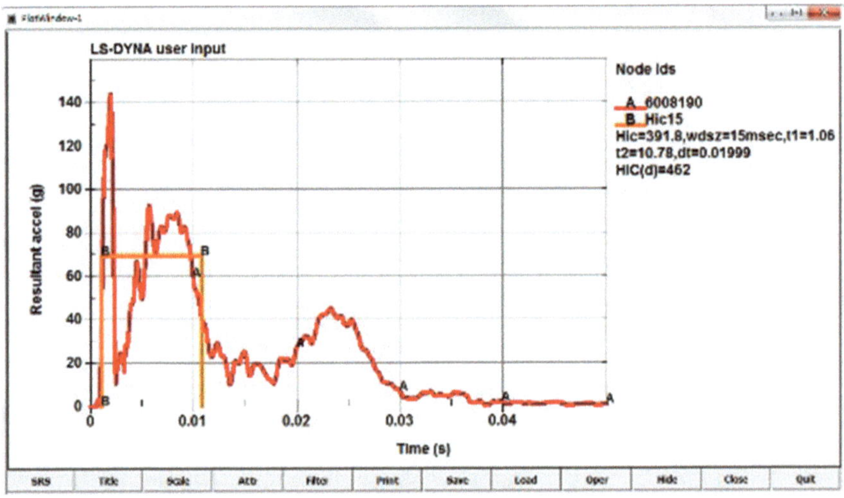

Fig. 6.37 History curve of acceleration against time

The simulation results are shown in Fig. 6.36.

(3) Click  → nodout* → File → **nodout** file. In the box, choose the node ID, choose the "**hic15**," then click "**plot**" to show the history curve of acceleration against time with the calculated HIC values. The curve is shown in Fig. 6.37.

6.2.2 *Modeling of Triple-Layered Laminated Glass*

The element in the single-layered windshield FE model should not be deleted, which results in the crack becoming invisible. The triple-layered windshield FE model is a more physically realistic model than the single-layered model and enables both glass and PVB failures. Thus, cracking patterns are expected to be realistic. The triple-layered FE model of laminated glass is proposed as follows.

6.2.2.1 FE Model Structure of Triple-Layered Laminated Glass

The triple-layered windshield FE model consists of two layers of fully integrated shell element (upper and lower) as shown in Fig. 6.2b, the thickness values of which are 2.55 and 2.1 mm, respectively. The interlayer of solid elements, with a thickness of 0.76 mm, represents the PVB material. In actual production, the laminated glass consists of two standard glasses sandwiched by PVB through pressing at high temperature and pressures. The upper and lower glasses connect with the interlayer PVB through node sharing, which simulates the adhesive strength of interlayer PVB and the upper/lower glass.

In FE, the default reference surface of the shell element is mid-surface, and the thickness extends up and down. In the meantime, the reference surface of the upper and lower shell elements are the lower layer surface and the upper layer surface for the thickness that extends up and down, respectively, thereby preventing parts of shell element from entering the PVB solid element.

6.2.2.2 Constitutive Description of Material for Triple-Layered Laminated Glass

To study the mechanical behavior of PVB in laminated glass, quasi-static and dynamic tensile experiments were performed. PVB exhibits viscoelastic material property under low-speed tension. Its response is influenced by loading speed. PVB demonstrates elasto-plastic material property under dynamic loading [11]. In addition, as Fig. 6.38 shows, within the vicinity of impact location, PVB had an irreversible plastic deformation (depression in the impact zone), and even the windshield is unloaded. Similar strategies were adopted in previous literature, where the elasto-plastic model was adopted to describe PVB [8, 12]. Thus, the material type MAT_123 "MAT_MODIFIED_PIECEWISE_LINEAR_PLASTICITY" (an elasto-plastic material model in LS-DYNA) is adopted for the material model to characterize the material properties of glass and PVB during the dynamic impact process.

For the MAT_123 material model, elasto-plastic mechanical properties can be expressed through bilinear stress–strain behavior, which can be characterized by defining the Young's modulus, failure strain, yield stress, and tangent modulus. Alternatively, when elastic-stage mechanical properties are expressed by defining

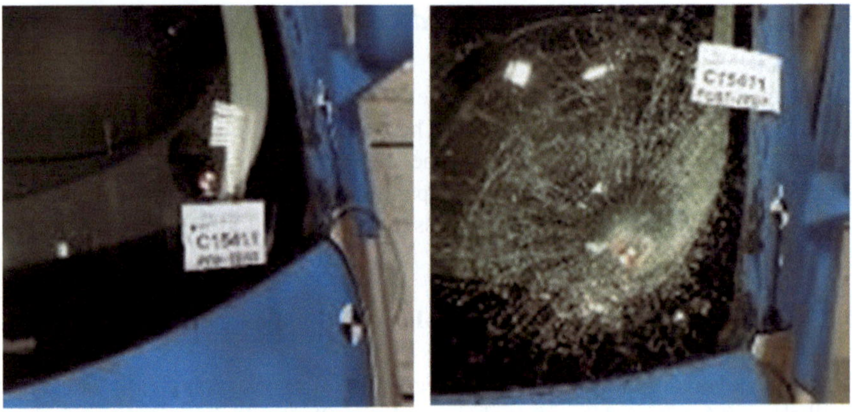

Fig. 6.38 Depression in the vicinity of impact location

the Young's modulus, yield stress, which is shown as a plastic stage stress–strain curve in Fig. 6.9b, can also be inputted to characterize the mechanical property in the plastic stage [3].

As mentioned earlier, the lower glass bears the tensile stress and fails first, whereas the upper glass bears the compression force and fails nearly immediately after the lower glass during the impact test. Thus, the mechanical properties of the upper and lower glasses are determined qualitatively by referring to all the dynamic compression and tensile test curves at different strain rates comprehensively [6], as shown in Figs. 6.10a, c.

As mentioned above, previous test and simulation study of windshield show that glass plastic failure strain is in the range of 0.01–0.1% [4–6, 8–10]. In the triple-layered model, an upper glass plastic failure strain of 0.04% is adopted as in the single-layered model. A smaller plastic failure strain of 0.024% is defined for the lower glass layer because the compression strength is greater than the tensile strength for glass. The local elements will be deleted when they achieve the defined plastic failure strain. The mechanical properties of the upper and lower glasses are also summarized and expressed in Table 6.3.

Figure 6.39 indicates that most of the distribution of the effective strain rate of PVB is in the range of 300/s–1500/s with 5 mm mesh size in the entire impact process. According to the dynamic tensile test of PVB in a previous study [7], the bilinear stress–strain curve with a strain rate of 1360/s is the most appropriate for defining the mechanical property of PVB. As Fig. 6.10d shows, for easy demonstration of PVB mechanical behavior at a the high strain rate, a pseudo yield stress σ where the material modulus changes abruptly and the corresponding pseudo yield strain ε are defined approximately [7]. Thus, the initial modulus is defined as the gradient of a secant line through the pseudo yielding point. The plastic stage stress–strain curve is used in this model to express plastic deformation. Table 6.3 presents the mechanical properties of PVB.

Table 6.3 Mechanical properties of the triple-layered structure model

	Triple-layered windshield FE model					
	Rho (kg/m^3)	*PRG*	*EG*(GPa)	*SYG* (MPa)	*ETG* (GPa)	*EFG*
Outer glass layer	2500	0.23	100	110	50	0.0004
Lower glass layer			68	16	60	0.00024
PVB	*Rho* (kg/m^3)	*EP* (MPa)	*PRP*	*LCSS*		
	1200	280	0.495	1360/s effective stress versus effective plastic strain curve		

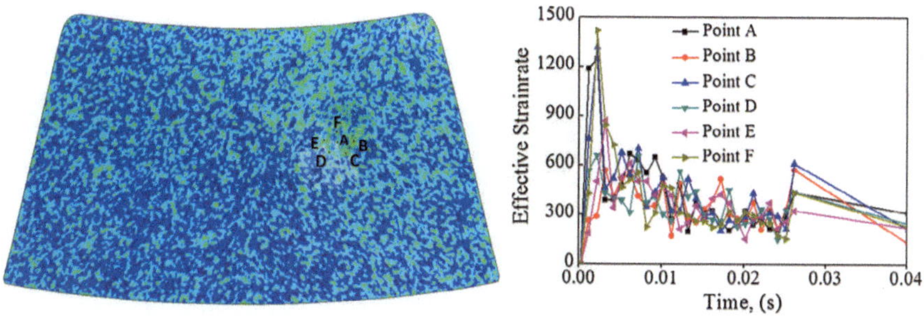

Fig. 6.39 Distribution of the effective strain rate of PVB in the triple-layered model

6.2.2.3 Tutorials on Numerical Modeling of Triple-Layered Laminated Glass

In this series of tutorials, the process of triple-layered numerical modeling of laminated glass on head impact will be introduced step by step. The main contents of the tutorials are as follows: editing the material, defining the element properties, adding materials and properties, and defining the boundary condition e.t. to the laminated glass and the contact between laminated glass and head.

1. **Creating and meshing the components**

   Creating and meshing the laminated glass (three-layered model)

(1) Create the solid element of glass model with a thickness of 0.76 mm (without the CAD processing in detail).

(2) Mesh the model with regular mesh by **3D** mesh in three-layered solid elements as shown in Figs. 6.40 and 6.41.

(3) Then, click "**faces**" on the interface of "**Tool**." The "**comps**" chooses PVB, and the "**tolerance**" is set as 0.001. Click the "**find face**" bottom to create face as shown in Fig. 6.42.

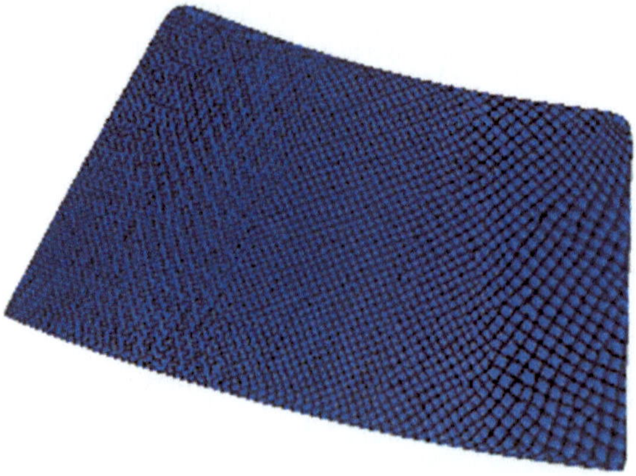

Fig. 6.40 FE model of laminated glass

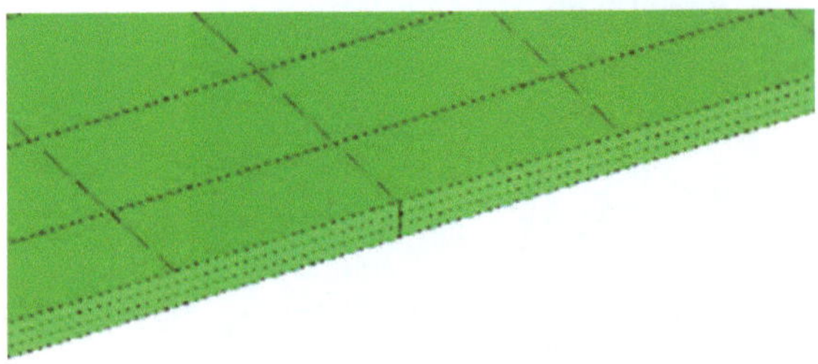

Fig. 6.41 3D mesh of interlayer PVB

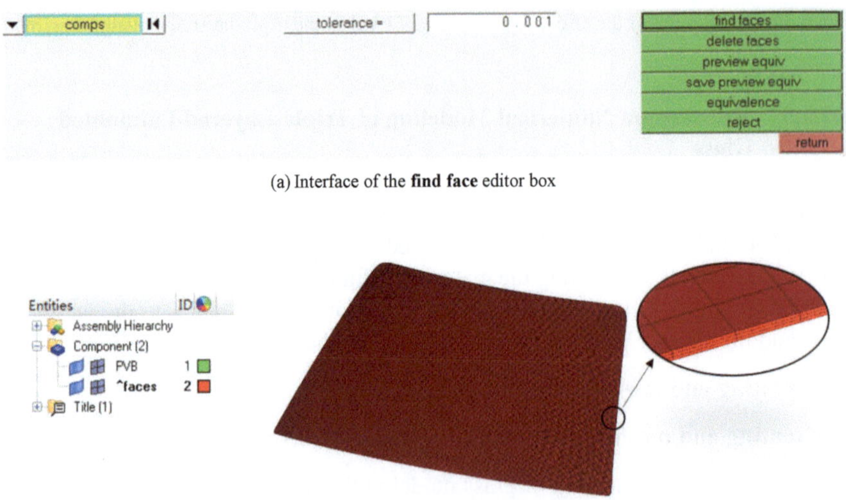

(a) Interface of the **find face** editor box

(b) Creating the face

Fig. 6.42 Find face on PVB

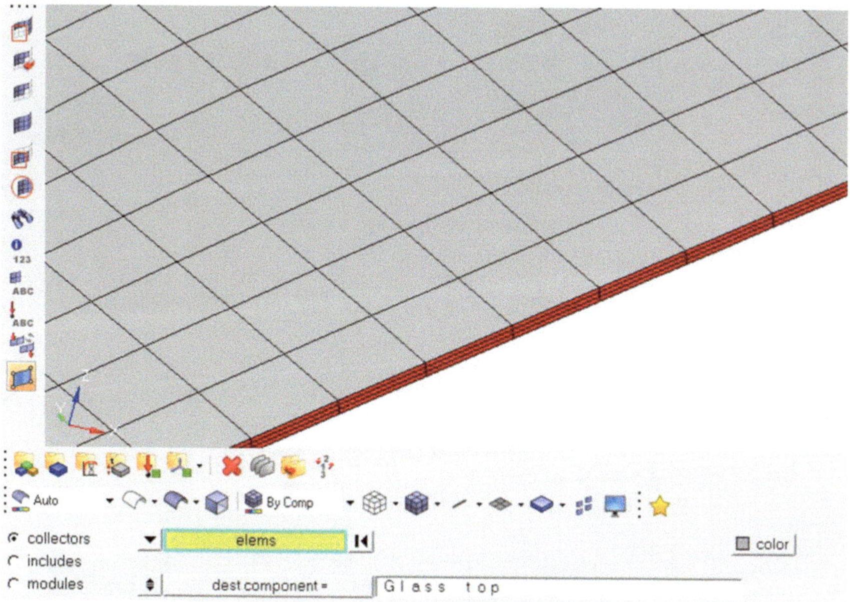

(a) Exact surface element on PVB face

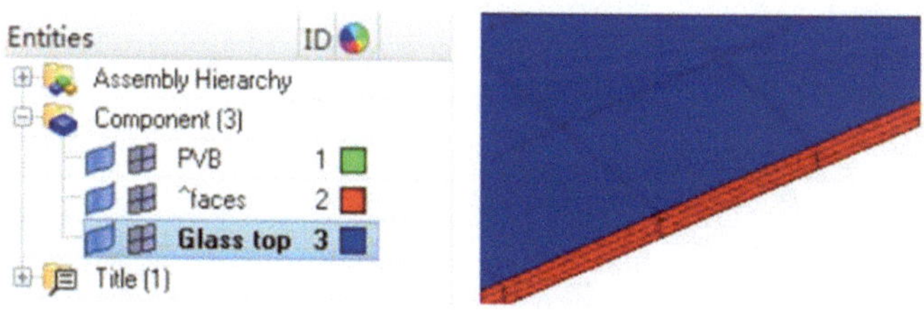

(b) Exact surface element on PVB face to **"Glass top"**

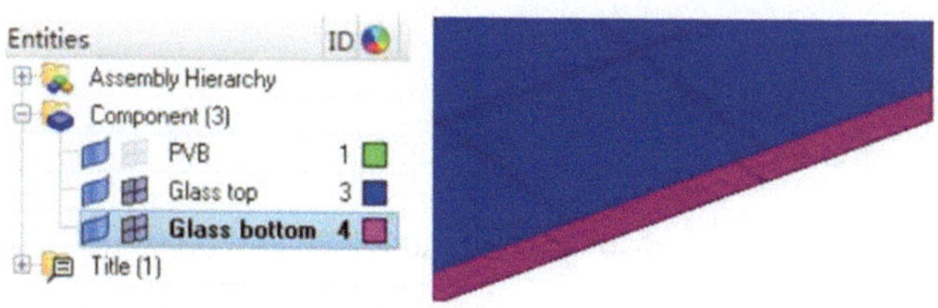

(c) Exact surface element on PVB face to **"Glass bottom"**

Fig. 6.43 Creating the glass top and bottom by exacting surface on the PVB face

(4) Create a new component and name it **"Glass top**." Then, click to exact surface on the face to **"Glass top"** to create the glass, as shown in Fig. 6.43.

1. **Defining materials**

Defining the material of laminated glass

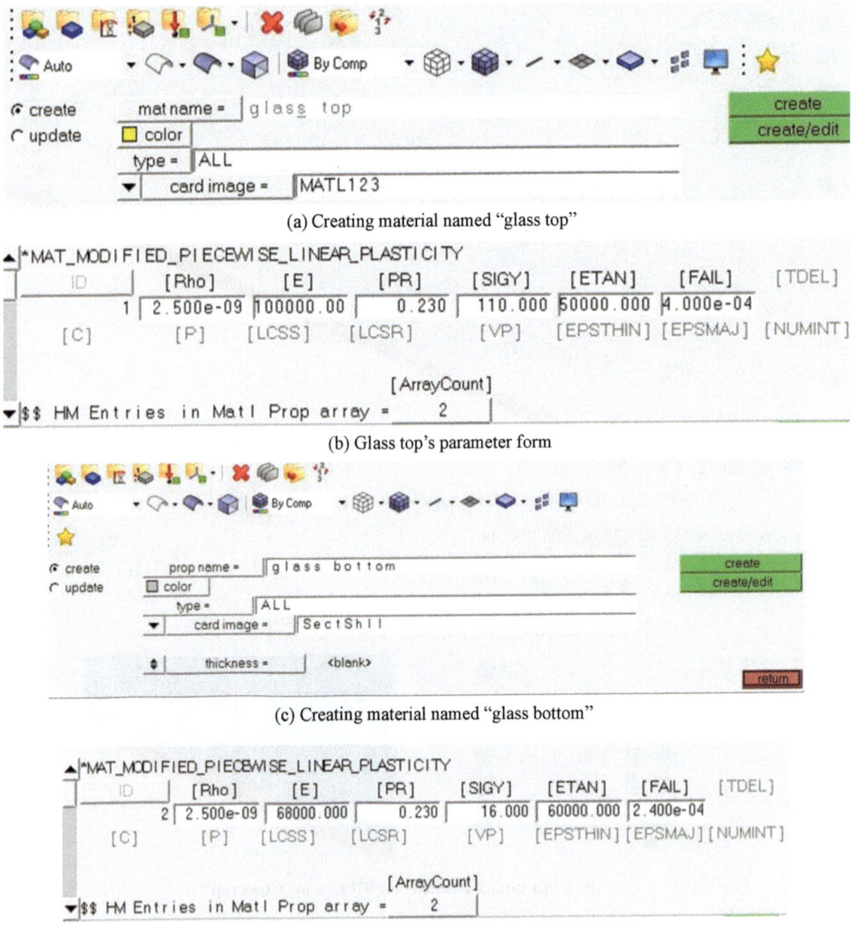

(a) Creating material named "glass top"

(b) Glass top's parameter form

(c) Creating material named "glass bottom"

(d) Glass bottom's parameter form

Fig. 6.44 Materials editor of glass

MAT_123 is chosen as the material of laminated glass and will be assigned to this component.

(1) Click the button. The dialog box of materials will then appear. Input the material's name as "glass top," "glass bottom," and "PVB," respectively. The "**type**" is defined as "**ALL**," edit the "**card image**," and choose the appropriate materials cards. Choose "MAT_123" for all the materials as shown in Figs. 6.43a and 6.44a, c. The "MAT_123" is Material Type 123, an elasto-plastic material with an arbitrary stress–strain curve and arbitrary strain rate dependency.
(2) Click the "**create/edit**" button and enter the corresponding values for the parameters of materials, as shown in Fig. 6.44b, c. Some important parameters are described as follows:

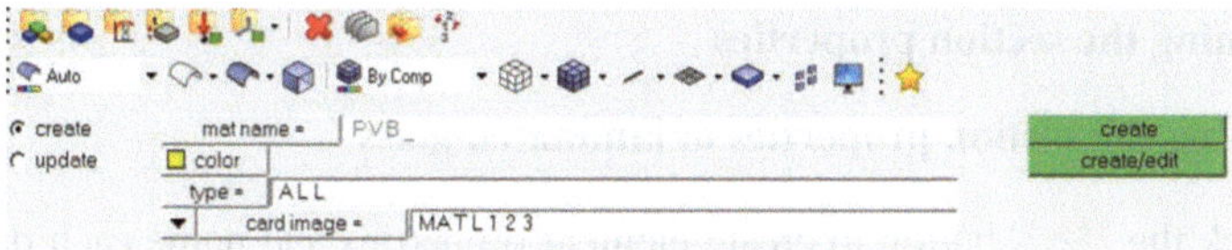

(a) Creating the material named "PVB"

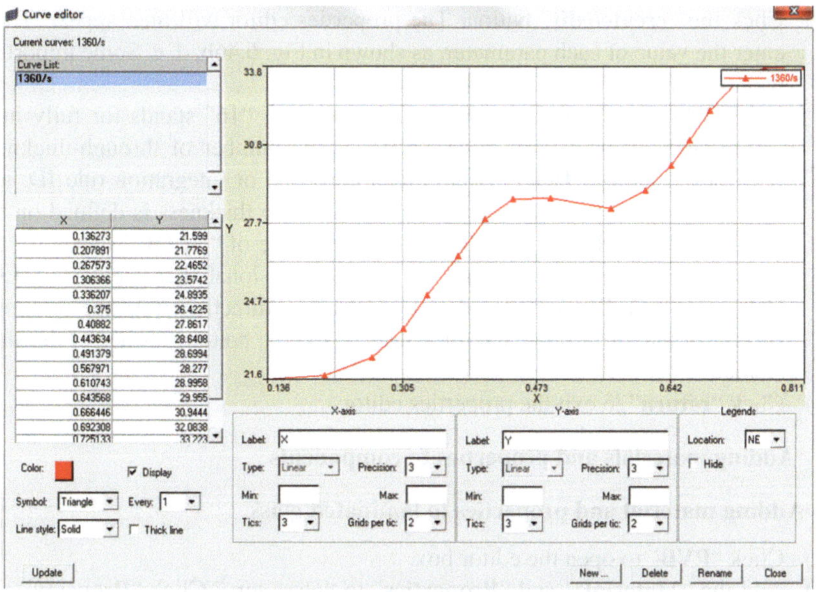

(b) Creating the 1,360/s stress and strain curve of PVB

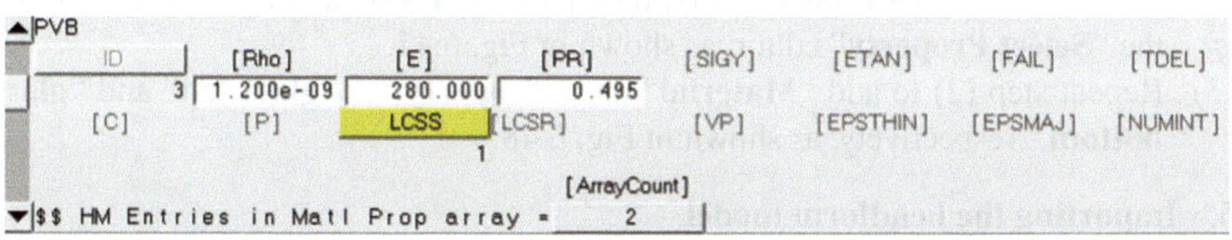

(C) PVB's parameter form

Fig. 6.45 Materials editor of PVB

For glass, "**Rho**" is the mass density, "**E**" is the Young's modulus, "**Nu**" is the Poisson's ratio, "**SYG**" is the yield stress, and "**ETG**" is the tangent modulus. For PVB, "**EP**" is the Young's modulus, "**PRP**" is the Poisson's ratio, "**SYP**" is the yield stress, and "**ETP**" is the tangent modulus.

(3) Then, click the "XYPlots" on the toolbars at the top to create the 1360/s effective stress versus effective plastic strain curve, as shown in Fig. 6.45b. The "**LCSS**" in the material editor box chooses the 1360/s curve.

(4) Click "**return**" to exit the materials editor.

3. **Defining the section properties**

Defining the section properties of laminated glass

(1) Click the  button to create element properties and name each one "glass top," "glass bottom," and "PVB," as shown in Fig. 6.46a, c, e.
(2) Click the "**create/edit**" button. The properties editor will then appear. Then, enter the value of each parameter, as shown in Fig. 6.46b, d, e. Some important parameters are described as follows:
"**ELFORM**" is the element formulation options, "16" stands for fully integrated shell element (very fast), "**NIP**" is the number of through-thickness integration points, "**IRID**" is the quadrature rule or integration rule ID, and "**T1**" is the shell thickness at node N1, unless the thickness is defined on the ***ELEMENT_SHELL_OPTION** card. The value of "**NLOC**" stands for the location of the reference surface for three-dimensional shell elements. "**1.0**," "**0.0**," and "**−1.0**" stands for the element offset direction from the reference surface of the top surface, the mid-surface, and the bottom surface of the shell element.
(3) Click "**return**" to exit the properties editor.

4. **Adding materials and properties to components**

Adding material and properties to laminated glass

(1) Click "**PVB**" to open the editor box.
(2) Add the "**Materials**" and "**Properties**" to "glass top." Click "**Property**" and open the "**Select Property**" box, then choose "**PVB**" and click "**OK**" to exit the "**Select Property**" editor, as shown in Fig. 6.47.
(3) Repeat step (2) to add "**Material**" and "**Property**" to "**glass top**" and "**glass bottom**," respectively, as shown in Fig. 6.48.

5. **Importing the headform model**

The process is the same as that of single-layered laminated glass modeling.

6. **Applying boundary and initial conditions to components**

The process is the same as that of single-layered laminated glass modeling.

7. **Defining the contact on the component**

The process is the same as that of single-layered laminated glass modeling. However, two other contact conditions between the head and glass top and bottom need to be created.

(1) Choose the "***CONTACT_AUTOMATIC_SURFACE_TO_SURFACE**" card to create contact between "**Glass top**" and "**Skin null shell**."
(2) In the contact dialog box, "**SSID**" chooses "**Glass top**" and "**MISD**" chooses "**Skin null shell**." Input the friction value of **FS**, **FD**, as shown in Fig. 6.49.

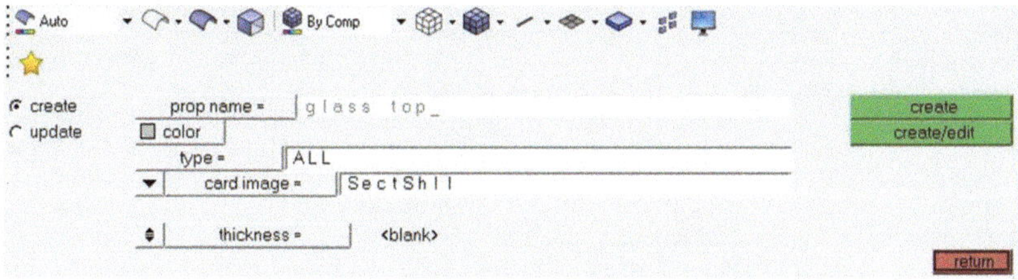

(a) Creating the element property named "glass top"

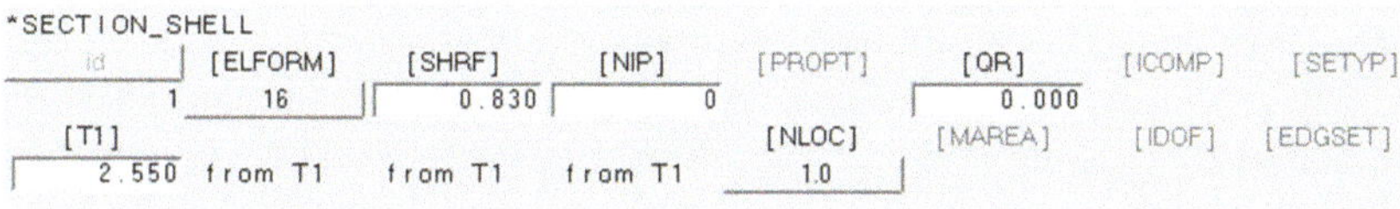

(b) Glass top's element property form

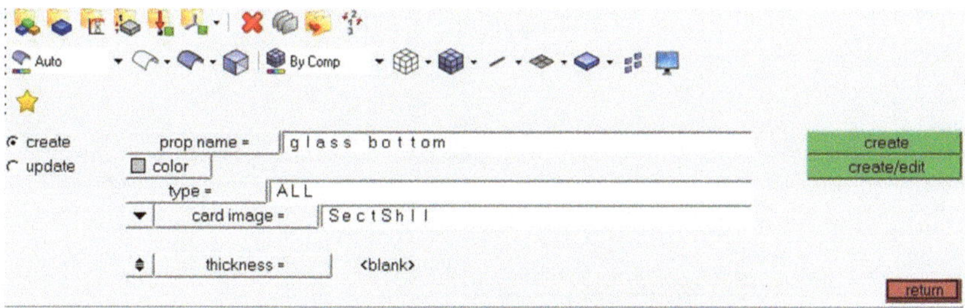

(c) Creating the element property named "glass bottom"

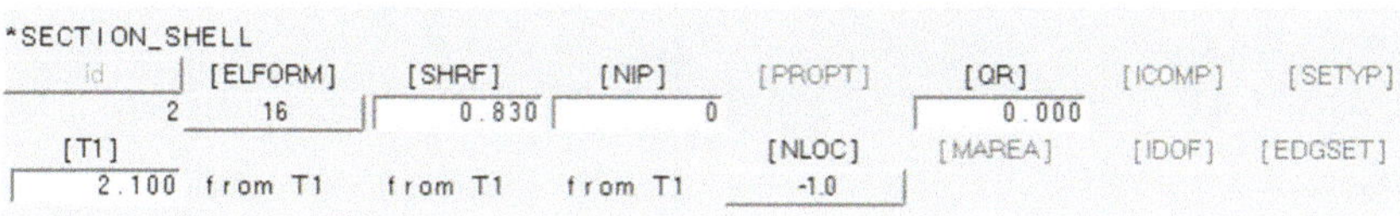

(d) Glass bottom's element property form

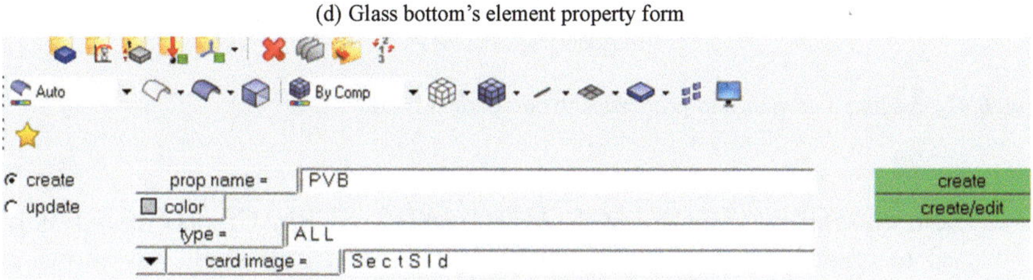

(e) Creating the element property named "PVB"

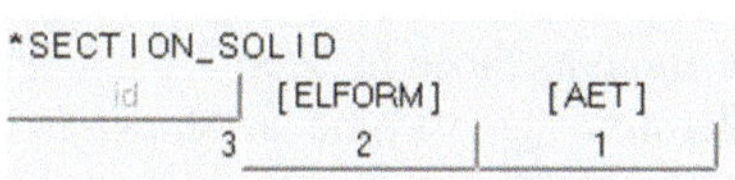

(f) PVB's element property form

Fig. 6.46 Defining the element properties

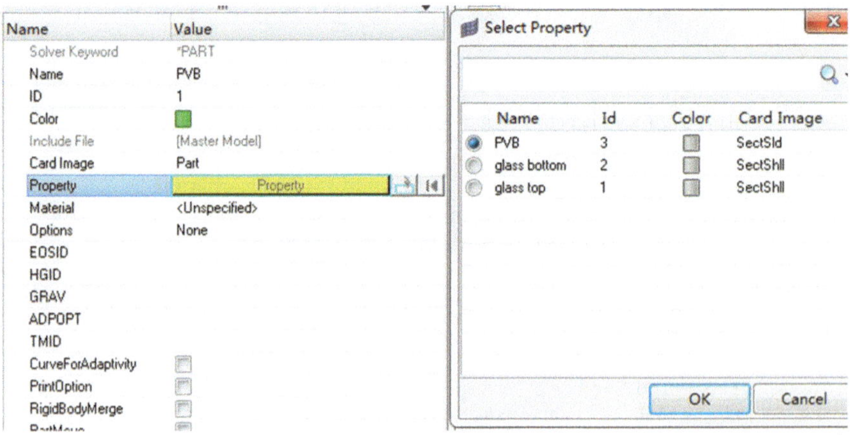

(a) Adding property to "PVB"

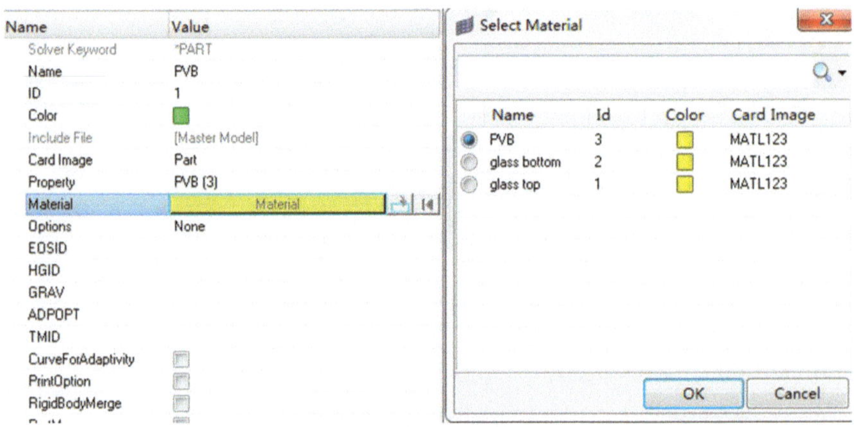

(b) Adding material to "PVB"

Fig. 6.47 Adding materials and properties to component PVB

(3) Then, choose the "*CONTACT_AUTOMATIC_SURFACE_TO_SURFACE" card to create contact between "**PVB**" and "**Skin null shell**," as shown in Fig. 6.50.

The next steps are the same as those in Sect. 7.1.2.3. The simulation result and the history curve of acceleration against time are shown in Figs. 6.51 and 6.52.

Name	Value
Solver Keyword	*PART
Name	Glass top
ID	3
Color	
Include File	[Master Model]
Card Image	Part
Property	glass top (1)
Material	glass top (1)
Options	None

(a) Glass top

Name	Value
Solver Keyword	*PART
Name	Glass bottom
ID	4
Color	
Include File	[Master Model]
Card Image	Part
Property	glass top (1)
Material	glass top (1)
Options	None

(b) Glass bottom

Fig. 6.48 Adding materials and properties to component glass top and bottom

6.3 Results and Discussion

6.3.1 Single-Layered FE Model

Figure 6.53 depicts the correlation of the linear acceleration curves of the headform between the numerical simulation and test results at five different points of the windshield for the single-layered windshield FE model. The first and second peaks in Phase I are determined accurately. The main behavior of the windshield in Phase II consists of a large deformation of PVB and crack propagation of glass along with the deformation of PVB. The simulation curves of the headform at five different impact points in Phase II correspond well with the test curves. The plastic strain contour plot for the laminated windshield model is also presented to illustrate the crack morphology in Fig. 6.53. The radial crack first appears after the initial impact at t_1 (0.18 ms) because of the large hoop stress. Subsequently, radial stress should be released after the dissipation of the hoop stress to generate the circular crack. The general profile of crack morphology evolution agrees qualitatively with previous observations in laminated glass sheets [1, 2, 13–19]. Therefore, the single-layered windshield FE model is acceptable in terms of the mechanical behavior of glass failure in Phase I and the large deformation of PVB in Phase II, which is reflected by the dynamic response of the headform impactor (linear acceleration curves) shown in Fig. 6.53.

6.3.2 Triple-Layered FE Model

Figure 6.54 illustrates a reasonably good correlation of the headform impactor between the numerical simulation and test results at five different points of the windshield for the triple-layered model. However, a delay occurs between glass failure and oscillations in the dynamic response of the headform impactor for the simulation

(a) "SSID" chooses "Glass top"

(b) "MISD" chooses "Skin null shell"

(c) Keyword of the contact

Fig. 6.49 Defining the contact between glass top and skin null shell

curves compared with the corresponding test curves and the numerical simulation results of the single-layered model, as shown in Fig. 6.58. In addition, to achieve a realistic fracture pattern of windshields, the two glass layers are allowed to fail. The deletion of the failed glass shell elements represents the cracks in terms of "damaged" elements. Crack propagation in Phase I, as shown in Fig. 6.55a, is consistent with the test results within the vicinity of the impact point in Fig. 6.55b, with a spiderweb crack. However, the final circular cracks far from the impact point during the simulation are fewer than the those in the real windshield fracture pattern, as shown in Fig. 6.55b.

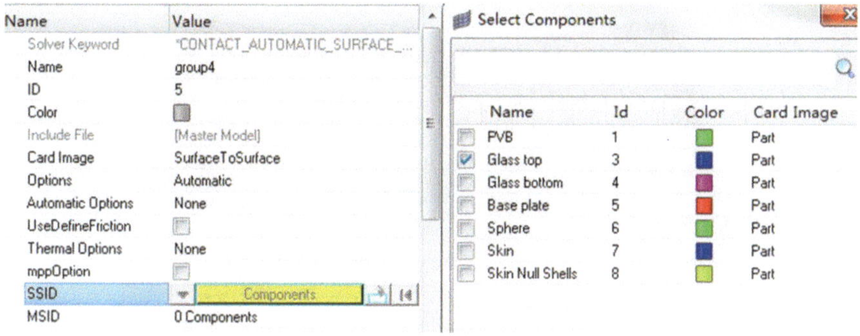

(a) "SSID" chooses "PVB"

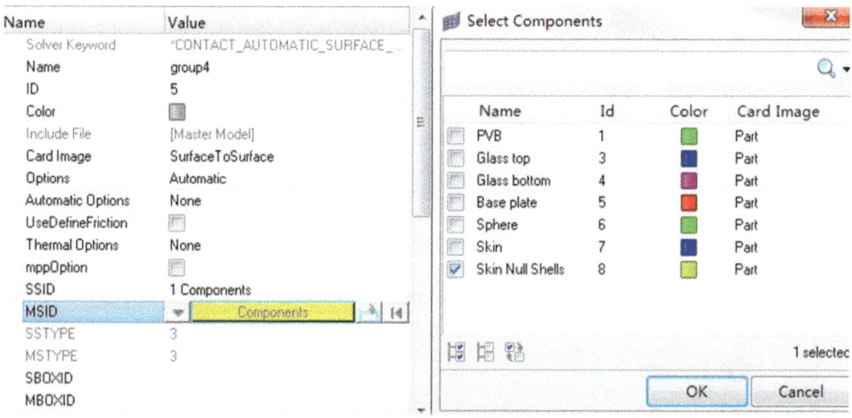

(b) "MISD" chooses "Skin null shell"

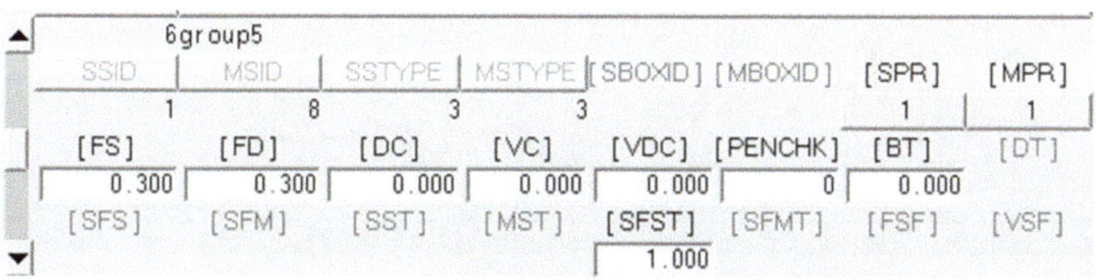

(c) Keyword of the contact

Fig. 6.50 Defining the contact between "PVB" and "Skin null shell"

6.3.3 *Influencing Factor Analysis of Simulation for the Two Models Above*

(1) Mesh sizes

Four mesh sizes were considered to investigate the mesh sensitivity of simulation results for the above two models. As indicated in Fig. 6.56a, the headform acceleration curves predicted by the single-layered model is sensitive to mesh size. The two peak

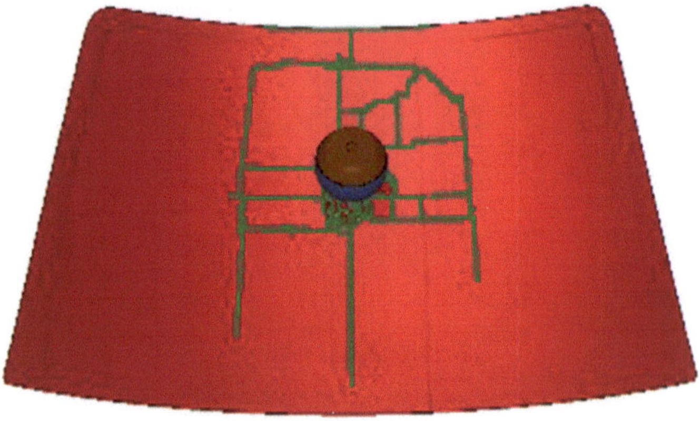

Fig. 6.51 Simulation results of the triple-layered model of laminated glass

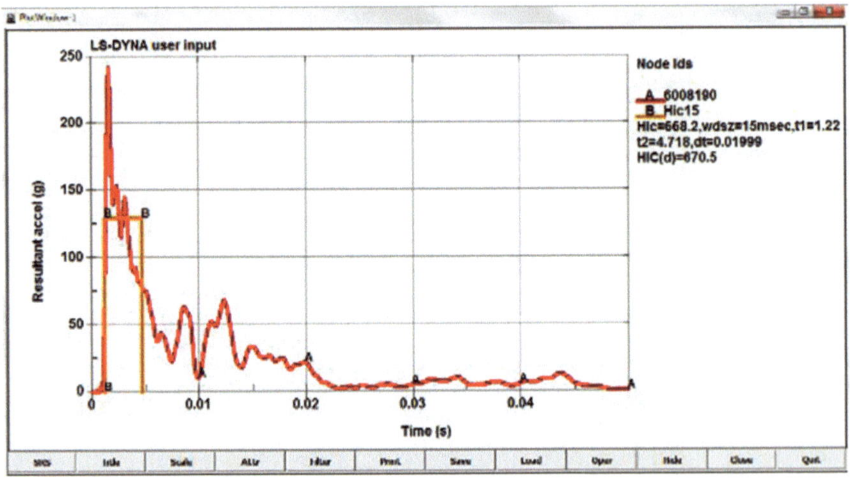

Fig. 6.52 History curve of acceleration against time

values decrease with mesh size, which indicates that a great mesh size corresponds to low stiffness. In addition, when the mesh size reached 15 or 20 mm, evaluating the headform acceleration curve further would obtain inaccurate results. Thus, the single-layered model with 5 mm mesh generally performs best. By contrast, the acceleration curve is not very sensitive to mesh size for the triple-layered model shown in Fig. 6.56b, and the model with 5 mm mesh still has the best agreement with the test acceleration curve. However, the model has very strong mesh dependency for the crack fracture (failure) pattern of impacted laminated windshield in the triple-layered model shown in Fig. 6.56c. The simulation results indicate that the crack

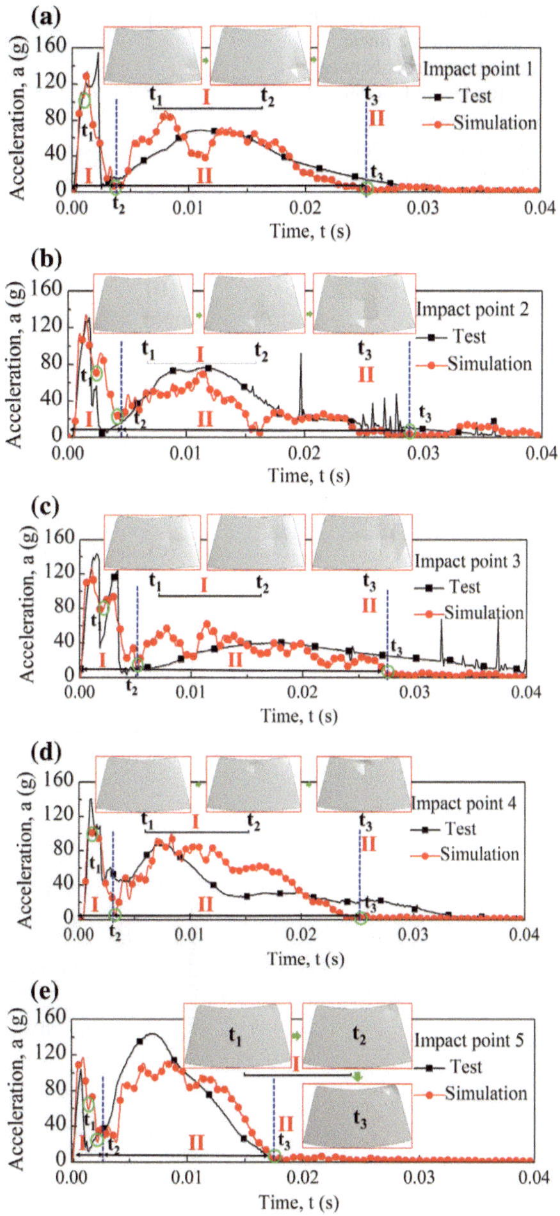

Fig. 6.53 Correlation of the dynamic response linear acceleration curves between the single-layered model and the test results for the headform impactor at five different impact points of the windshield

area is bigger and the number of circular cracks is greater for the 5 mm mesh than other mesh sizes in the triple-layered model. Therefore, the triple-layered model with 5 mm mesh has the best agreement with the test results in terms of headform acceleration curve and fracture pattern.

Fig. 6.54 Correlation of the dynamic response linear acceleration curves between the triple-layered model and the test results for the headform impactor at five different impact points of the windshield

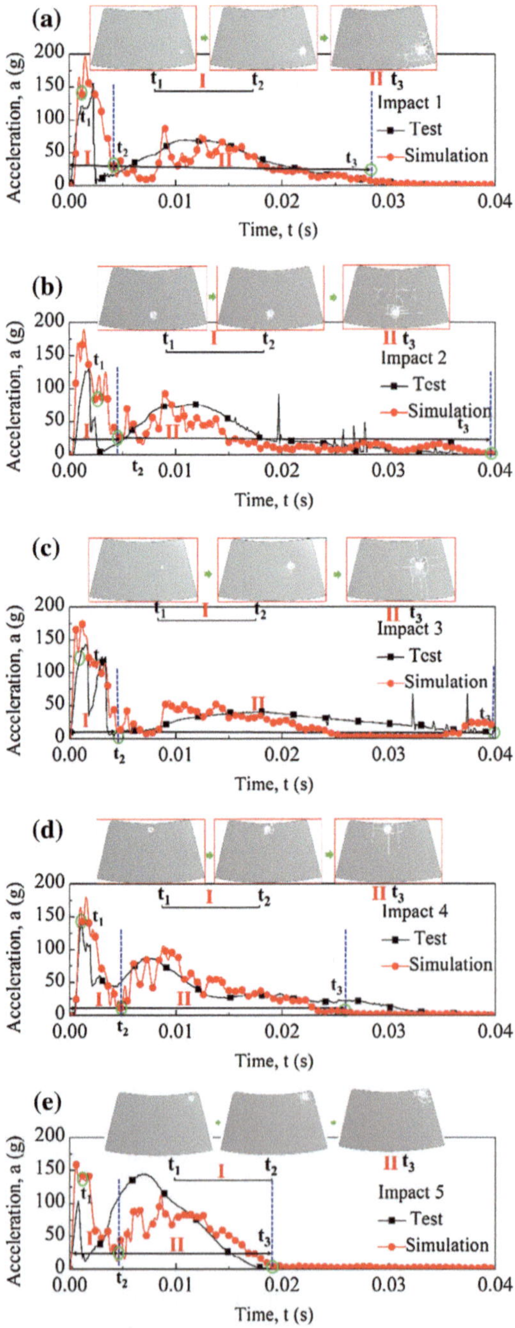

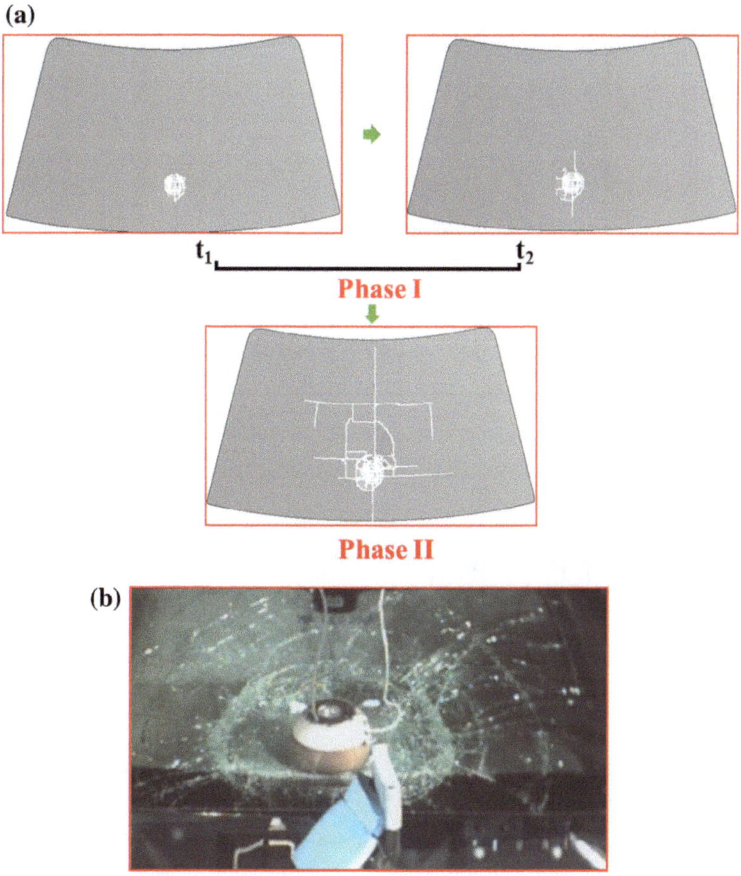

Fig. 6.55 Crack propagation of the windshield at impact point 2 for the triple-layered model. **a** Crack propagation process of impact point 2 in the numerical simulation. **b** Crack fracture pattern of impact point 2 in the test

(2) Element deletion criteria

In addition to mesh size, the element deletion criteria may also be sensitive to simulation results in terms of the headform acceleration curve and fracture pattern in the triple-layered model (no element deletion exists in the single-layered model). Thus, instead of using previous element deletion criteria (where elements fail when they achieve plastic failure strain), another element deletion criterion (where elements fail when they achieve principal stress at failure) was adopted to analyze the sensitivity of element deletion criteria for the acceleration curve and crack fracture pattern. The results are illustrated in Fig. 6.57. Figure 6.4a shows that the crack fracture pattern of the impacted laminated windshield obtained by the model whose element failed when they achieved principal failure stress is more realistic with a more circular crack and

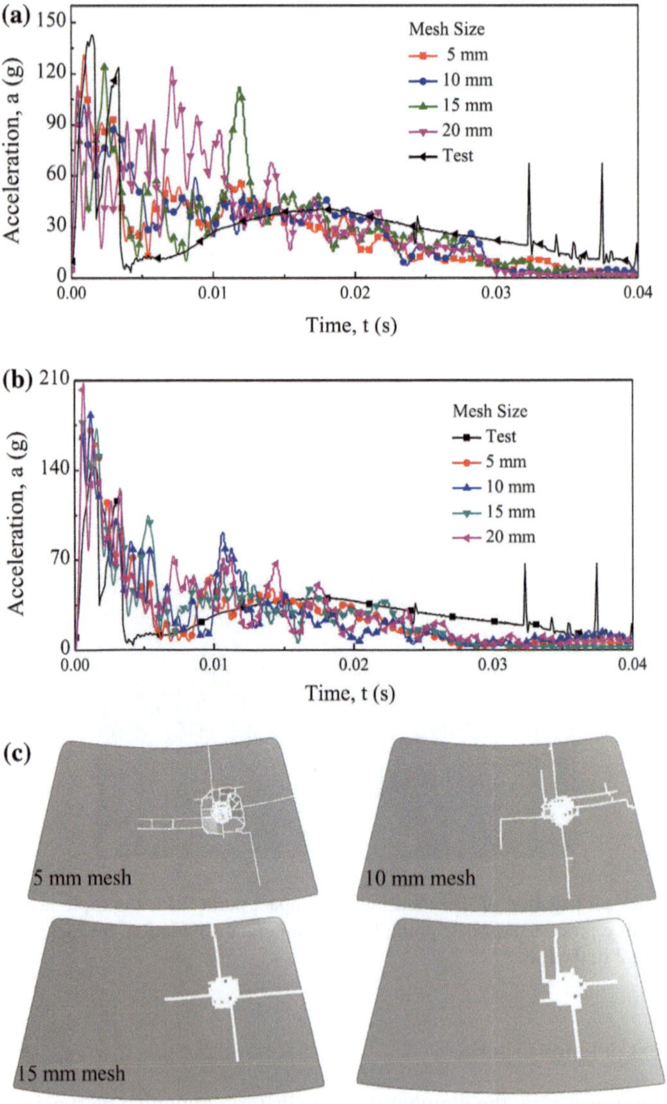

Fig. 6.56 Mesh size sensitivity analysis for the single-layered and triple-layered models. **a** Mesh size sensitivity analysis of headform acceleration curve obtained from single-layered model. **b** Mesh size sensitivity analysis of headform acceleration curve obtained from triple-layered model. **c** Mesh dependency for crack fracture (failure) pattern of impacted laminated windshield in the triple-layered model

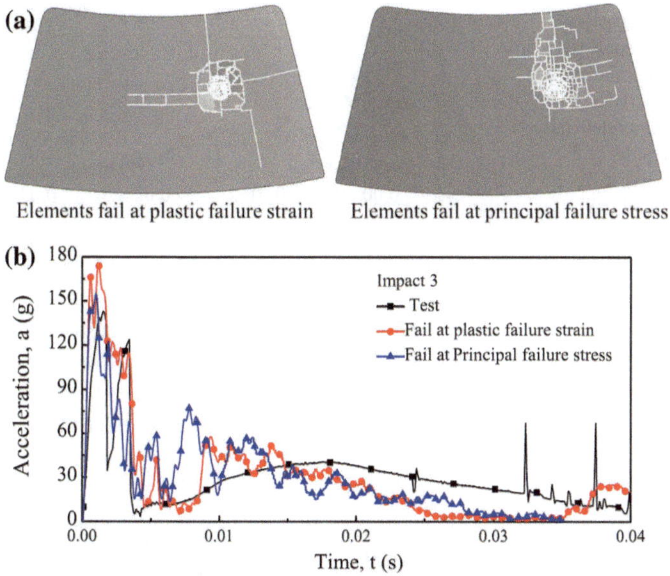

Fig. 6.57 Effect of different element deletion criteria on simulation results. **a** Element deletion criteria effect on fracture pattern of laminated windshield in triple-layered model. **b** Element deletion criteria effect on headform acceleration curve in triple-layered model

a larger crack area. However, the corresponding headform acceleration–time curve is more oscillating (thereby indicating instability in numerical calculation) than previous results, as shown in Fig. 6.57b. Thus, the new element deletion criterion is not considered in this study.

6.3.4 Comparison of Results Between the Two FE Models

A comparison between the simulation results of the triple-layered model and those of the single-layered model in Fig. 6.58 shows obvious oscillations in the dynamic response of the headform impactor. A delay of glass failure occurs in the triple-layered model, as shown in the comparison with the test curve and single-layered model in Fig. 6.58. The oscillations verify that the deletion of the glass-failed elements for the triple-layered model influences the dynamic response of the headform due to the rougher contact face caused by the deleted glass elements compared with that of the single-layered model and the realistic test scenario when residual failed glass splinters are still stuck on the windshield. Furthermore, the structure of the shell element layer in the single-layered model more effectively prevents the delay of failure of the two glass layer than the triple-layered model that separates the glass and PVB layers. This result proves that the structure of the windshield FE model

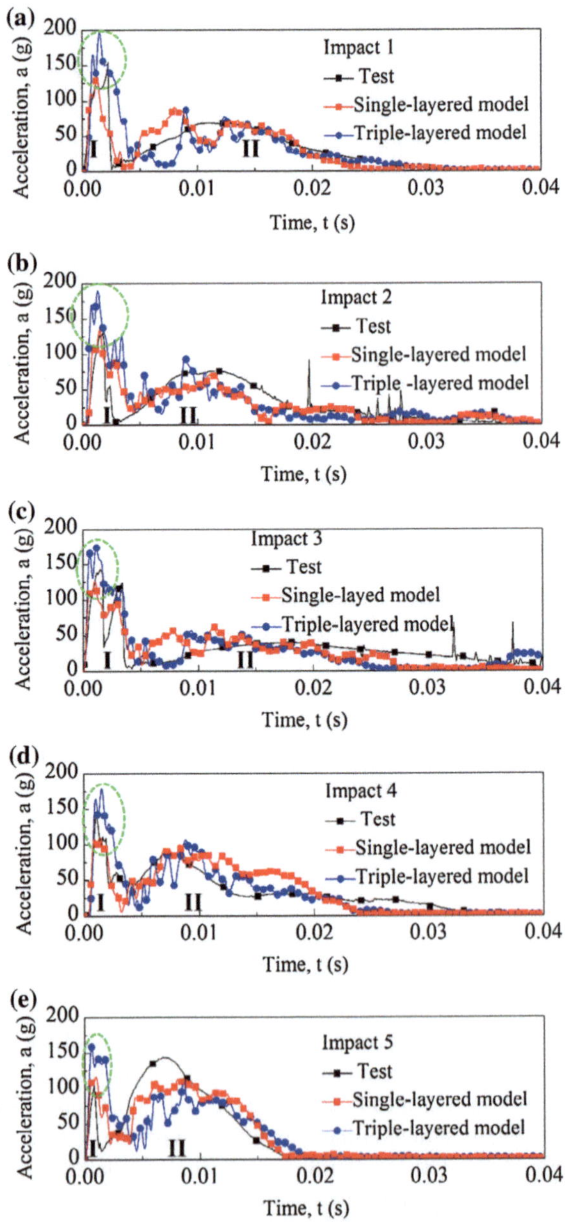

Fig. 6.58 Comparison of the dynamic response linear acceleration curve between the single-layered and triple-layered models based on the test results of the headform at five different impact points of the windshield

also influences glass fracture and the dynamic response of the headform impactor. Defining the glass and PVB layers individually in the triple-layered model may delay glass failure.

For the final fracture pattern of the windshield between the two FE models, the triple-layered model clearly performs better than the single-layered model, as shown in Fig. 6.59. First, crack propagation is visible in the triple-layered model but invisible in the single-layered model because of the absence of failed elements. Second, the deformation trace of the shell element layer in the single-layered model consists of a few radial and circular cracks, particularly at impact points 1 and 2, which is inconsistent with the test results with a spiderweb crack. For the triple-layered model, crack propagation corresponds well with the test results, particularly for a spiderweb crack. However, circular cracks are fewer than those in the test results because the deletion of an increasing number of failed glass elements discontinues the propagation of radial stress on the glass layer to a certain extent, which prevents the production of more circular cracks. This fracture pattern is quantitatively consistent with the work of Zang [20] and is described using DEM.

6.3.5 Application to a New Windshield

The validation test based on a new vehicle with a laminated windshield with a different geometric size is performed, as shown in Fig. 6.60a. The complicated structure and strong stiffness around the lower area of the windshield, which may make the lower area more dangerous in pedestrian–vehicle crash accidents, may result in a complex dynamic response of the headform impactor. The impact point located on the lower left corner of the windshield is selected in this test.

Figure 6.60b shows the corresponding cutoff vehicle FE model for numerical simulation and physical dimensions of the windshield used in the test. The windshield consists of a 2.1 mm upper glass layer, a 0.76 mm middle layer, and a 1.6 mm lower layer. First, a single-layered windshield FE model is established based on this cutoff vehicle FE model. Figure 6.60d shows the correlation results between the test and the numerical simulation. Results show that the curve of the contact force between the headform impactor and the windshield corresponds well with the test results as time progresses. This single-layered windshield FE model can enable the headform impactor achieve an accurate dynamic response that is consistent with the impact on windshields in reality. In addition, Fig. 6.60d shows that Phase II of the contact force curve exhibits a steep rise because of the instrument panel positioned behind the lower area of the windshield, as shown in Fig. 6.59. When the upper and lower glass layers fail sufficiently, the PVB layer and the instrument panel (IP) begin to bear the main loading from the headform. Once the headform comes in contact with the IP, the contact force suddenly increases because of the high stiffness of the IP. It will continue to increase with the movement of the headform until the headform impactor rebounds or the IP fails completely.

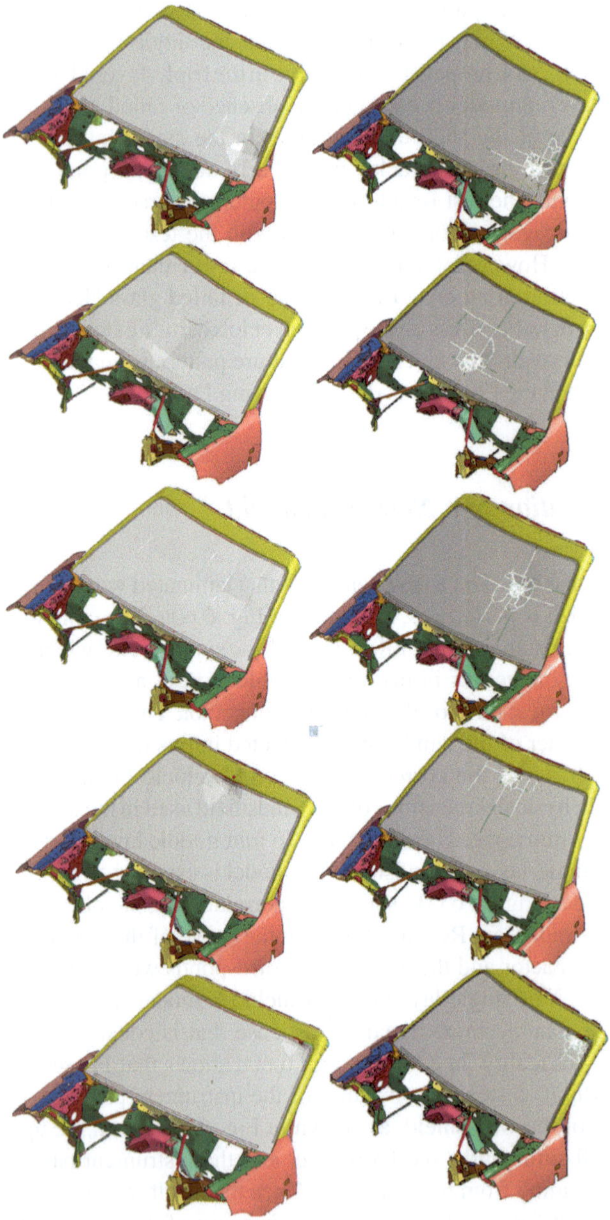

Fig. 6.59 Final fracture pattern of the windshield for the single-layered and triple-layered FE models

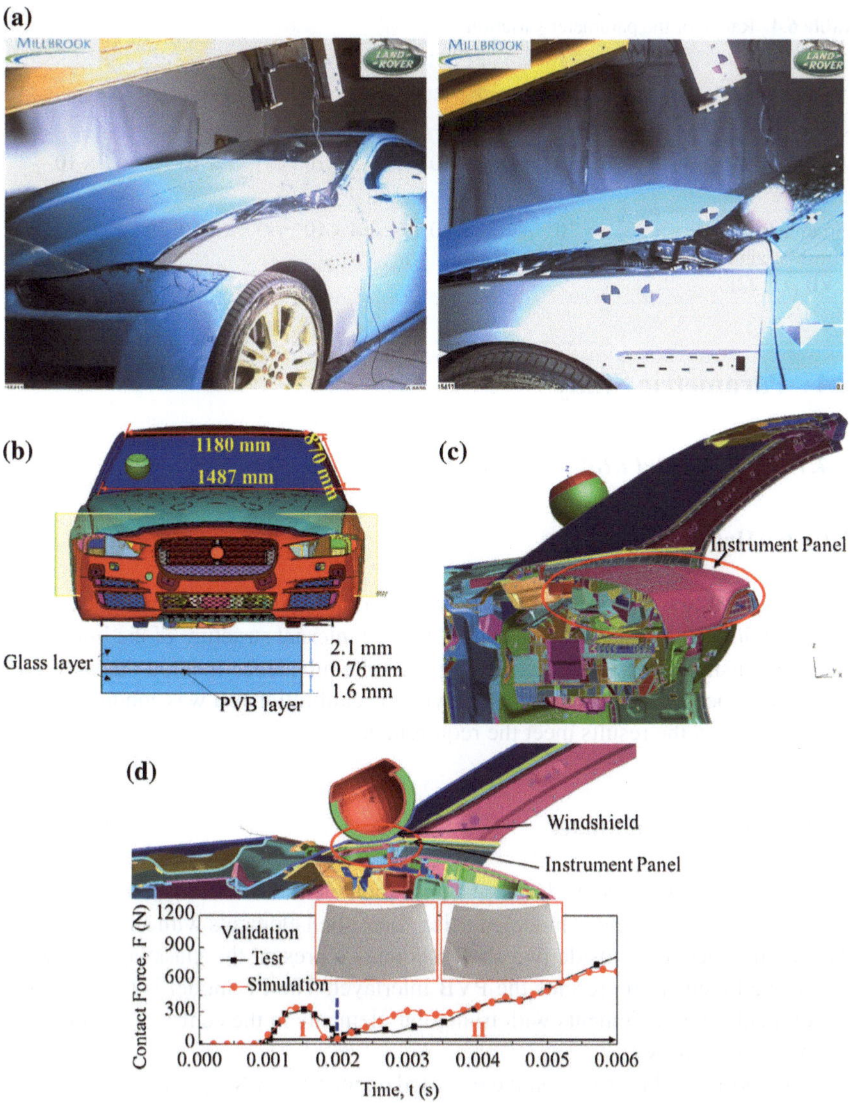

Fig. 6.60 Headform impact test and numerical simulation for validation. **a** Headform impact test. **b** FE model. **c** Front view of the cutoff vehicle FE model and the physical dimensions of the windshield. **d** Side view of the cutoff vehicle FE model and the correlation of the dynamic response linear acceleration curves between the numerical simulation and test results

Table 6.4 Range of the parameter variation

	1	2	3	4	5
Young's modulus of glass (GPa)	0.7	7	70	350	700
Plastic failure strain of glass	5×10^{-3}	1×10^{-3}	5×10^{-4}	1×10^{-4}	5×10^{-5}
Strain rate of stress–strain curve of PVB [11, 22]	4×10^{-3}/s	2×10^{-3}/s	4×10^{-2}/s	8×10^{-2}/s	118/s

6.4 Parametric Study

6.4.1 FE Model and Method

6.4.1.1 Head Model

The FE model of the pedestrian headform impactor is built according to the requirement of Global Technical Regulations [21]. The internal structure of the impactor is simplified using the mass and centroid position equivalence method during the FE modeling process and then calibrated. Dynamic calibration test was conducted using LS-DYNA, and the results meet the requirements.

6.4.1.2 Windshield Model

The windshield FE model is developed with shell elements. The layered structure of the laminated windshield is modeled using three shell elements with shared nodes at their boundaries; the outside two shell elements represent the glass layers, and the inside shell element represents the PVB interlayer. The FE model is meshed using 2 mm quadrilateral elements with triangular elements in the central region through a convergence study.

A series of simulations is conducted based on the LS-DYNA platform for different conditions to obtain the parameters' influence on the windshield model. The contact form between the head model and the windshield apply surface-to-surface contact with a friction coefficient of 0.3 [1, 2]. First, we vary the glass Young's modulus, glass plastic failure strain, and PVB stress–strain curve according to the data listed in Table 6.4. Other conditions are kept at their reference values, which are shown in Table 6.5. Afterwards, four boundary conditions, i.e., four edges clamped (FC), four edges simply supported (FS), left and right edges clamped while upper and lower edges simply supported (LR), and left and right edges simply supported while upper and lower edges are clamped (UD), are set with other simulation conditions at their reference values.

Table 6.5 Parameters used in the windshield FE model [1, 23]

Layer	Variable	Reference parameters
Glass layer	Density ρ_{glass}	2500 kg/m^3
	Young's modulus E_{glass}	70 GPa
	Poisson's ratio υ_{glass}	0.227
	Yield stress σ_{glass}	0.143 GPa
	Plastic strain to failure ε_{glass}	0.001
	Thickness t_{glass}	0.002 m
PVB layer	Density ρ_{glass}	2000 kg/m^3
	Poisson's ratio υ_{glass}	0.435
	Strain rate of strain–stress curve	118/s
Global settings	Boundary condition	Clamped

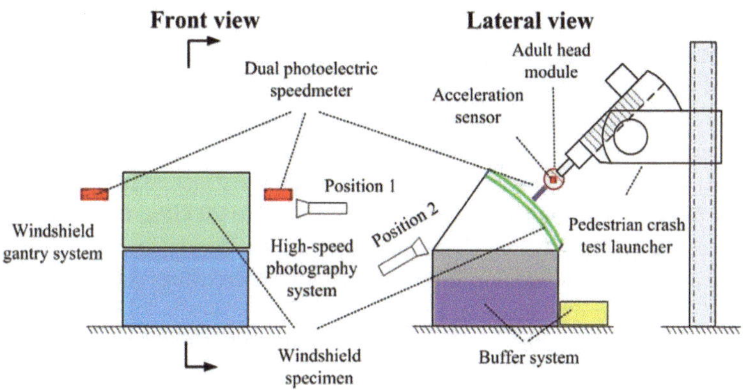

Fig. 6.61 Schematic of the experimental system

6.4.1.3 Model Verification

Further simulations are conducted to verify the windshield FE model with the dynamic experiments of PVB laminated windshield under headform impact. The head module impact windshield experimental system consists of seven parts, namely, pedestrian crash test launcher, windshield gantry system, high-speed photography system, dual photoelectric speedometer, windshield specimen, adult head module, acceleration sensor, and buffer system, as shown in Fig. 6.61.

The headform impact velocities and angles are set to 6.6–11.2 m/s and 60°–90°, respectively, according to real-world vehicle–pedestrian accidents in the traffic accident cases database of Tsinghua University. Eight experiments are conducted, as shown in Table 6.6. Simulations that use the most accurate windshield model that was obtained are conducted according to the settings in Table 6.2 to verify the windshield FE model.

Table 6.6 Test parameter settings

No.	Impact angle (°)	Impact velocity (m/s)
1	90	6.6
2	90	8.1
3	90	8.8
4	90	11.2
5	90	8.1
6	80	8.2
7	71	8.2
8	59	8.2

6.4.2 Results and Discussion

6.4.2.1 Effect of the Young's Modulus of Glass

Figure 6.62 shows the acceleration–time curves of different Young's moduli of glass compared with the corresponding test results. When the Young's modulus of glass is set to 700 MPa and 7 GPa, the windshield does not crack and the acceleration trends are different from the experimental results. Once the Young's modulus of glass rises beyond 70 GPa, the cracks initiate and the crack intensity increases with the Young's modulus. When the Young's modulus of glass is set to 70 GPa, the simulation and experimental results of the acceleration curve become more similar. Regardless of cracks, head model acceleration increases with the increasing Young's modulus of glass.

As the Young's modulus of glass increases, the windshield becomes stiff, and the peak acceleration increases correspondingly. The yield stress and plastic failure strain are fixed. Thus, the failure strain becomes smaller and cracks generate easily with the increasing yield stress of glass.

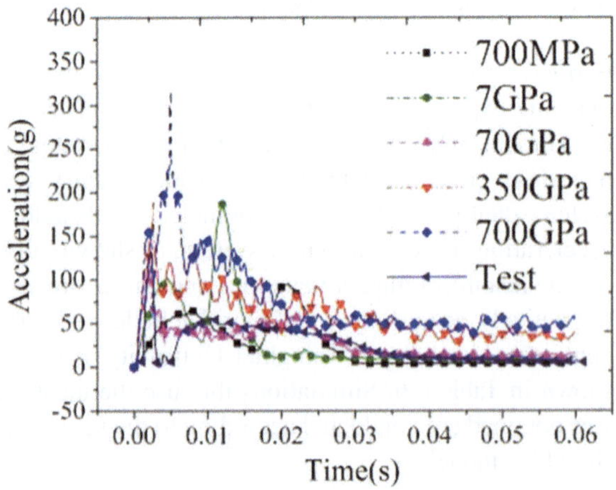

Fig. 6.62 Acceleration–time curves of different Young's moduli of glass compared with the corresponding test results

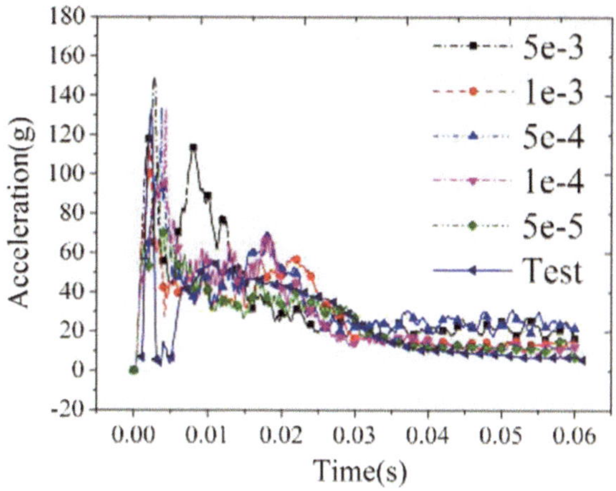

Fig. 6.63 Acceleration–time curves of different plastic failure strains compared with the corresponding test results

6.4.2.2 Effect of Plastic Failure Strain

Figure 6.63 shows the acceleration–time curves of different plastic failure strains compared with the corresponding test results. As the plastic failure strain increases, the acceleration does not change obviously. As the plastic failure strain decreases, the cracks intensify. When the plastic failure strain is set to 1×10^{-3} and 5×10^{-4}, simulation and experimental results achieve better agreement.

As a result of the fixed yield stress and Young's modulus, the failure stress is reduced as the plastic failure strain decreases. The cracks intensify as crack initiation occurs easily.

6.4.2.3 Effect of PVB Stress–Strain Curve

Figure 6.64 shows the acceleration–time curves of different PVB stress–strain curves compared with the corresponding test results. Compared with the application of dynamic stress–strain curve, the second acceleration peak is not obvious when the quasi-static stress–strain curves are employed. When we apply quasi-static stress–strain curves with different strain rates, the obtained acceleration curves are not very different. Simulation and experimental results of acceleration curve exhibit better agreement when we apply the dynamic stress–strain curve.

From the quasi-static and dynamic tensile test results of the specimens, we can see that the PVB materials exhibit different tensile characteristics in dynamic and quasi-static loads [22]. The impact process itself is a dynamic process. Therefore, the results obtained by the application of the dynamic tensile test curve are more in line with the actual situation.

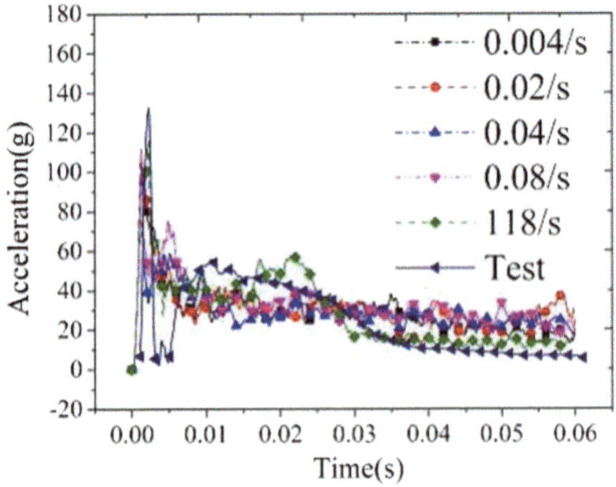

Fig. 6.64 Acceleration–time curves of different glass PVB stress–strain curves compared with the corresponding test results

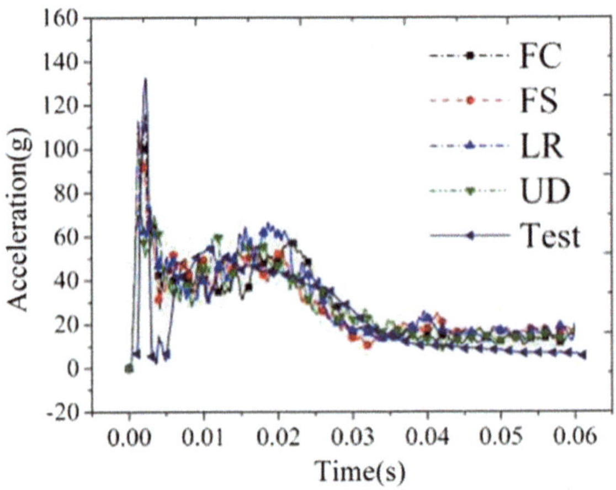

Fig. 6.65 Acceleration–time curves under different boundary conditions compared with the corresponding test results

6.4.2.4 Effect of Boundary Condition

Figure 6.65 shows the acceleration–time curves under different boundary conditions compared with the corresponding test results. With FC and FS boundary conditions, the obtained acceleration curves are almost the same. However, the second acceleration peak is not stable with LR and UD boundary conditions. This finding shows that the influence of boundary conditions is only minimal. Thus, the FC boundary condition is selected in line with the real situation.

6.4.3 Model Verification

Figure 6.66 shows the obtained acceleration–time curves from the simulations according to Table 6.6 compared with the corresponding results for further verification. The peak values and the entire time course of the simulation fit well with the corresponding impact test results with different impact velocities and impact angles.

Figure 6.67 shows a comparison of crack propagation patterns for the simulation results and the test results with a velocity of 8 m/s at vertical impact. The crack propagation patterns obtained from the impact test are captured by high-speed cameras [24]. In the beginning of the impact, radial cracks first appear on the windshield panel to release the hoop stress because the hoop stress is much larger than the radial stress [2]. Circular cracks begin to appear approximately within 1 s and expand from the inside outward, consistent with the initial propagation direction of the stress wave as presented by literature findings [25, 26]. The simulation results are in good agreement with the experimental results.

6.5 Summary

In this chapter, we investigate the mechanical behavior of laminated windshield subjected to headform impact through FE numerical simulation.

First, we establish two windshield FE models, namely, the single-layered and triple-layered models, to investigate the mechanical behavior of laminated windshield subjected to headform impact. MAT_032 ("MAT_LAMINATED_GLASS") material model in LS-DYNA is chosen to describe the constitutive relationship of single-layered laminated glass, which can describe the multilayered structure with only one shell element through the property card (Integration_shell card) with various properties associated with different layers (with their own integration points). The element in the single-layered windshield FE model is not allowed to be deleted because doing so would result in the cracks becoming invisible. Thus, the triple-layered FE model of laminated glass is proposed, which is a more physically realistic model than the single-layered model, supports both glass and PVB failures, and has cracking patterns that are expected to be realistic. Then, MAT_123 "MAT_MODIFIED_PIECEWISE_LINEAR_PLASTICITY" (an elasto-plastic material model in LS-DYNA) is adopted as the material model to characterize the material properties of glass and PVB during the dynamic impact process. The upper glass plastic failure strain of 0.04% is adopted. A smaller plastic failure strain of 0.024% is defined for the lower glass layer because the compression strength is greater than the tensile strength for glass. The local elements will be deleted when they achieve the defined plastic failure strain. A strain rate of 1360/s is most appropriate for defining the mechanical property of PVB according to the dynamic tensile test of PVB in a previous study. To validate the accuracy of the two FE models,

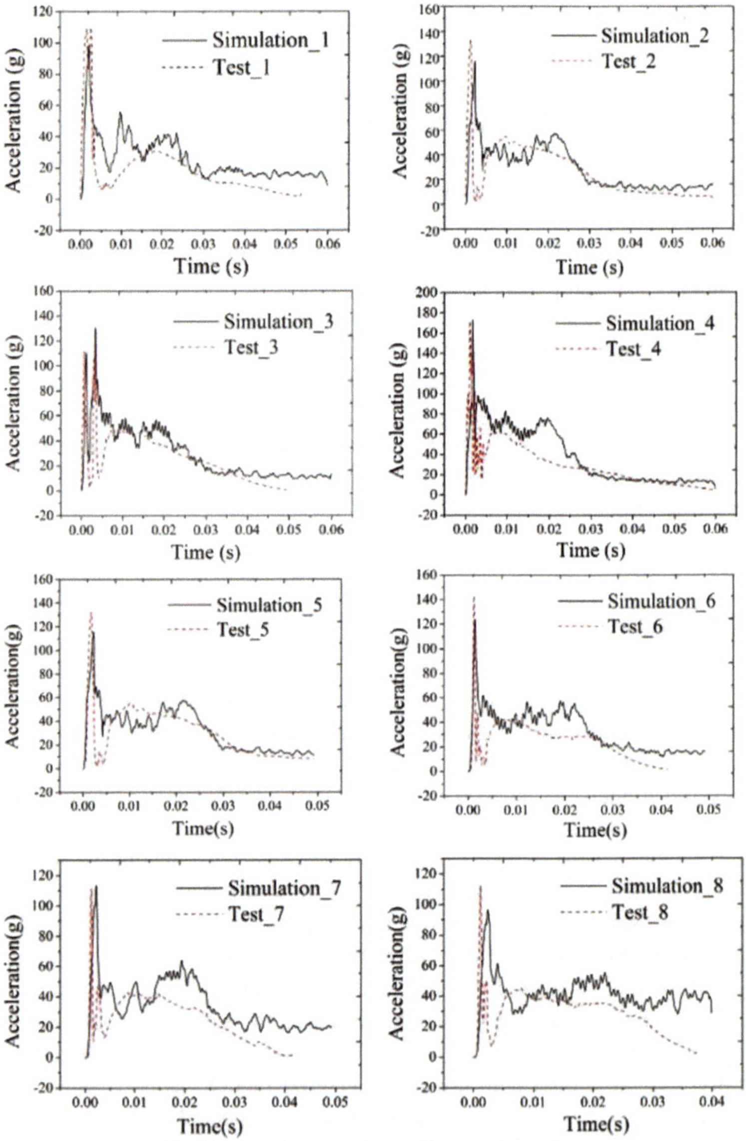

Fig. 6.66 Acceleration–time curves of different impact velocities and angles compared with the corresponding test results for further verification

impact tests are conducted at five different impact points on the windshield selected according to the pedestrian testing protocol of Euro NCAP.

The modeling steps for single-layered and triple-layered laminated glass are described in detail, and the simulation results are compared with the test results

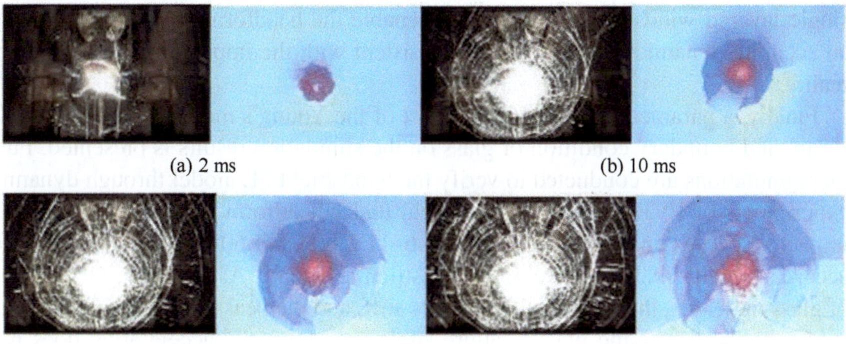

(a) 2 ms (b) 10 ms

(c) 15 ms (d) 20 ms

Fig. 6.67 Comparison of crack propagation patterns for the simulation results and test results with a velocity of 8 m/s at vertical impact

at five impact points. Results show that the single-layered windshield FE model is acceptable in terms of the mechanical behavior of glass failure in Phase I and the large deformation of PVB in Phase II. The results produced by the triple-layered model also show a reasonably good correlation of the headform impactor between the numerical simulation and test. The crack propagation in Phase I is consistent with the test results within the vicinity of the impact point with a spiderweb crack.

The influencing factors of simulation for the two models are also discussed. The headform acceleration curves predicted by the single-layered model is sensitive to mesh size, and a large mesh size corresponds to low stiffness. The single-layered model with 5 mm mesh generally performs best. However, the acceleration curve is not very sensitive to the mesh size for the triple-layered model, and the triple-layered model with 5 mm mesh has the best agreement with the test results in terms of headform acceleration curve and fracture pattern. The effect of element deletion criteria on the simulation results is also studied. Results show that the crack fracture pattern of an impacted laminated windshield obtained by the model whose element failed when they achieved principal failure stress is more realistic, with a more circular crack and a larger crack area than that where the elements failed when they achieved plastic failure strain.

A comparison of the simulation results of the two FE models shows that a delay occurred for glass failure in the triple-layered model because the structure of this windshield FE model described glass and PVB individually. However, the triple-layered model clearly performs better than the single-layered model for the final fracture pattern of the windshield between the two FE models. In other words, crack propagation corresponds well with the test results by the triple-layered FE model of laminated glass, and the single-layered FE model of laminated glass can better depict the acceleration–time behavior of laminated glass. The single-layered FE model of laminated windshield with different thickness is applied on another new cutoff vehicle FE model to verify the accuracy of the windshield FE model. Results show that the single-layered windshield FE model can enable the headform impactor to achieve an accurate dynamic response that is consistent with the impact on windshields in reality.

Finally, a parametric study on the effect of the Young's modulus, plastic failure strain, and boundary condition of glass on the simulation results is presented. Further simulations are conducted to verify the windshield FE model through dynamic experiments of PVB laminated windshield under headform impact; the headform impact velocities and angles are set as 6.6–11.2 m/s and 60°–90°, respectively, in accordance with real-world vehicle–pedestrian accidents. As the Young's modulus of glass increases, the windshield becomes stiff, and the peak acceleration increases correspondingly. As the plastic failure strain increases, the acceleration does not change obviously. As the plastic failure strain decreases, the cracks intensify. The impact process of PVB itself is a dynamic process. Thus, the results obtained by the application of dynamic tensile test curve are more in line with the actual situation. The boundary conditions of laminated glass have only minimal influence on the acceleration–time curves. In addition, the acceleration–time curves of the simulation results according to test parameter settings are obtained and compared with the corresponding results for further verification. The peak values and the entire time course of the simulation fit well with the corresponding impact test results with different impact velocities and impact angles. Moreover, the crack propagation patterns for the simulation results and the test results with the velocity of 8 m/s at vertical impact are in good agreement, that is, circular crack appeared after the radial crack first appeared in both cases.

References

1. J. Xu, Y. Li, X. Chen, Y. Yan, D. Ge, M. Zhu, B. Liu, Numerical study of PVB laminated windshield cracking upon human head impact. Cmc Comput. Mat. Continua **18**, 183–211 (2010)
2. J. Xu, Y. Li, X. Chen, Y. Yan, D. Ge, M. Zhu, B. Liu, Characteristics of windshield cracking upon low-speed impact: numerical simulation based on the extended finite element method. Comput. Mater. Sci. **48**, 582–588 (2010)
3. LS-DYNA Keyword User's Manual, Version 971, Livermore Software Technology Corporation (2007)
4. Y. Peng, J. Yang, C. Deck, R. Willinger, Finite element modeling of crash test behavior for windshield laminated glass. Int. J. Impact Eng **57**, 27–35 (2013)
5. J. Grünert, R. Hardy, S. Joonekindt, N. Dijk, F. Swartjes, F. Bayer, Constitutive laws for complete components windscreen and other materials. APROSYS report "AP-SP3-012" (2004)
6. X. Zhang, Y. Zou, H. Hao, X. Li, G. Ma, K. Liu, Y. Zou, X. Li, G. Ma, K. Liu, Laboratory test on dynamic material properties of annealed float glass. Int. J. Protective Struct. **3**, 407–430 (2012)
7. X. Zhang, H. Hao, Y. Shi, J. Cui, The mechanical properties of polyvinyl butyral (PVB) at high strain rates. Constr. Build. Mater. **93**, 404–415 (2015)
8. B. Liu, T. Xu, X. Xu, Y. Wang, Y. Sun, Y. Li, Energy absorption mechanism of polyvinyl butyral laminated windshield subjected to head impact: experiment and numerical simulation. Int. J. Impact Eng. (2015)

9. J. Xu, Y. Li, B. Liu, M. Zhu, D. Ge, Experimental study on mechanical behavior of PVB laminated glass under quasi-static and dynamic loadings. Compos. B Eng. **42**, 302–308 (2011)
10. Y. Peng, C. Deck, J. Yang, R. Willinger, Modeling and validation of windscreen laminated glass behavior during fracture 541–544 (2012)
11. R. Iwasaki, C. Sato, J.L. Latailladeand, P. Viot, Experimental study on the interface fracture toughness of PVB (polyvinyl butyral)/glass at high strain rates. Int. J. Crashworthiness **12**, 293–298 (2007)
12. X. Xu, B. Liu, Y. Wang, Y. Li, J. Xu, A numerical model on PVB laminated windshield subjected to headform low-speed impact. J. Phys. Conf. Ser. (2013)
13. X. Xu, J. Chen, J. Xu, Y. Li, X. Yao, Modeling and experimental studies of crack propagation in laminated glass sheets. Sae Int. J. Mat. Manuf. **7**, 328–336 (2014)
14. J. Chen, J. Xu, B. Liu, X. Yao, Y. Li, Quantity effect of radial cracks on the cracking propagation behavior and the crack morphology. PLoS ONE **9**, e98196 (2014)
15. J. Chen, J. Xu, X. Yao, X. Xu, B. Liu, Y. Li, Different driving mechanisms of in-plane cracking on two brittle layers of laminated glass. Int. J. Impact Eng. **69**, 80–85 (2014)
16. J. Chen, J. Xu, X. Yao, B. Liu, X. Xu, Y. Zhang, Y. Li, Experimental investigation on the radial and circular crack propagation of PVB laminated glass subject to dynamic out-of-plane loading. Eng. Fract. Mech. **112–113**, 26–40 (2013)
17. J. Xu, Y. Li, Crack analysis in PVB laminated windshield impacted by pedestrian head in traffic accident. Int. J. Crashworthiness **14**, 63–71 (2009)
18. J. Xu, Y. Sun, B. Liu, M. Zhu, X. Yao, Y. Yan, Y. Li, X. Chen, Experimental and macroscopic investigation of dynamic crack patterns in PVB laminated glass sheets subject to light-weight impact. Eng. Fail. Anal. **18**, 1605–1612 (2011)
19. S. Zhao, L.R. Dharani, X. Liang, L. Chai, S.D. Barbat, Crack initiation in laminated automotive glazing subjected to simulated head impact. Int. J. Crashworthiness **10**, 229–236 (2005)
20. M. Zang, Z. Song, Z. Yang, Finite element model of windscreen for pedestrian protection. J. South China Univ. Technol. **42**, 143–148 (2014)
21. ECE/TRANS/180/Add.9, Global technical regulation No. 9 (2008)
22. B. Liu, Y. Sun, Y. Li, Y. Wang, D. Ge, J. Xu, Systematic experimental study on mechanical behavior of PVB (polyvinyl butyral) material under various loading conditions. Polym. Eng. Sci. **52**, 1137–1147 (2012)
23. J. Xu, Y.T. Sun, Y. Wang, B.H. Liu, Y.B. Li, R. Willinger, *Evaluation of Pedestrian's Head Injury Subject to PVB Laminated Windshield Impact* (2011)
24. B. Liu, Experimental research on energy absorption characteristic of PVB laminated windshield subjected to headform low-speed impact. J. Mech. Eng. **49**, 128 (2013)
25. J. Xu, Y.B. Li, Study of damage in windshield glazing subject to impact by a pedestrian's head. Proc. Inst. Mech. Eng. Part D: J. Automobile Eng. **223**, 77–84 (2009)
26. J. Xu, Y. Li, G. Lu, W. Zhou, Reconstruction model of vehicle impact speed in pedestrian–vehicle accident. Int. J. Impact Eng. **36**, 783–788 (2009)

Chapter 7
Numerical Simulation Based on XFEM

Cracks, including radial and circumferential cracks, will appear on the surface of the glass material after the head impacts on the windshield. The fracture characteristics of a windshield are widely recognized as among the most important factors in automotive crashworthiness. The crack profiles (e.g., length and pattern) on the windshield material provide critical information regarding impact speed (which is extremely useful in accident reconstruction), vehicle crashworthiness, and insights into improving pedestrian and passenger protection. This situation has motivated us to study the propagation characteristics of a crack induced by low-speed head impact on a windshield glass material. In previous studies on impact on windshield glass, automotive engineering researchers frequently avoid the explicit simulation of crack propagation due to the intrinsic complexity of the numerical analysis of cracking.

The current study aims to investigate the crack propagation characteristics (including radial and circumferential cracks) when a model windshield undergoes low-speed impact. The numerical investigation of multiple crack propagation is based on XFEM.

7.1 Fundamentals of XFEM

XFEM incorporates a discontinuous displacement field across the crack that is facing away from the crack tip in the following form:

$$u^h = \sum_{i=1}^{n} N_i(x)\left(u_i + b_i H(x) + \sum_{l=1}^{4}\sum_{j=1}^{2} c_{il} F_l^i(x) \right), \tag{7.1}$$

where n is the number of nodes in the mesh, $N_i(x)$ is the shape function of node i, and u_i are the classical degrees of freedom (DOFs) of node i. b_i and c_{il} are the DOFs associated with the Heaviside jump function $H(x)$, with a value of 1 above and below the crack. The crack tip function $F_l^i(x)$ is

J. Xu and Y. Li, *Impact Behavior and Pedestrian Protection of Automotive Laminated Windshield*, https://doi.org/10.1007/978-981-13-2441-3_7

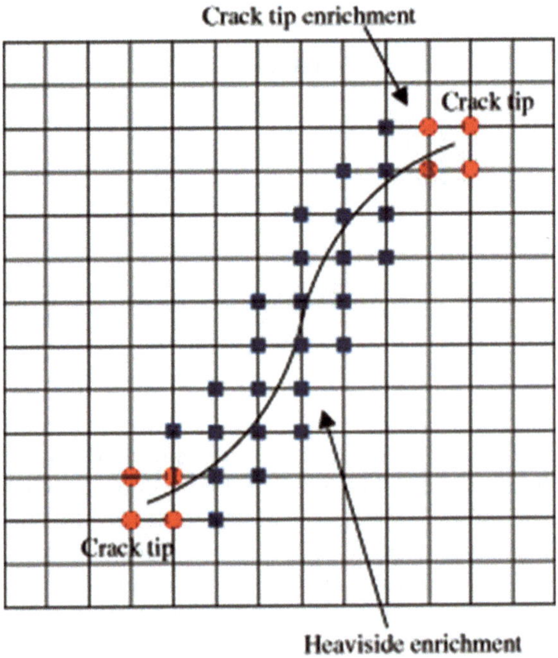

Fig. 7.1 Schematic of a uniform mesh that surrounds a crack used in XFEM simulation. The blue square nodes denote Heaviside function enrichment, whereas the red circular nodes indicate crack tip enrichment. (The reader is referred to the web version of this article for the interpretation of references to color in this figure legend)

$$F_l(r,\theta) = \left(\sqrt{r}\sin\left(\frac{\theta}{2}\right), \sqrt{r}\cos\left(\frac{\theta}{2}\right), \sqrt{r}\sin\left(\frac{\theta}{2}\right)\sin\theta, \sqrt{r}\cos\left(\frac{\theta}{2}\right)\sin\theta\right), \tag{7.2}$$

where (r, θ) are the local polar coordinates at the crack tip.

Figure 7.1 illustrates the enriched nodes near a crack tip, where $H(x)$ is discontinuous across the crack surface. Under general mixed-mode loadings, the asymptotic near-tip hoop and shear stress components are

$$\begin{cases} \sigma_{\theta\theta} = \frac{1}{4\sqrt{2\pi r}}(k_1(3\cos(\theta/2) + \cos(3\theta/2) + k_{II}(-3\sin(\theta/2) - 3\sin(3\theta/2)) \\ \sigma_{\theta\theta} = \frac{1}{4\sqrt{2\pi r}}(k_1(\sin(\theta/2) + \sin(3\theta/2) + k_{II}(\cos(\theta/2) - 3\cos(3\theta/2)) \\ \sigma_{\theta\theta} = \frac{1}{4\sqrt{2\pi r}}(k_1(5\cos(\theta/2) - \cos(3\theta/2) + k_{II}(5\sin(\theta/2) - 3\sin(3\theta/2)) \end{cases} \tag{7.3}$$

For fracture propagation, we use the popular criterion for maximum local hoop stress, where the crack propagation direction is determined from Eq. (7.3) as

$$\theta_c = arc\cos\left(\frac{3k_{II}^2 + \sqrt{K_I^4 + 8k_I^2k_{II}^2}}{K_I^2 + 9k_{II}^2}\right) \tag{7.4}$$

where θ_c is measured with respect to a local polar coordinate system with its origin at the crack tip and aligned with the direction of the existing crack. Numerical simulations based on the preceding XFEM algorithm is performed using the commercial code ABAQUS.

7.2 Model and Methods

7.2.1 Model Setup

A featureless spherical headform is used to model the human head. The model head has a radius R of 90 mm and a mass of 4.5 kg. Notably, the effective head weights are changeable in different accident cases because the body may also get involved due to inertia, thereby causing the effective impact mass to become larger than the pure head mass, particularly during high-speed impact. Thus, the effective head mass M is also varied from 4.5 to 90 kg by changing the density (from 1500 to 30,000 kg/m^3) while maintaining the volume in this study.

On the basis of 151 pedestrian–vehicle accident cases from NTADTU, the impact speed of most vehicles involved in an accident falls within the range of 40–80 km/h, i.e., 11.1–22.2 m/s, when low-speed impact is considered. We focus on normal impact and assume that impact occurs at the center of the plate. We neglect a small windshield curvature and focus on a flat rectangular plate with dimensions of a $\times$ b = 700 mm $\times$ 700 mm and a thickness of t = 4.76 mm, as shown in Fig. 7.2.

In terms of material properties, the typical material parameters of glass are assigned to the plate, as follows: density ρ = 2500 kg/m^3, Young's modulus

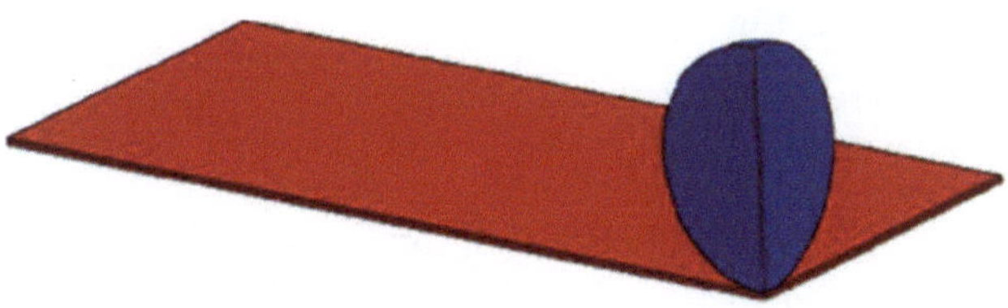

(a) Global view of the quarter model of head impact on a model windshield plate

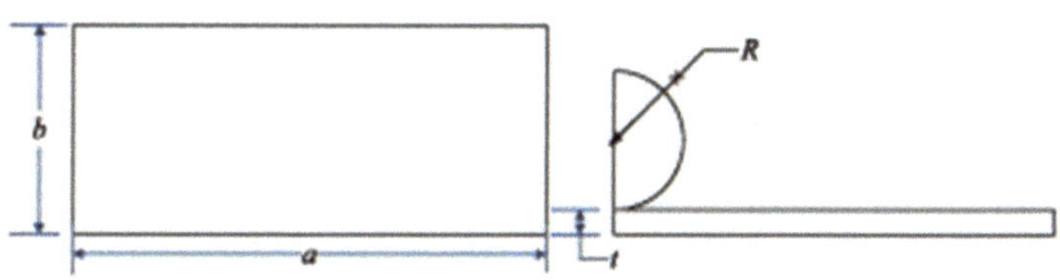

(b) Dimensions of the featureless headform and windshield plate in the quarter model

Fig. 7.2 Computational model setup

$E = 70$ GPa, Poisson's ratio $\nu = 0.22$, mode I energy release rate $G_{\mathrm{I}} = 10$ J/m^2, and modes II and III energy release rate $G_{\mathrm{II}} = G_{\mathrm{III}} = 50$ J/m^2. The properties are assumed to be rate-independent. To utilize symmetry, only one quarter of the model is required for simulation. The boundary condition of the windshield is assumed to be fixed. The plate and head are modeled with eight-node linear brick elements. The contact between the head and plate follows Coulomb's friction law with a coefficient of $f = 0.1$ (a small slip exists between the head and windshield according to our accident scene investigation).

7.2.2 Computational Method

Suppose that no crack or damage occurs. We first compute the deformation response with the response to impact. For a given effective head mass M and impact speed v, the maximum displacement D is computed and presented in a dimensionless form, as shown in Fig. 7.3. For the ranges of impact speed and head mass considered in this study, D/t can be fitted as a dimensionless function of v/v_0 and M/M_0, as follows:

$$D/t = \left(135.3(v/v_0)^{0.7115} - 0.0305^{518.5(v/v_0)}\right)(M/M_0)^{0.2669} \tag{7.5}$$

where $v_0 = \sqrt{E/\rho}$ and $M_0 = abt\rho$.

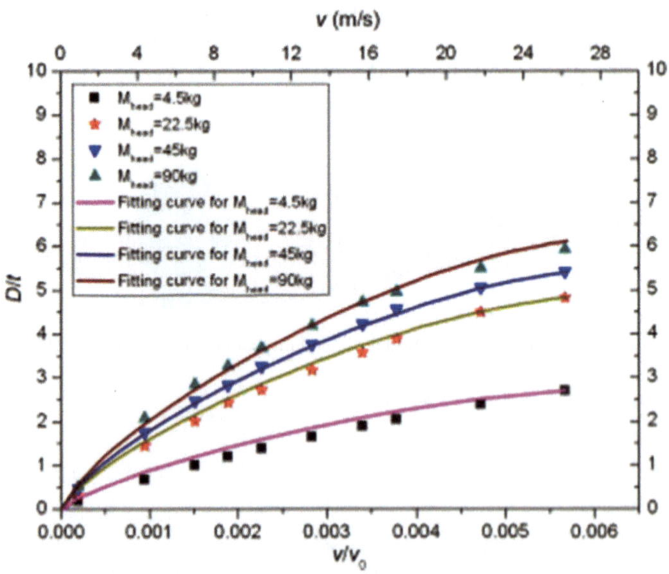

Fig. 7.3 Relationship between initial impact velocity and displacement under different head masses. The lines denote curve fitting, Eq. (7.5)

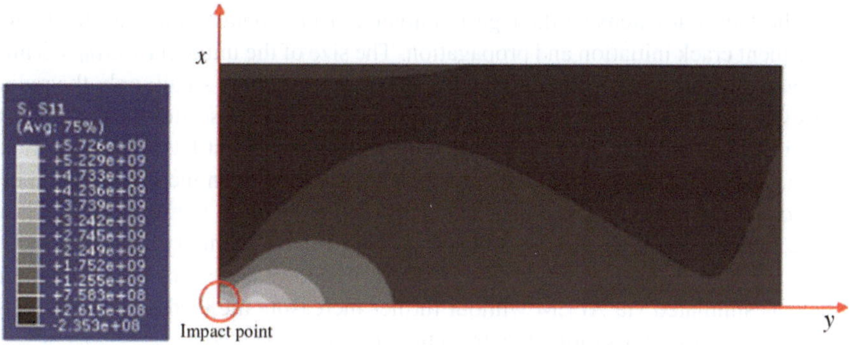

(a) Contour plot of the radial stress field (unit: Pa)

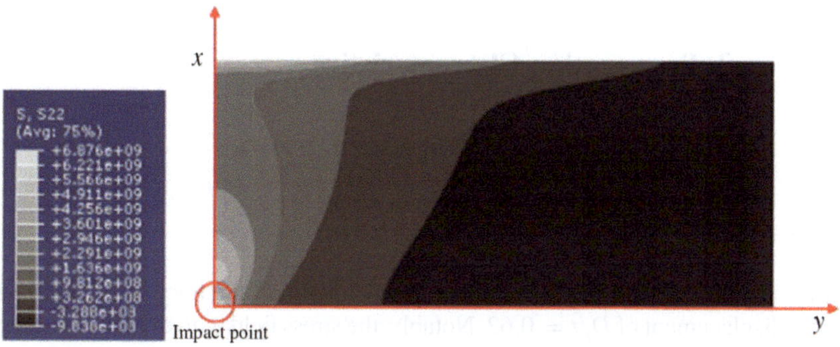

(b) Contour plot of the hoop stress field (unit: Pa)

Fig. 7.4 Comparison between the radial and hoop stress fields that result fully from impact deformation. Cracks should form to relieve prominent stress. A radial crack is preferred to propagate before a circumferential crack due to its more prominent hoop stress field

Impact speed is relatively low (considerably lower than wave speed); thus, our simulation shows that the dynamic effect is relatively negligible for the range of parameters investigated in this study. That is, for a given head mass, the computed stress field from a dynamic impact (with incident speed v) is extremely close to that of quasi-static indentation (with a maximum indentation depth of D/t) as long as they induce the same deformation curvature of the plate, i.e., when v and D/t satisfy Eq. (7.5). Figure 7.4 provides an example of the stress field. This computational method enables us to simplify the dynamic impact problem into a quasi-static indentation problem, where the normalized indentation depth D/t becomes the governing variable to indicate the effective "impact condition."

Our previous study showed that circumferential cracks always grow after the radial crack has developed in windshield fracture, except near the vicinity of impact where extensive damage can be found [1]. This result is consistent with Fig. 7.4, in which the hoop stress field is more prominent than the radial stress field. Thus, crack simulation can be divided into two steps. First, we embed a number of initial flaws into the center of the plate (i.e., the impact location). This step is consistent with the fact that extensive damage is initiated at the contact zone and facilitates subsequent crack initiation and propagation. The size of the initial flaw is $a_0 = 3$ mm with variable directions. When the initial flaws are extremely small, only the radial flaws will grow to become radial cracks upon impact to release strain energy. For a given indentation depth D, a radial crack will grow gradually until its reaches a certain length. The final length of a radial crack also increases with an increase in D. In the second step, new circumferential flaws with size $a_0 = 3$ mm are introduced based on the radially cracked specimen (the sensitivity of flaw geometry and distribution is discussed in the succeeding section). Then, the development of circumferential cracks is simulated via XFEM without further increasing the load and by adopting the residual stress and strain fields from the first step.

7.3 Crack Propagation Characteristics

7.3.1 Radial Crack Propagation Characteristics

7.3.1.1 Overall Crack Feature

In Fig. 7.5, the evolved radial crack is shown along with the hoop stress field for a typical displacement of $D/t = 0.62$. Notably, the stress field with the fully developed crack is already considerably more relaxed than its counterpart without crack (refer to the difference between Figs. 7.4 and 7.5 for an example). Moreover, the current hoop stress field is specified for the global coordinate, whereas the local hoop stress determines crack trajectory during crack propagation. When the local maximum hoop stress exceeds a critical threshold (referred to as damage stress σ_d), the crack grows incrementally toward a local principal direction.

In the beginning stage of impact, the radial crack grows with an increase in indentation depth D/t given that the growth of the radial crack can release more energy than that of the circumferential crack (due to the more prominent hoop stress field, as shown Fig. 7.4). Noticeably, the crack growth direction keeps changing slightly with a growth direction of approximately 22°. This phenomenon is mainly caused by the rectangular shape of the windshield plate, which makes the overall strain in the y-direction greater than that in the x-direction. When the crack grows further, the boundary constraint changes and the crack direction approaches approximately 45°.

During its growing process, the crack strives to release as much energy as possible through mode I fracture. The released strain energy E_{strain} is computed as the difference between the strain energies of two specimens: one cracked and one without crack (at the same D/t). In Fig. 7.6, E_{strain} is normalized via the initial impact kinetic energy of the head $E_{kinetic}$, with M = 4.5 kg, and the impact speed can be deduced using Eq. (7.5). The released strain energy is relatively lower compared with the overall system energy, and it increases as D/t increases. Nevertheless, when it is normalized by the incident kinetic energy, which also increases as D/t increases, the dimensionless number $E_{strain}/E_{kinetic}$ fluctuates to a certain extent.

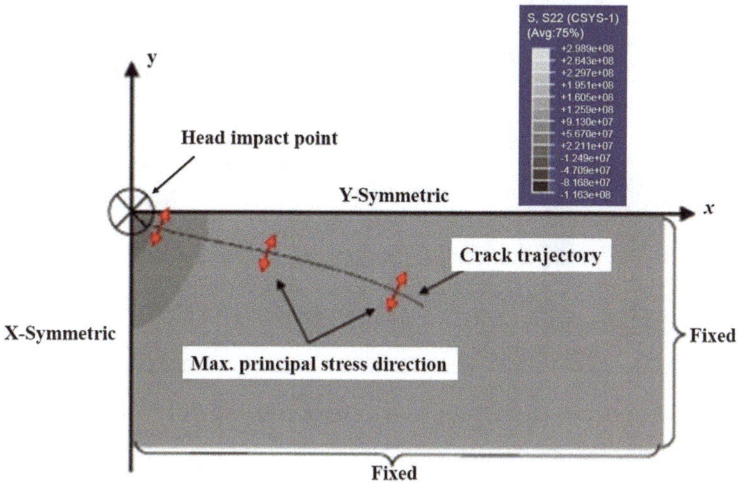

Fig. 7.5 View of the crack trajectory and hoop stress field after the radial crack has fully developed at $D/t = 0.62$

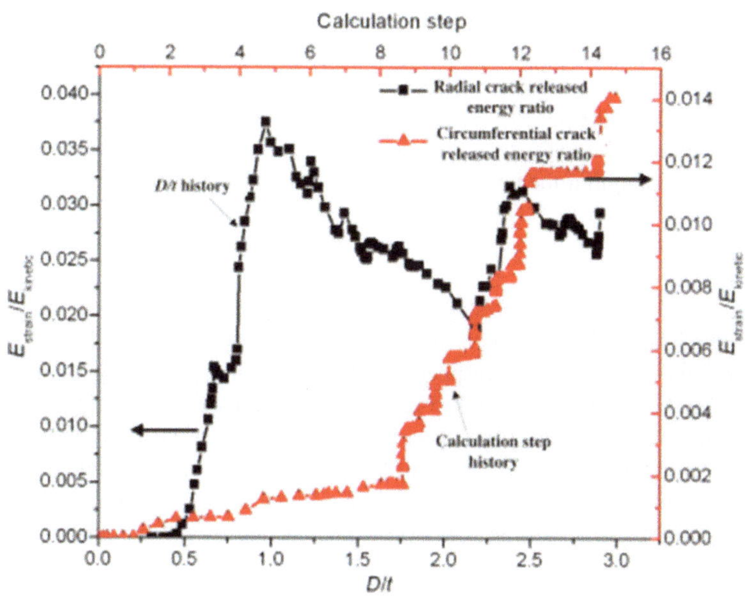

Fig. 7.6 Normalized released strain energy during the propagation of radial and circumferential cracks. The released strain energy of a radial crack is a function of D/t, or equivalently, the impact speed deduced through Eq. (7.5). It is normalized by the incident kinetic energy with $M = 4.5$ kg. The circumferential crack propagates at $D/t = 2.89$ and is normalized by the corresponding incident kinetic energy

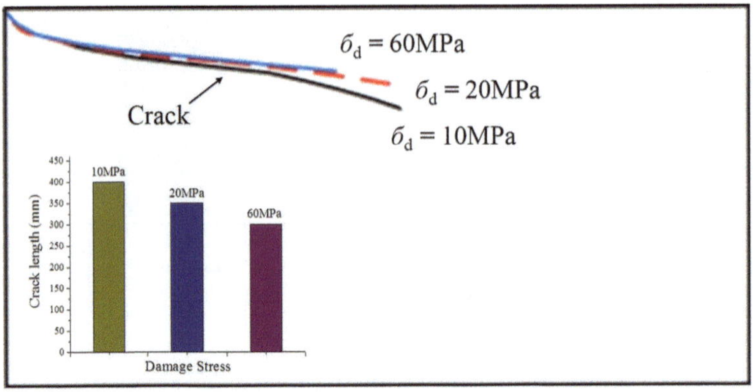

Fig. 7.7 Effect of different damage stresses on the radial crack trajectory at $D/t = 0.62$

7.3.1.2 Sensitivity of the Initial Flaw and Fracture Criterion

The damage stresses σ_d of different soda-lime glass specimens vary due to their intrinsic stochastic flaws, which typically range from 10 to 60 MPa. The effect of σ_d on crack evolution under quasi-static indentation is shown in Fig. 7.7. The results indicate that as the critical stress of crack initiation σ_d increases, the crack angle becomes smaller (the boundary constraint is more prominent) and the crack becomes shorter. That is, the damage stress threshold can be roughly estimated according to the crack path and length if the impact condition is known. Subsequently, $\sigma_d = 10$ MPa is used.

Mode I energy release rate is another critical parameter that governs the brittleness of a material. Consider typical candidate materials for a windshield, such as soda-lime glass and PMMA. The former is brittle, whereas the later has a certain degree of ductility. The mode I energy release rates of these materials are approximately $G_{\mathrm{I}} = 10$ J/m^2 and $G_{\mathrm{I}} = 400$ J/m^2, respectively. Without losing generality, we normalize crack length C with plate size *a*. In Fig. 7.8, the effect on the resulting crack length is examined by varying G_{I} from 10 to 400 J/m^2. A numerical study demonstrates that if the value of G_{I} is within the brittle material domain, then the effect on crack length is minimal. Otherwise, if the windshield is made from a ductile material, then the crack length will be shorter under the same plate deformation. Subsequently, $G_{\mathrm{I}} = 10$ J/m^2 is used.

The characteristics of the initial flaw have several minor effects on crack propagation. First, we maintain the initial flaw length at 3 mm and set the flaw angle θ_i as 30°, 45°, and 60° with respect to the *x*-axis. The final crack pattern is shown in Fig. 7.9. Although θ_i exerts influences on the initial crack trajectory, all crack paths eventually converge with similar lengths, thereby indicating the robustness of the current model. The initial flaw length exerts negligible effect on the crack propagation pattern and crack length (provided that the initial flaw is short). When the initial flaw length is set as 0.5, 1, 3, and 5 mm, the lengths of the resulting radial cracks

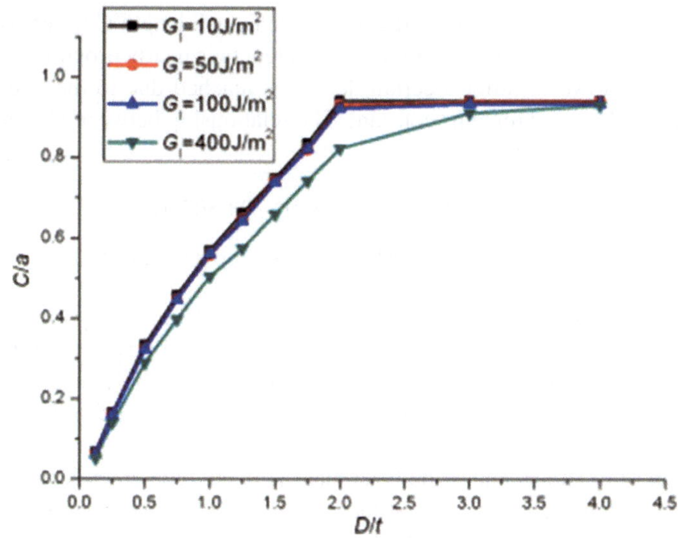

Fig. 7.8 Effect of mode I energy release rate on radial crack length

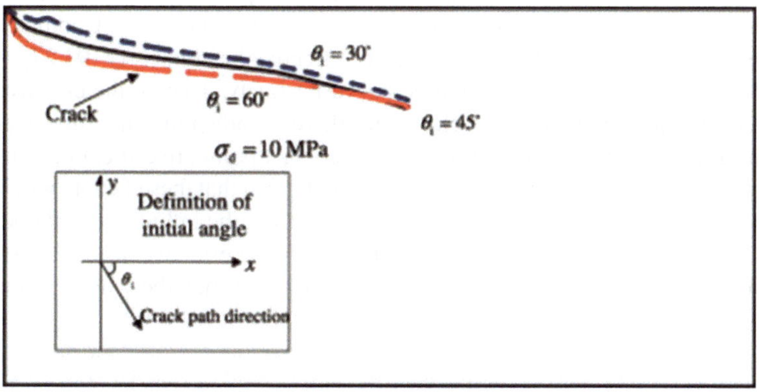

Fig. 7.9 Effect of the initial flaw angle on the radial crack pattern at $D/t = 0.62$

in the four scenarios are the same. If the initial flaw is 10 mm long, then the crack propagation pattern slightly differs. Therefore, the 3 mm-long initial flaw used in the current study is validated and is determined to be a relatively insensitive parameter in the present XFEM simulation.

7.3.1.3 Radial Crack Length

The length of a radial crack can be represented as a function of the quasi-indentation depth D/t, which, in turn, is related to impact speed v for a given effective head mass [Eq. (7.6)]. Figure 7.8 shows the normalized crack length C/a versus the normalized head displacement D/t. Initially, crack length is nearly proportional to head displacement. However, after a certain length is reached due to the fixed boundary, crack length nearly remains constant. The relationship between C/a and D/t is nonlinear and may be fitted as

$$C/a = -0.4046 + 0.9231(D/t) - 0.05074(D/t)^3 + 0.03803\exp(D/t) - 0.4243\sqrt{(D/t)}\ln(D/t) \qquad (7.6)$$

when Eqs. (7.5) and (7.6) are combined, a new method can be derived for directly estimating impact speed based on the knowledge of radial crack length (assuming that the effective head mass is known) using the Newton–Raphson method. This new method is useful for traffic accident reconstruction, as illustrated in the succeeding section.

Equation (7.6) shows that if the head mass is fixed, then crack length increases nonlinearly with impact speed. To estimate the influence of the effective head mass, Figs. 7.3 and 7.8 can be combined, along with Eqs. (7.5) and (7.6). For example, if the impact speed is fixed at 5 m/s, as shown in Fig. 7.3 or Eq. (7.5), then the variation in crack length (as a function of the corresponding *D*/*t*) can be obtained at different effective head masses from either Fig. 7.8 or Eq. (7.6). Compared with that for the crack length of $M = 4.5$ kg, the effective head mass is 193.27, 241.35, and 294.23% for $M = 22.5$, 45, and 90 kg, respectively. That is, the effective head mass considerably influences crack length, thereby indicating that the shape of a vehicle front end and the pedestrian–vehicle contact point are critical in estimating impact speed during a low-speed impact accident given that these factors will largely determine the pedestrian revolution angle. In addition, the influence of M decreases when M is high, which implies that the influence of the effective head mass starts to decline if the effective head mass is considerably larger than the windshield mass.

7.3.2 *Circumferential Crack Propagation Characteristics*

On the basis of radial crack geometry, we introduce circumferential flaws with a length of $a_0 = 3$ mm along the radial crack and a spacing of 50 mm (spacing is a relatively minor factor of the final circumferential crack pattern) into the second step. The stress field calculated from the end of first step is transferred to the second step. The first step has relaxed most of the hoop stresses; hence, the remaining excessive radial stress should be relaxed by developing circumferential cracks. The most efficient means to release radial stress is to develop a circumferential crack with a maximum radius, i.e., in the same order of C. This finding is verified in Fig. 7.10; that is, cracks also grow near the contact zone because of high stress. This result is consistent with the observation that intensive cracks are identified within the contact area, crack density is lower outside, and the largest circumferential crack nearly

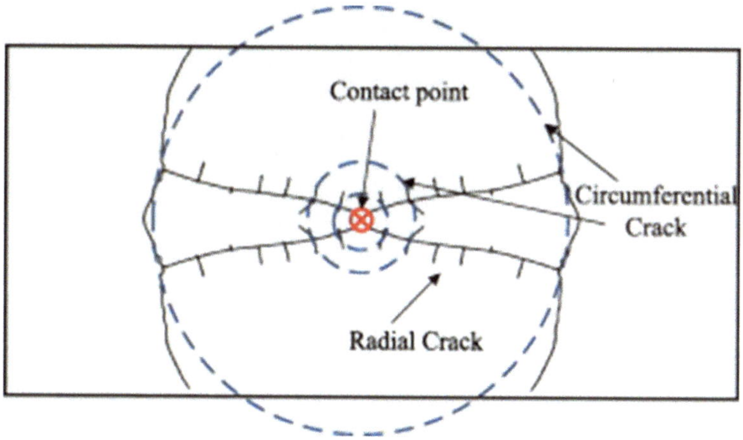

Fig. 7.10 Final crack pattern (including radial and circumferential cracks) on the model windshield plate subjected to head impact at $D/t = 0.62$

bounds the radial cracks. The strain energy released by the circumferential crack is shown in Fig. 7.6, where the total released energy is less than 50% that of the radial crack. This evidence strongly supports the argument that a crack is prone to grow in a radial manner rather than in a circumferential manner given that more energy can be released in the former. As shown in Fig. 7.6, less than approximately 5% of the total kinetic impact energy is consumed by radial and circumferential cracks, and thus, automotive windshield can still be improved for enhanced energy mitigation efficiency and pedestrian protection (e.g., the use of transparent energy absorption materials.

7.4 Parametric Study

7.4.1 Computational Model

7.4.1.1 Head Model

In real-life pedestrian–vehicle accidents, the interaction between pedestrian and vehicle can be divided into three phases [2]: (i) pedestrian–vehicle contact, (ii) impact between the windshield/engine hood and pedestrian, and (iii) pedestrian–vehicle separation (either by flipping over or sliding down), as illustrated in Fig. 7.11. The head is the most vulnerable (and probable) part that can get involved during impact (phases ii and iii) [3]. A featureless headform (Fig. 7.12) with $R_{\mathrm{Head}} = 90$ mm and a reference mass of 4.5 kg is commonly used in windshield testing and evaluation in the automotive industry [4]. Although the same model head radius is adopted in the

Fig. 7.11 Illustration of the pedestrian–vehicle accident process

(a) Pedestrian–vehicle contact phase

(b) Impact between windshield/engine hood and human head

(c) Pedestrian–vehicle separation

current study, we determine that the effective head mass depends on the vehicle's impact speed [5], pedestrian–vehicle contact location, and front-end shape of the vehicle (Fig. 7.11); thus, the effective head mass M_{Head} is varied from 4.5 to 90 kg by fixing the size of the head and changing its density. The head material is consistent with both the featureless headform standard and Ref. [6]. $E_{\text{Head}} = 6.5$ GPa and $\nu_{\text{Head}} = 0.26$, without any plastic deformation and damage.

7.4.1.2 Windshield Model

The in-plane dimension (a × b) of the reference windshield is standard (1400 mm × 600 mm) and without curvature. As shown in Fig. 7.12, a real windshield may have a curvature with respect to the z-direction. The curvature and dimension of a windshield may vary within moderate ranges in different automobiles, and thus, their effects are explored in detail in the subsequent paragraph.

The reference thickness of the PVB interlayer is typically $t_{PVB_0} = 0.76$ mm for passenger cars. In this study, thickness is varied up to 3.04 mm to determine

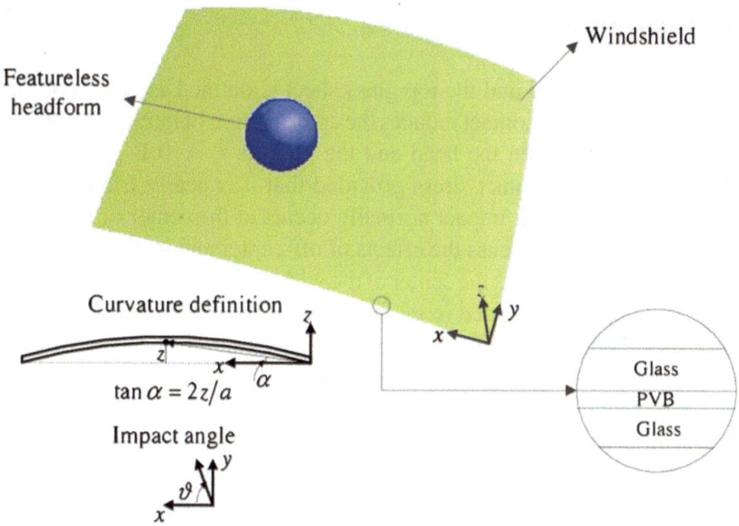

Fig. 7.12 Setup of computational model

whether changing this parameter may lead to different crashworthiness. The density of the PVB material is $\rho_{\mathrm{PVB}} = 1100\ \mathrm{kg/m^3}$ [7]. The reference stress–strain relationship of the PVB material is obtained from our previous experiment [8] (refer to section "Effect of modified PVB material property" for details). Although PVB is a viscoelastic material, we ignore the viscoelasticity of PVB according to the suggestions in [8] due to the small viscoelasticity effect. The physical properties of PVB may vary with respect to age, temperature, and working environment. Here, we also moderately vary the mechanical properties to explore their influences on cracking characteristics.

The in-plane dimensions of the two glass layers are the same as those of PVB except for thickness. We adopt the typical thickness of the windshield, $t_{\mathrm{Glass}} = 2$ mm for each glass sheet. The thickness of the glass layer seldom changes in different types of vehicles; thus, t_{Glass} remains a constant in this study. The typical material parameters of soda–lime glass are density $\rho_{\mathrm{Glass}} = 2500\ \mathrm{kg/m^3}$, Young's modulus $E_{\mathrm{Glass}} = 70$ GPa, Poisson's ratio $\nu_{\mathrm{Glass}} = 0.22$, mode I energy release rate $G_{\mathrm{I}} = 10\ \mathrm{J/m^2}$, and modes II and III energy release rates $G_{\mathrm{II}} = G_{\mathrm{III}} = 50\ \mathrm{J/m^2}$. These properties are assumed to be rate-independent in [9] and also fixed in the present study (the effects of the energy release rate, fracture criterion, and initial flaw are discussed in Ref. [10]). The size and distribution of the initial flaws on the glass exert minimal effects on the final crack pattern and length as long as they are sufficiently small relative to the windshield dimension [3].

7.4.1.3 Model Setup

Contact between the head and the top glass sheet is defined as "hard contact." The impact stress field of hard contact induces the crack pattern of interest. The Coulomb's friction coefficient between the head and the glass is $f = 0.1$, generally a minor factor for the resulting contact stress provided that it is nearly frictionless [11]. In most cases, we assume that impact normally occurs at the center of the windshield, and thus, we also briefly discuss the effects of off-center and inclined impacts. If the impact is normal and occurs at the center of the windshield, then only 1.4 of the system should be modeled and analyzed (e.g., the result of the crack pattern in Fig. 7.14 of this paper shows only 1.4 of the windshield). Otherwise, for general off-center and inclined impacts, the entire windshield (e.g., that in Fig. 7.12) should be considered. The head and the windshield are meshed with eight-node linear brick elements, and a mesh convergence study is conducted to determine mesh density, i.e., 1 mm × 1 mm. For most parts of this study, the boundary condition of the windshield is fixed (clamped), which is qualitatively consistent with the frame setup used in practice [12]. A parametric study is also conducted under different boundary conditions to investigate the influence of boundary condition on cracking properties.

Census data from NTADTU indicate that most impact speeds involved in vehicle accidents are between 40 and 80 km/h, i.e., 11.1–22.2 m/s, which is the low-speed impact range used in the current study.

7.4.2 *Computational Method*

Our previous study [10] showed that the dynamic effect is relatively minor on the range of parameters investigated in this study. Thus, we may simplify the low-speed dynamic impact cracking problem into a quasi-static indentation fracture problem and use the quasi-static indentation stress field to initiate and propagate radial and circumferential cracks.

Dynamic FEM simulations (without cracking) are first conducted under different impact conditions to obtain the equivalent quasi-static indentation conditions. We initially focus on normal and center head impacts (off-center and inclined impacts may be analyzed in a similar manner). In addition, we maintain the other simulation conditions as "regular" reference parameters, i.e., the windshield dimension is standard without curvature, the boundary condition is clamped, and PVB properties are retained at their reference values (the reference parameters are listed in Table 7.1).

For different values of the effective head mass M, impact speed v, and PVB interlayer thickness t_{PVB}, the maximum impact displacement of the head D is computed and presented in a dimensionless form. Finally, D/t ($t = t_{\mathrm{PVB}} + t_{\mathrm{Glass}}$ is the total thickness of the windshield) can be fitted as a dimensionless function of v/v_0, M/M_0, and $t_{\mathrm{PVB}}/t_{\mathrm{PVB}_0}$ as

Table 7.1 Values of the parameters used in the reference windshield model

Parts	Variables	Reference parameters
Head	Mass M_{Head}	4.5 kg
	Radius R_{Head}	90 mm
	Density ρ_{Head}	1412 kg/m^3
	Poisson's ratio ν_{Head}	0.26
	Young's modulus E_{Head}	6.5 GPa
Glass	Dimension $a \times b$	1400 mm × 700 mm
	Thickness t_{Glass}	2 mm (each layer)
	Density ρ_{Glass}	2500 kg/m^3
	Young's modulus E_{Glass}	70 GPa
	Poisson's ratio G_{Glass}	0.22
PVB	Dimension $a \times b$	1400 mm × 700 mm
	Thickness t_{PVB}	0.76 mm
	Density ρ_{PVB}	1100 kg/m^3
Crack setting	Mode I energy release rate G_{I}	10 J/m^2
	Mode II energy release rate G_{II}	50 J/m^2
	Mode III energy release rate G_{III}	50 J/m^2
Global settings	Windshield curvature α	0°
	Boundary conditions	Clamped

$$
\begin{aligned}
D/t = {} & \left[(v/v_0)^{-0.8799} + (M/M_0)^{-3.334} + 0.6679\right]^{-3.338} \\
& + \left[(M/M_0)^{3.357} + \left(t_{PVB}/t_{PVB_0}\right)^{2.194} + 1.070\right]^{0.07599} \\
& + \left[(v/v_0)^{0.09454} + \left(t_{PVB}/t_{PVB_0}\right)^{-0.03154} - 1.082\right]^{6.131} - 1.424
\end{aligned}
\tag{7.7}
$$

for the range of parameters used in this study, where $v_0 = \sqrt{G_{\text{I}}/\rho t}$ is the reference velocity, and $M_0 = 9.1$ kg is the mass of the reference PVB laminated windshield with the aforementioned "regular" parameters. The correlation coefficient of fitting is $R = 0.999$ (Fig. 7.13). Subsequently, the normalized displacement D/t can be used to represent the combined (and coupled) effect of the effective impact mass, center and normal impact speeds, and PVB interlayer thickness. A quasi-static indentation simulation is then conducted (to reach a particular D/t of the "head" indenter) via XFEM, and the resulting stress field is the driving force of the concurrent crack propagation.

For impact-induced windshield fracture, circumferential cracks always grow after radial cracks have developed [1, 13] (except near the vicinity of impact where extensive damage can be found) because the hoop stress field is more prominent than the radial stress field. Therefore, we divide crack simulation based on XFEM into two

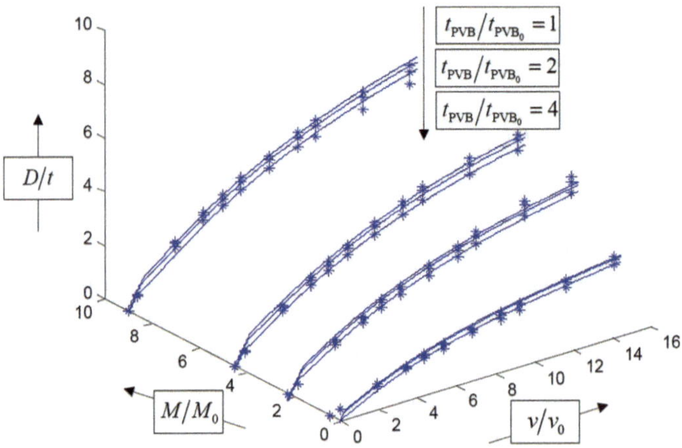

Fig. 7.13 Relationship of head displacement to initial impact velocity, PVB interlayer thickness, and head mass (all parameters are normalized). The lines denote the curve fitting based on Eq. (7.7)

steps. First, a radial crack initiates from the center of the windshield (with the aid of random initial flaws). Second, circumferential crack growth is calculated. Additional details and the robustness of such approach are provided in [10]. Thereafter, a systematic parametric study is performed by considering several major factors that can affect crack shape, angle, and length, which are the main parameters that characterize the crack pattern.

7.4.3 *Results and Discussions*

7.4.3.1 Effects of PVB Interlayer Material Property

Effect of PVB Interlayer Thickness

The invention of the PVB laminated windshield has contributed considerably to the automotive industry because this type of windshield has saved many human lives and has significantly reduced injuries. In the current study, crack patterns and crack lengths are investigated and compared in windshield models with and without a PVB interlayer. In particular, crack length may be an effective indicator of PVB's energy absorption capability.

In Fig. 7.14, radial crack characteristics are compared for reference windshield models with or without a PVB interlayer at $D/t = 0.62$. Here, the crack length C is normalized by the characteristic plate length as C/a, and the crack angle is defined in Fig. 7.14. As shown in this figure, the radial crack is prone to grow near the longer border (with a crack angle q of approximately 22°) in a pure glass windshield (with thickness t =4.76 mm, which is the same as the overall thickness

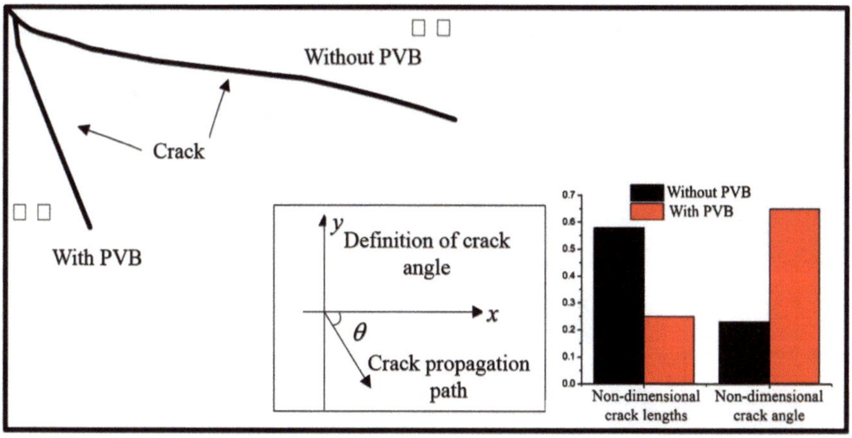

Fig. 7.14 Different crack patterns in the reference pure glass windshield and PVB laminated windshield. The values of the input parameters of the reference models used in XFEM simulation are listed in Table 7.1. The normalized steady-state crack angle (defined in the inset) and length under the two situations are compared at $D/t=0{:}62$

of the reference PVB laminated windshield). By contrast, the radial crack tends to propagate near the shorter border (with a crack angle of approximately 59°) in a PVB laminated windshield at the same D/t. Moreover, crack length is shorter in a laminated windshield.

From Fig. 7.13 and Eq. (7.7), D/t is expected to decrease as the thickness of the PVB interlayer increases under the same impact speed v because a higher amount of impact energy is mitigated by a thicker PVB interlayer. That is, a higher v should be imposed on a windshield with a thicker PVB interlayer to reach the same D/t. Consequently, we convert the relation between C/a and D/t into that between C/a and v/v_0 in Fig. 7.15 by using Eq. (7.7) (for different PVB interlayer thickness values and with the other parameters fixed at their reference values). The results demonstrate that crack length is shorter in a windshield with a thicker PVB interlayer under the same v/v_0. This result is consistent with the fact that the cracks on the windshield of commercial trucks or buses is shorter (at nearly the same impact speed), which may be largely attributed to the thicker PVB interlayer (in cases where windshield size is similar to that of passenger cars) [14]. However, the gain (reduction in crack length) becomes less significant as the PVB interlayer becomes extremely thick, thereby implying that energy absorption enhancement and crack length reduction may not be achieved by simply thickening the PVB interlayer. Therefore, we further explore the possibility of modifying the material properties of PVB.

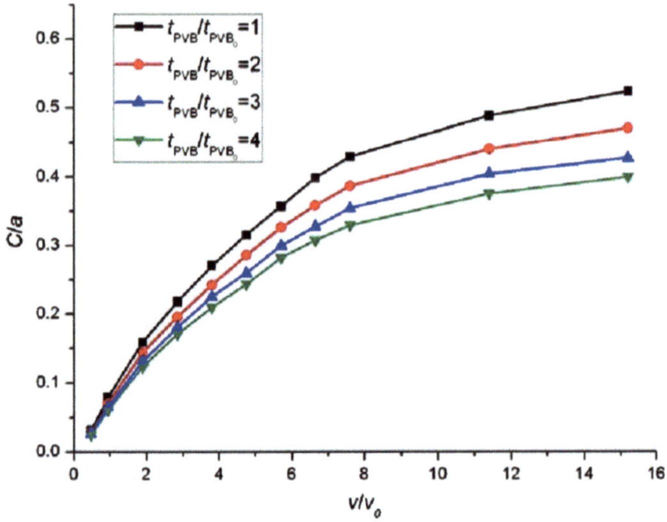

Fig. 7.15 Relations between the normalized crack length and normalized initial impact speed as the normalized thickness of the PVB interlayer varies from 1 to 4. The reference values listed in Table 7.1 are adopted for the other parameters used in the simulation

Effect of Modified PVB Material Property

In practice, interlayer material property may be modified to increase pedestrian protection and further understand its contribution to windshield crack formation (which may help in accident reconstruction).

As shown in our previous experiment [8, 15], the constitutive behavior of PVB is similar to that of rubber and its uniaxial stress–strain curve can be approximated as $\sigma_{\mathrm{PVB}}(\mathrm{MPa}) = 428.9(\varepsilon_{\mathrm{PVB}})^{1.832}$, $0 \leq \varepsilon_{\mathrm{PVB}} \leq 0.525$, where σ_{PVB} and ν_{PVB} denote stress and strain, respectively. Subsequently, we generalize the stress–strain relation as $\sigma_{\mathrm{PVB}} = m\kappa(\varepsilon_{\mathrm{PVB}})^{n\psi}$, where κ =428.9 MPa, ψ =1.832, and m and n are the scale factors. We can perturb the mechanical property of the PVB interlayer (at least theoretically to "strengthen" the PVB layer) by varying the scale factors m and n.

In Fig. 7.16, we vary m and n from 1 to 3 with several different combinations to investigate the influences on C/a. The results show that C/a becomes longer when m is smaller or when n is larger (i.e., when the interlayer material is "compliant"). That is, in addition to thickening the PVB layer, increasing the stiffness of the interlayer may also shorten cracks, thereby implying improved pedestrian/passenger protection in traffic accidents.

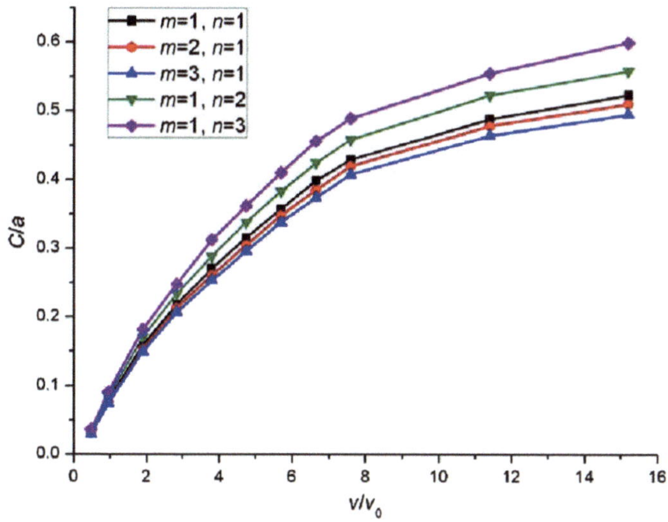

Fig. 7.16 Relations between the normalized crack length and normalized initial impact speed as the interlayer material scale factors *m* and *n* are varied. The reference values listed in Table 7.1 are adopted for the other parameters used in the simulation

7.4.3.2 Effects of Windshield Geometric Parameters

In this section, we separately vary the curvature, length/width aspect ratio, and dimension of the windshield. The results will help to illustrate clearly the relationship between these geometric factors and the crack parameters (e.g., the angle of the longest radial crack).

Effect of Windshield Curvature

The main advantage of windshield curvature design is to keep the entire vehicle under good aerodynamic condition. We examine whether curvature is influential to crack length to further assess its benefit from a mechanical point of view. Here, α is defined as the curvature parameter in Fig. 7.12, which can be normalized by the reference value $\alpha_0 = \pi/2$. In general, crack length tends to become shorter with an increase in α under the same impact condition. The magnitude of the stress field upon impact is basically reduced with the enhancement of curvature due to the reduction in head displacement D/t at the same impact speed, which will favor a shorter crack.

Figure 7.17 shows the normalized crack length at different curvature parameters (all the other parameters retain their reference values provided in Table 7.1 except for curvature α). A short crack length at a large windshield curvature implies improved PVB energy dissipation capability and fracture resistance at the same head displacement. The steady-state crack propagation angle (defined in Fig. 7.14 and normalized

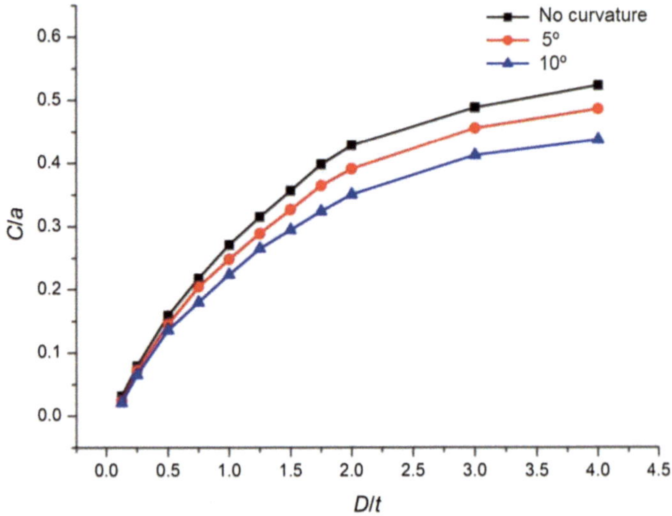

Fig. 7.17 Relations between the normalized crack length and normalized head displacement into the windshield as the windshield curvature angle is changed from 0° to 10° (see Fig. 7.2 for the definition of curvature angle). The reference values listed in Table 7.1 are adopted for the other parameters used in the simulation

by the reference value $\theta_0 = \pi/2$) also changes considerably during curvature variation, as shown in Fig. 7.18 (with $D/t = 1.21$).

The results indicate that a large windshield curvature causes radial crack propagation to incline toward the longer border. From the fitting of the XFEM analysis results, the relationship between normalized α and θ can be derived as (for the current ranges of parameters)

$$\theta/\theta_0 = e^{-1.398-85.63(\alpha/\alpha_0)^{1.5}} - 0.777(\alpha/\alpha_0)^{0.5}, \tag{7.8}$$

$$\theta/\theta_0 = 2.08\left(4.03 + (\alpha/\alpha_0)^{1.06}\right)^{0.5} - 0.530\left(2.89 + (\alpha/\alpha_0)^{0.470}\right)^{1.5} - 0.932, \tag{7.9}$$

where Eq. (7.8) is for pure glass, whereas Eq. (7.9) is for PVB laminated glass.

Effect of the Length/Width Aspect Ratio of the Windshield

The aspect ratio of the windshield imposes an effect on crack patterns through the boundary constraint. This ratio is changed by fixing the total area ($a \times b$) while simultaneously varying a and b. Numerical simulations are divided into two sets: pure glass windshield and PVB laminated windshield. All the other parameters have fixed reference values (see Table 7.1) and $D/t = 1.21$. The variation in crack angle as a function of aspect ratio χ ($\chi = a/b$) is shown in Fig. 7.19.

For a pure glass windshield, the relation between θ and χ can exhibit the following form (for the current ranges of parameters):

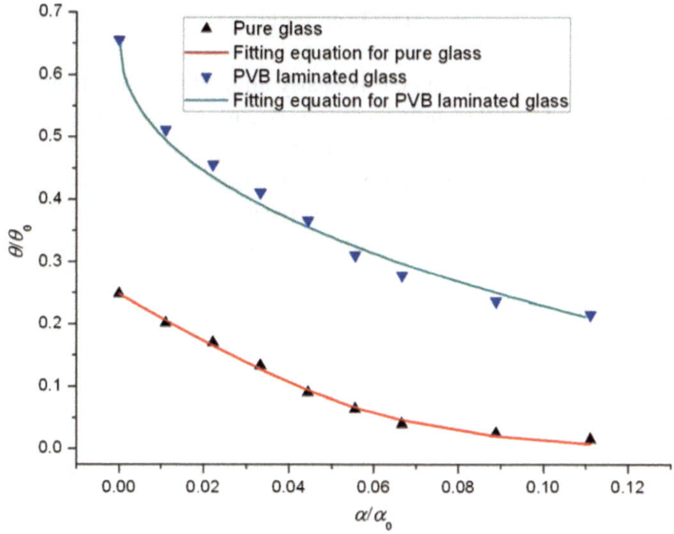

Fig. 7.18 Relations between the normalized crack angle and windshield curvature for PVB laminated windshield and pure glass windshield. Solid lines denote the fitted results. The reference values listed in Table 7.1 are adopted for the other parameters used in the simulation

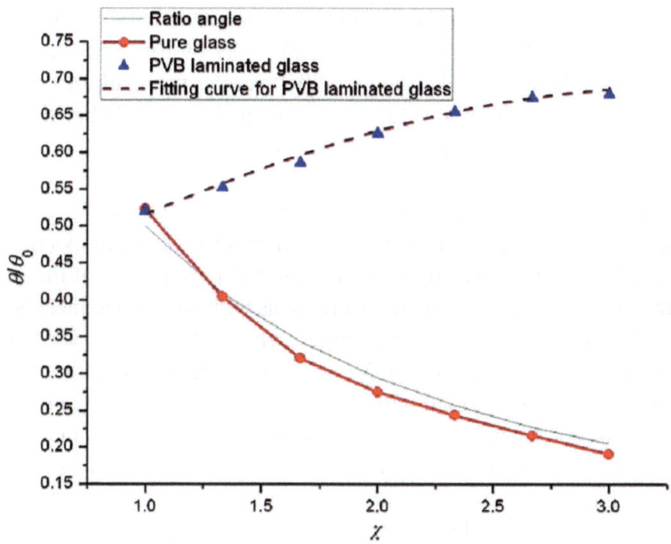

Fig. 7.19 Steady-state crack angle as a function of windshield aspect ratio for PVB laminated windshield and pure glass panel. The reference values listed in Table 7.1 are adopted for the other parameters used in the simulation. Here, the ratio angle is denoted by *a*/*b*

$$\theta = 0.9934\,\arctan(1/\chi), \tag{7.10}$$

which indicates that the crack angle θ is nearly aligned with the diagonal direction of the windshield and that the angle increases as aspect ratio decreases.

By contrast, the crack angle increases as χ increases in a PVB laminated windshield. As shown in Fig. 7.19, the variation in the normalized crack angle in a PVB laminated windshield occurs within a relatively small region (when the aspect ratio of the windshield is changed), particularly when the aspect ratio is within the range for common windshields (where χ typically varies between 2 and 2.5, for which the computed crack angle differs by approximately 7.46%), which may be valuable for accident reconstruction. Similarly, the relation between θ/θ_0 and χ for a PVB laminated windshield can present the following form (for the current ranges of parameters):

$$(\theta/\theta_0) = 2.08\left(1.00 + (\chi)^{1.31}\right)^{0.5} - 0.530\left(1.00 + (\chi)^{0.747}\right)^{1.5} - 0.932\cdot \tag{7.11}$$

Effect of Windshield Panel Size

Apart from windshield aspect ratio, windshield dimension also varies in different automobiles. In most modern passenger cars (the focus of this study), the ratio of the head dimension to the windshield (λ) is approximately 0.03 based on the definition of $\lambda = \frac{\text{Projection area of head}}{\text{Projection area of windshield}}$. We vary $\lambda = 0.005 \sim 0{:}1$ by fixing the head dimension and aspect ratio of the windshield. Figure 7.20 illustrates the resulting crack length as a function of v/v_0 under different values of λ by using Eq. (7.7). When the windshield is small, the normalized crack length will be long under the same impact speed. An impact can essentially result in a local stress concentration, which is the driving force of crack propagation. Hence, a prominent impact stress zone will be relatively small if the windshield is large, which will lead to short normalized cracks (although the absolute length of a crack will remain long due to the large D/t at the same v/v_0). Notably, the effect of λ will gradually decrease as windshield size increases. Census data from NTADTU indicate that passenger cars have $0.028 \leq \lambda \leq \leftarrow 0.034$, whereas shuttles and buses have $0.005 \leq \lambda \leq \leftarrow 0.025$. Therefore, the variation in crack length is relatively small within the range of λ for passengers. An accident investigator may not regard windshield dimension as a dominant factor during the reconstruction of a pedestrian–vehicle accident based on crack analysis. However, researchers should be cautious in using crack information to reconstruct accidents with buses and should consider the effect of windshield size.

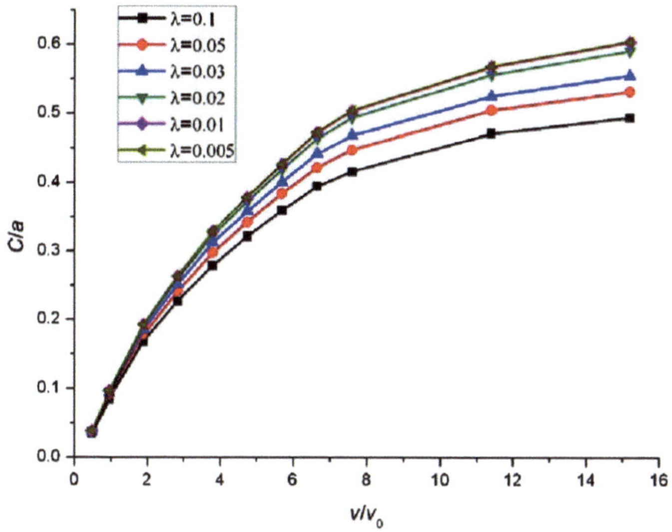

Fig. 7.20 Influence of windshield size on the normalized crack length when the normalized size parameter changes from 0.005 to 0.1 and as impact velocity is varied. The reference values listed in Table 7.1 are adopted for the other parameters used in the simulation

7.4.3.3 Effects of Boundary Conditions

Different Supporting Means and Materials

In actual vehicles, the windshield is typically supported and surrounded by a rubber material (polyurethane). Reference [16] provides valuable stress–strain data for the three-point bending test of polyurethane, including failure strength and failure strain. In this section, the edge of the reference windshield is bounded by a polyurethane material (similar to a saddle bar found in real automobiles).

To simplify the problem, we proportionally perturb the original stress of polyurethane σ_0 to a new value σ while assuming that the corresponding strain remains constant.

The radial crack pattern for different boundary constraints can be calculated by following the same numerical approach (while keeping all the other governing parameters at their reference values provided in Table 7.1). With a softer supporting material (i.e., a smaller scale factor σ/σ_0), or equivalently, a looser boundary condition, the maximum displacement of the head during impact will be larger under the same initial impact speed. We plot the relation between v/v_0 and C/a in Fig. 7.21 by using Eq. (7.7). The currently widely used polyurethane material I windshield saddle bar provides a boundary constraint that is "stronger" than that under the simply supported condition but "weaker" than the clamped constraint. Similar results are found in [17]. The normalized crack length can be up to 5% shorter in the simply supported case

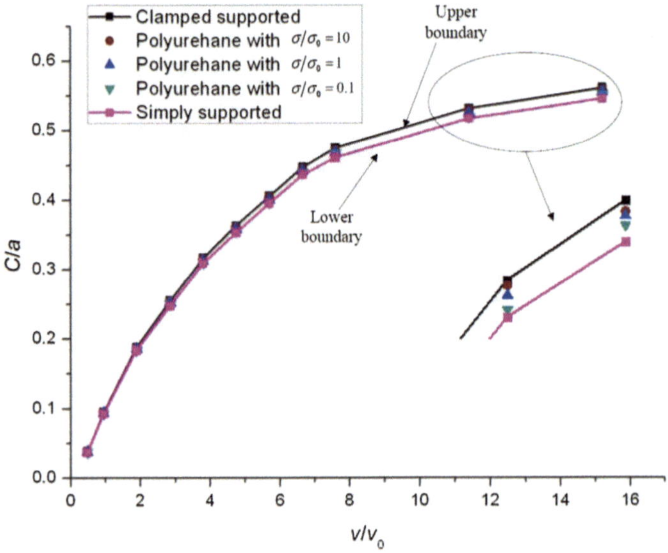

Fig. 7.21 Influence of saddle bar (boundary support) material on the normalized crack length as impact speed is varied. The reference values listed in Table 7.1 are adopted for the other parameters used in the simulation. The clamped and simply supported boundary conditions represent the upper and lower bounds, respectively, in the present simulation (in terms of crack length)

than in the clamped case under the same impact speed. Thus, boundary condition has a relatively insignificant influence on crack length. Accident investigators may ignore the effect of boundary conditions on different vehicle types.

Effect of Off-Center Impact

Off-center impact is extremely common in actual accidents; however, no previous effort to evaluate cracks on a windshield induced by off-center impact has yet been conducted. In this section, we assume that the off-center impact is still located along the x-axis and denote s as the distance measured from the center of the windshield to the impact location. A normalized offset distance is defined as $\mu = 2s/\mu$.

In Fig. 7.22, an example of an XFEM simulation of an off-center impact-induced crack (with the reference parameters listed in Table 7.1) is illustrated at $D/t = 1.21$ and $\mu = 0.6$. In general, crack propagation is mainly affected by the local stress field; thus, the windshield plate may be divided into four subsections (with respect to the off-center impact location). The crack extension in each subsection approximately follows that of a center impact on a windshield with four times the subsection dimension. For the cracks simulated in Fig. 7.22, each radial crack branch grows roughly according to the unique geometric characteristics (i.e., aspect ratio and size) of the corresponding subsection. Crack lengths and angles differ in the four

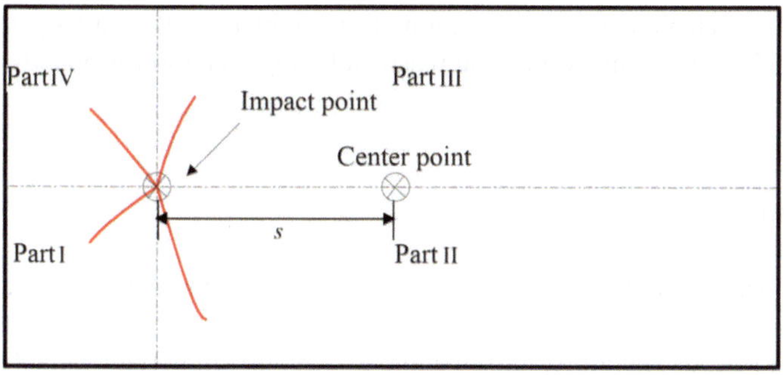

Fig. 7.22 Radial crack pattern simulated using XFEM at $D/t = 1.21$ and with the off-center parameter $m = 0.6$. The reference values listed in Table 7.1 are adopted for the other parameters used in the simulation. When the impact is off-center, crack propagation is no longer symmetric. The crack in each subsection grows in a manner that is approximately consistent with that on a fourfold-sized windshield upon center impact

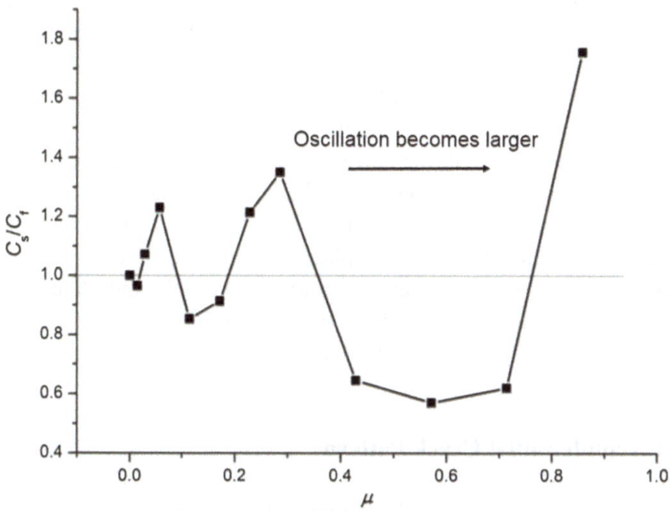

Fig. 7.23 Oscillation in C_s/C_f becomes larger when the impact is more off-center

subsections; hence, we let C_s be the simulated crack length (under the off-center impact) in a given subsection and C_f be the crack length that results from a center impact of the twofold-sized subsection. Consequently, the average C_s/C_f is obtained for all four subsections.

We vary μ within relatively large ranges, and Fig. 7.23 shows that the difference between C_s and C_f will increase as the impact point moves farther away from the center point.

If the impact point is not too far from the center of the windshield, then an accident investigator may divide the windshield into four subsections and analyze the cracking behavior in any subsection. The summary of the investigations of all the four subsections should provide reasonable upper and lower bounds of impact speed.

Inclined Impact

In actual pedestrian–vehicle accidents, the vehicle's front-end shape, pedestrian's height, and vehicle impact speed will exert influences on the impact angle of the pedestrian's head. Therefore, impact angle ϑ is another important variable for crack patterns on a windshield. We assume that an inclined impact occurs at the center of the windshield and initially compute x, y, and z deformation responses (without cracking) under different impact angles at a fixed total impact speed. The x-o-z plane is provided as an example. Impact angle varies from $\vartheta = 90°$to $\vartheta = 15°$ ($\vartheta = 0°$ is not included because no displacement will occur in the z direction). Impact speed in the y direction is maintained at 0, whereas the total impact speed of the x and z directions ($\vec{v} = \vec{v}_x + \vec{v}_z$) is retained at the same value. We investigate two scenarios with $|\vec{v}| = 10$ m/s and 20 m/s.

We first check the normalized displacements in the x, y, and z directions according to different impact angles at varying impact speeds (Fig. 7.24). Evidently, displacement increases with velocity in the corresponding direction. Figure 7.25 shows the normalized crack lengths at various normalized crack angles (ϑ/ϑ_0, where $\vartheta_0 = 90°$). When the impact angle is large, crack length is short; that is, crack length is mainly affected by the resulting head displacement in the z direction. Notably, cracks grow asymmetrically in the inclined scenario and we only show the longest crack branch. That is, if a radial crack exhibits strong asymmetry during a center impact, then it may be induced by an inclined impact.

7.4.3.4 Circumferential Crack Pattern

In the previous sections, radial crack characteristics under the influences of different geometries, materials, impact conditions, and boundary constraints are discussed. In this section, we investigate the second stage of crack propagation, i.e., the circumferential fracture (after the radial cracks have developed). As mentioned in Sect. 7.4.2, the stress field is retrieved by the end of the first step of radial crack propagation to enable the initialization of circumferential cracks (which grow out of small closely spaced random defects along the previously developed radial crack without applying additional external loads; see [10] for details).

Figure 7.26 shows a typical complete crack pattern (radial and circumferential) on a PV laminated windshield (with the reference parameters provided in Table 7.1) that has been subjected to normal and center impacts of the human head ($D/t = 1.21$). The largest circumferential crack initiates from the outmost tip of the radial crack, and a smaller closed circumferential crack is formed near the impact crater. In addition, a few less developed partial circumferential cracks are found between the two

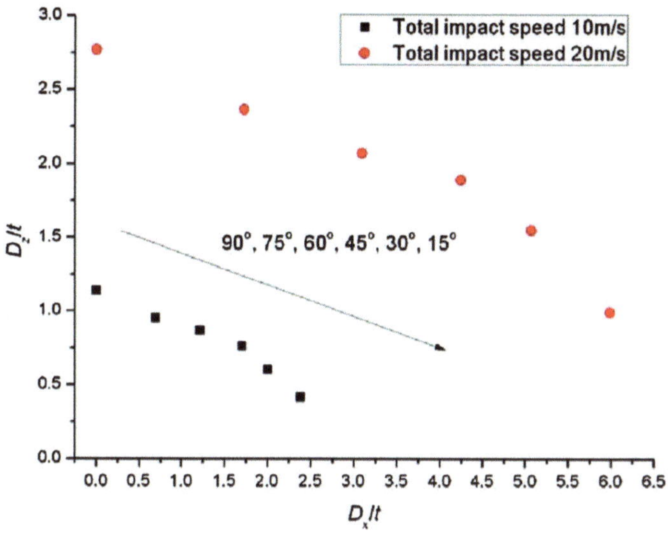

Fig. 7.24 Normalized displacement components in the x and z directions under different impact angles when the total impact speed is either 10 m/s or 20 m/s. The reference values listed in Table 7.1 are adopted for the other parameters used in the simulation

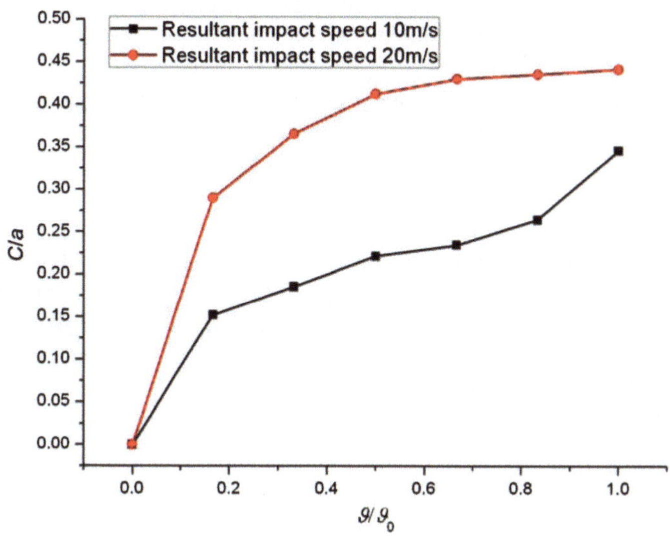

Fig. 7.25 Normalized crack length as a function of the inclined impact angle at two different total impact speeds. The reference values listed in Table 7.1 are adopted for the other parameters used in the simulation

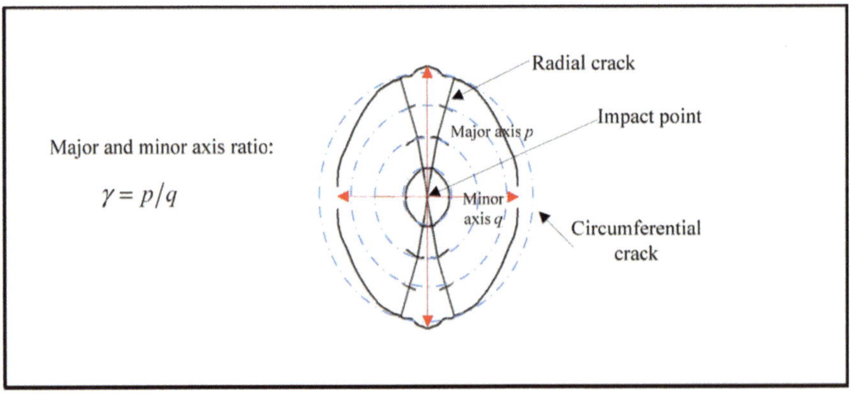

Fig. 7.26 Complete radial and circumferential crack pattern in a PVB laminated windshield at *D*/*t* = 1.21 during XFEM simulation (with the reference parameters provided in Table 7.1). *g* is defined as the ratio between the major (*y*) and minor (*x*) axes of the elliptical circumferential crack

fully developed cracks. Notably, "spiderweb" cracks are occasionally observed on real windshields [4]. These cracks include many radial cracks and a few circumferential cracks the cross one another probably due to multiple contacts, which are not considered in the present study. The radial crack referred to in this study corresponds to the longest one in actual windshield cracking, thereby presenting good agreement with real-life accidents [10].

The aspect ratio γ between the major and minor axes of the elliptical circumferential crack (Fig. 7.26) is an effective indicator for the shape of the crack pattern. Figure 7.27 presents the results of the effects of different variables on aspect ratio γ (in each figure, only one parameter in the x-axis is changed while the others are fixed at their reference values). Under the conditions of a compliant PVB layer, loose boundary constraint, or large windshield curvature, aspect ratio γ tends toward 1 (i.e., approaching a circle). Essentially, the shape of the circumferential crack is related to the prior radial crack growth to a certain extent, and thus, γ depends on radial crack angle (θ/θ_0) and length (C/a) to a certain degree. Nevertheless, the trends shown in Fig. 7.27 indicate that the effects of various material and system parameters are mixed (coupled) due to the complicated interaction between the hoop and radial stress fields during the two-stage crack propagation process. Thus, the shape of the circumferential crack is not determined solely by the previous radial crack parameters.

7.4.3.5 Governing Parameters of the Crack Pattern on the Windshield

When the preceding results are combined, crack pattern characteristics (including radial crack length and angle, and circumferential crack number and aspect ratio) can

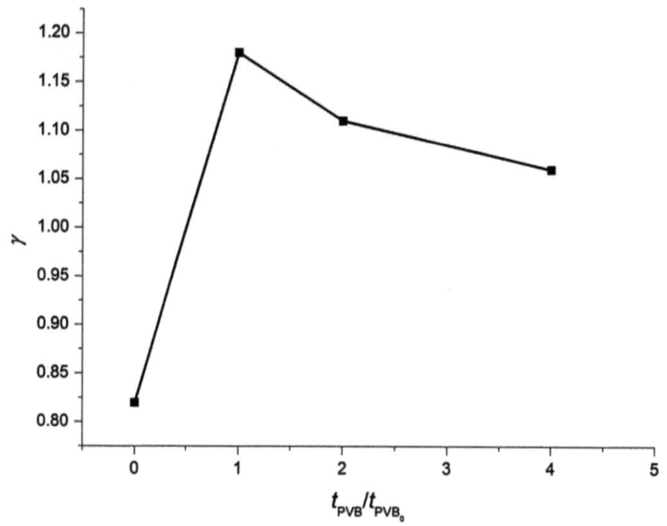

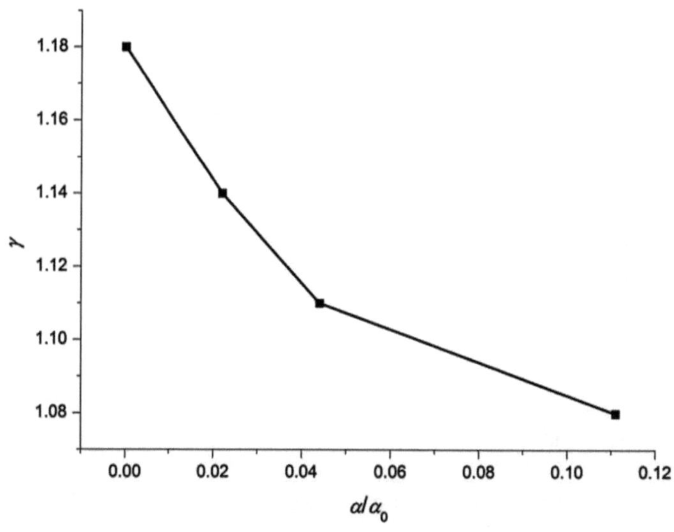

Fig. 7.27 Influences of the normalized PVB interlayer thickness, windshield curvature, supporting material stiffness, and interlayer material stiffness on the circumferential crack ratio g. The reference values listed in Table 7.1 are adopted for the other parameters used in the simulation

be expressed as functions of several main geometric, material, and boundary factors, which are approximated in a dimensionless functional form as

$$\left[C/a, \theta/\theta_0, \gamma\right] = F[D/t, \alpha/\alpha_0, \chi, \lambda, \vartheta/\vartheta_0, \mu, \sigma/\sigma_0, (m, n)], \tag{7.12}$$

where D/t is expressed in terms of M/M_0, v/v_0, and $t_{PVB} = t_0$ [Eq. (7.7)].

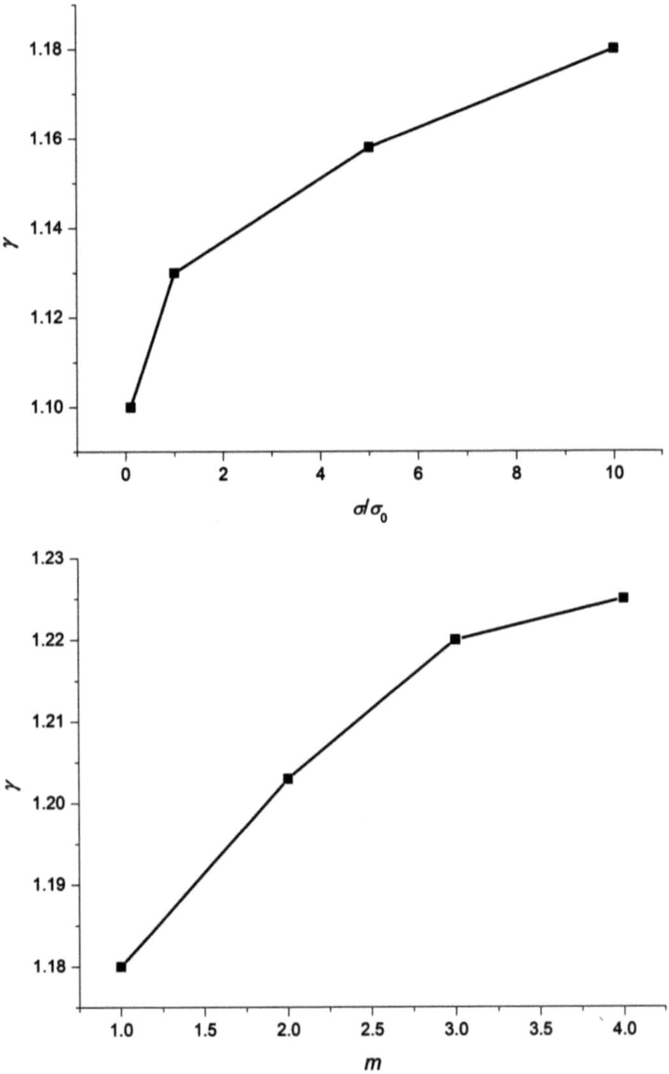

Fig. 7.27 (continued)

If we assume that the automobile industry uses the same PVB interlayer material with the same boundary condition (and if the effect of the boundary condition is indeed minimal, as indicated in Section "Different supporting means and materials"), then Eq. (7.12) can be simplified as follows for center impact:

$$\left[C/a, \theta/\theta_0, \gamma\right] = F^{'}[D/t, \alpha/\alpha_0, \chi, \lambda, \vartheta/\vartheta_0] \tag{7.13}$$

The variables at the left side of Eq. (7.12) can effectively describe the entire crack pattern, whereas the variables at the right side denote the governing impact, material, geometric conditions, and parameters. Subsequently, we fit the three crack pattern parameters separately.

(1) The crack length $[C/a] = f[D/t, \alpha/\alpha_0, \chi, \lambda, \vartheta/\vartheta_0]$, which is affected by all the variables at the right side, can be expressed as follows (for the ranges of parameters used in this study):

$$\begin{aligned} C/a = {} & -0.198\left[(D/t)^{2.56} + (\alpha/\alpha_0)^{0.551} + (\chi)^{-0.0806} + (\lambda)^{7.81} + (\vartheta/\vartheta_0)^{34.7}\right]^{0.5} \\ & + 0.141\left[(D/t)^{0.911} + (\alpha/\alpha_0)^{10.26} + (\chi)^{-0.0306} + (\lambda)^{1.23} + (\vartheta/\vartheta_0)^{0.765}\right]^{1.5} \\ & - 0.118 \end{aligned} \quad (7.13)$$

where D/t can be obtained from Eq. (7.7). The correlation coefficient is $R = 0.960$.

(2) For the crack angle $[\theta/\theta_0] = g[D/t, \alpha/\alpha_0, \chi, \lambda, \vartheta/\vartheta_0]$, we assume that the steady-state crack propagation direction is essentially insensitive to the depth of head displacement on the windshield, i.e., $[\theta/\theta_0] = g[\alpha/\alpha_0, \chi, \lambda, \vartheta/\vartheta_0]$. Through nonlinear curve fitting, we can obtain

$$\begin{aligned} \theta/\theta_0 = {} & 2.08\left[(\alpha/\alpha_0)^{1.06} + (\chi)^{1.31} + (\lambda)^{1.76} + (\vartheta/\vartheta_0)^{-0.0262}\right]^{0.5} \\ & - 0.530\left[(\alpha/\alpha_0)^{0.470} + (\chi)^{0.747} + (\lambda)^{1.99} + (\vartheta/\vartheta_0)^{-0.00615}\right]^{1.5} - 0.932. \end{aligned} \quad (7.14)$$

The correlation coefficient is $R = 0.9999$.

(3) The circumferential crack aspect ratio is also assumed to be independent of D/t, and thus, $[\gamma] = h[\alpha/\alpha_0, \chi, \lambda, \vartheta/\vartheta_0]$, which can be fitted by

$$\begin{aligned} \gamma = {} & -0.114\left[(\alpha/\alpha_0)^{0.203} + (\chi)^{-24.8} + (\lambda)^{0.0846} + (\vartheta/\vartheta_0)^{0.476}\right]^{0.5} \\ & - 0.716\left[(\alpha/\alpha_0)^{2.48} + (\chi)^{0.00707} + (\lambda)^{-0.00182} + (\vartheta/\vartheta_0)^{-0.0110}\right]^{1.5} + 5.05. \end{aligned} \quad (7.15)$$

The correlation coefficient is $R = 0.9968$.

Equations (7.13) to (7.15) indicate the general form and extent of influences from several governing variables. In practice, on the basis of the crack pattern on a PVB laminated windshield, the inverse analyses of Eqs. (7.13) to (7.15) may lead to the determination of relevant information, including impact speed, windshield material, and geometric properties, which may be helpful to accident investigators and vehicle crashworthiness designers.

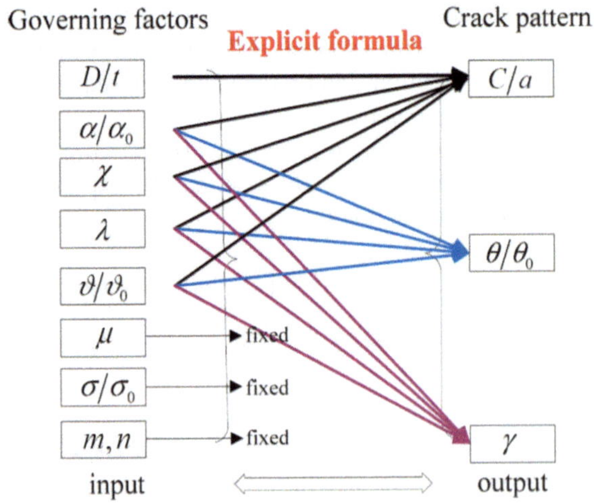

Fig. 7.28 Bridging the main materials and system parameters with the crack pattern on a PVB laminated windshield

7.4.3.6 Concluding Remarks

The crack pattern formed on a windshield upon head impact has long been a popular problem but remains unsolved until now. This issue is critical for accident reconstruction, vehicle crashworthiness analysis, and pedestrian and passenger protection. We conduct an extensive study on PVB laminated windshield crack patterns (including radial and circumferential cracks) using XFEM simulations. The effects of the most prominent materials, geometries, impact conditions, and parameters are explored.

Explicit relations are established between crack characteristics and prominent material/system parameters based on systematic numerical simulations. In summary, among the various parameters considered in this study, we propose a map that bridges the most influential parameters and crack pattern in Fig. 7.28. The map is established based on the extensive XFEM simulations performed in this study. Furthermore, the qualitative impact on crack geometry as each governing factor is varied is summarized in Table 7.2. Figure 7.28 and Table 7.2 can provide an overall qualitative view of the numerical results of this study [the quantitative results are related to Eqs. (7.13) to (7.15)].

The results may provide accident investigators and vehicle safety designers with fundamental insights into the general relations between governing materials/system variables and crack patterns, particularly the effects of a single variable and a combination of several variables.

Table 7.2 Summary of the qualitative relationship between the governing variables and crack pattern parameters

		Governing material/system variables								
		D/t	α/α_0	χ	λ	ϑ/ϑ_0	μ	σ/σ_0	m	n
Crack pattern parameter	C/a		↑↓	↑↓	↑↓	↑↑	–	↑↑	↑↓	↑↑
	θ/θ_0	–	↑↓	↑↑	↑↑	↑↓	–	–	–	–
	γ	–	↑↓	↑↑	↑↓	↑↑	–	↑↑	↑↑	↑↓

Note "↑ ↑" denotes the scenario in which an increase in one variable will cause an increase in the other variable (and vice versa). "↑ ↓" denotes the scenario in which an increase in one variable will cause a decrease in the other variable (and vice versa). "–" denotes the scenario in which the relationship between two variables is either oscillating (such as in cases that involve μ) or extremely weak (one variable will nearly remain constant while the other variable changes)

7.5 Summary

In this chapter, we investigate the crack propagation characteristics when a model windshield experiences low-speed impact. Compared with other numerical techniques, XFEM is unarguably more efficient become it does not require remeshing during crack propagation. Hence, XFEM is adopted to numerically investigate multiple crack propagation on PVB laminated glass that is subjected to quasi-static indentation or low-speed impact (below 30 m/s).

With the relatively minor dynamic effects, the impact problem is translated to a quasi-static indentation problem, where impact speed and effective head mass are correlated with the displacement of the windshield plate. Radial crack initially grows along a particular angle and its length depends on plate displacement and boundary constraint due to the more prominent hoop stress field. Subsequently, circumferential cracks will develop near the contact zone and at the perimeter of the previous radial crack to release excessive radial stress. The effects of prominent factors, i.e., impact speed and effective head mass, are explored along with other variables, such as initial flaw, material damage stress, and fracture criterion.

Then, explicit formulae are established based on numerical simulations among impact speed, effective head mass, plate displacement, and crack length. Critical accident information, such as impact speed or damage stress, can be deduced from the crack pattern characteristics. A qualitative bridge can be established between the numerical simulation result and real-life accident via the crack growth mechanism.

Finally, a parametric study on the effects of the PVB interlayer material property, windshield geometric parameters, and boundary conditions of the windshield on the crack propagation characteristics is conducted within the same XFEM simulation framework. Explicit relations are established between the crack characteristics and prominent material/system parameters based on systematic numerical simulations. Among the various parameters considered in this study, we propose a map that bridges the most influential parameters and the crack pattern mentioned in this chapter. This map is established based on the extensive XFEM simulations performed in this study. Furthermore, the qualitative impact on the crack geometry as each governing factor is varied is summarized in Table 7.2 in this chapter. Figure 7.18 and Table 7.2 can provide an overall qualitative view of the numerical results in this chapter. The results may provide accident investigators and vehicle safety designers with fundamental insights into the general relations between governing material/system variables and crack patterns, particularly the effects of a single variable and a combination of several variables.

References

1. J. Xu, Y. Li, G. Lu, W. Zhou, Reconstruction model of vehicle impact speed in pedestrian–vehicle accident. Int. J. Impact Eng **36**, 783–788 (2009)
2. L.Y.U. Xu Jun, An overview on the technique of accident reconstruction for pedestrian-vehicle crash. Automotive Engineering, **31,** 1029–1033 (2009)
3. J. Yao, J. Yang, D. Otte, Investigation of head injuries by reconstructions of real-world vehicle-versus-adult-pedestrian accidents. Saf. Sci. **46**, 1103–1114 (2008)
4. S. Zhao, L.R. Dharani, X. Liang, L. Chai, S.D. Barbat, Crack initiation in laminated automotive glazing subjected to simulated head impact. Int. J. Crashworthiness **10**, 229–236 (2005)
5. D.P. Wood, Impact and movement of pedestrians in frontal collisions with vehicles, in *ARCHIVE Proceedings of the Institution of Mechanical Engineers Part D Transport Engineering 1984–1988*, vols. 198–202, 202, 101–110 (1988)
6. S. Zhao, L.R. Dharani, L. Chai, S.D. Barbat, Dynamic response of laminated automotive glazing impacted by spherical featureless headform. Int. J. Crashworthiness **11**, 105–114 (2006)
7. S. Zhao, L.R. Dharani, L. Chai, S.D. Barbat, Analysis of damage in laminated automotive glazing subjected to simulated head impact. Eng. Fail. Anal. **13**, 582–597 (2006)
8. J. Xu, Y. Li, D. Ge, B. Liu, M. Zhu, Experimental investigation on constitutive behavior of PVB under impact loading. Int. J. Impact Eng. **38**, 106–114 (2011)
9. W. Holand, G.H. Beall, Glass Ceramic Technology (2012)
10. J. Xu, Y. Li, X. Chen, Y. Yan, D. Ge, M. Zhu, B. Liu, Characteristics of windshield cracking upon low-speed impact: numerical simulation based on the extended finite element method. Comput. Mater. Sci. **48**, 582–588 (2010)
11. N. Ogasawara, N. Chiba, X. Chen, Limit analysis-based approach to determine the material plastic properties with conical indentation. J. Mater. Res. **21**, 947–958 (2006)
12. X. Yuan, S. Li, Analysis of rectangular thin plate vibration under different support conditions. Aeroengine, 31 (2005) 39–43
13. J. Xu, Y. Li, Crack analysis in PVB laminated windshield impacted by pedestrian head in traffic accident. Int. J. Crashworthiness **14**, 63–71 (2009)
14. J. Xu, Y.B. Li, Study of damage in windshield glazing subject to impact by a pedestrian's head. Proc. Inst. Mech. Eng Part D: J. Automobile Eng. **223**, 77–84 (2009)
15. J. Xu, Y. Li, B. Liu, M. Zhu, D. Ge, Experimental study on mechanical behavior of PVB laminated glass under quasi-static and dynamic loadings. Compos. B Eng. **42**, 302–308 (2011)
16. L. Mkrtchyan, M. Maier, Fibre reinforced polyurethane foam to increase passive safety of vehicle occupants. Plast. Rubber Compos. **36**, 445–454 (2013)
17. S.M.L.X. Yuan, Analysis of rectangular thin plate vibration under different support conditions. Aeroengine **31**, 39–43 (2005)

Chapter 8
Pedestrian Protection and Energy Dissipation

The windshield plays an important role in accidents. When a pedestrian is hit by a car, impact usually occurs on the windshield, which accounts for approximately 50% of pedestrian deaths during accidents [1], typically due to head impact with the windshield [2]. Notably, pedestrians comprise the largest group of motor vehicle deaths and account for approximately 24.78% of total traffic accident fatalities (2010). The percentage is even higher in Korea, i.e., approximately 40% of all traffic-related fatalities in 2004 [3].

When a vehicle crashes into another vehicle with sharp deceleration, the passenger may be injured by bumping into the windshield [4, 5]; however, such incident can be decreased through the widespread use of three-point/two-point seat belts. Unfortunately, with regard to pedestrian protection during accidents, effective modification and improvement on the energy absorption capability of windshields remain minimal. By contrast, the progressive failure of the energy absorption design of bumpers and engine hoods has been effectively addressed [6–8]. As one of the hardest and most dangerous of the vehicle front, the windshield, which is a primary cause of pedestrian injury, has gained increasing concern from automotive and mechanical researchers.

PVB has two major advantages over monolithic glass, namely, energy absorption and fragment glutinosity, which indicates that less fragments will spin on the ground. Therefore, PVB is the preferred interlayer material for manufacturing laminated windshields. The investigation of the energy absorption capability of PVB laminated windshields can provide useful guidance for pedestrian protection during pedestrian–vehicle accidents and for automobile design. A series of experiments, including compression tests and dynamic experiments, has been conducted to study the mechanical behavior of PVB laminated windshields. However, studies on the energy absorption capability of PVB laminated windshields subjected to headform impact are still rare [9].

Accordingly, several works on the pedestrian protection and energy absorption capabilities of PVB laminated windshields based on experiments and simulations are introduced in the succeeding sections.

J. Xu and Y. Li, *Impact Behavior and Pedestrian Protection of Automotive Laminated Windshield*, https://doi.org/10.1007/978-981-13-2441-3_8

8.1 Experiments on PVB Laminated Windshields Impacted by a Headform

8.1.1 Experiment 1

(1) Experimental setup

Yang et al. [10] designed an experimental system with a dropping impact hammer to conduct an impact experiment. Figure 8.1 shows the diagram of the experimental system. The headform was located in the vertical direction above the windshield. An accelerometer placed at the center of the headform was used to measure the acceleration of the headform. A high-speed camera was installed parallel to the impact direction to observe the crack propagation process, while a video camera was set to record the movement of the headform and the deformation of the windshield. The boundary conditions of the experiment are provided in Table 8.1.

(2) Experimental results

Figures 8.2, 8.3, and 8.4 show the movement of the headform as it impacts the windshield at 18 km/h, and the resultant acceleration of the headform at 18 and 20 km/h, respectively. The headform–windshield impact process is divided into three parts: collision, rolling, and flying off [10].

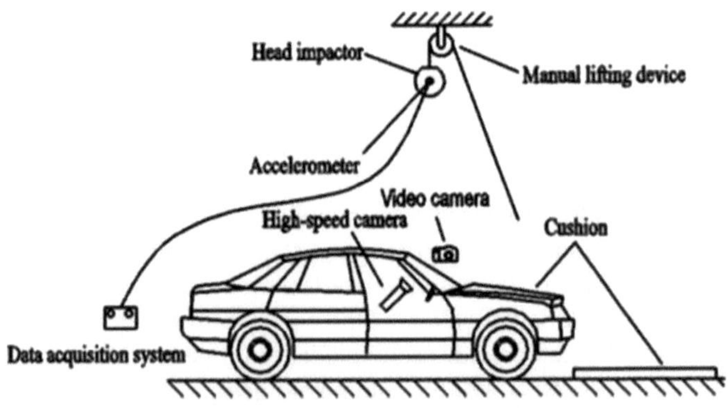

Fig. 8.1 Diagram of the experimental system [10]

Table 8.1 Boundary conditions of each experiment [10]

No.	Impact speed (km/h)	Impact location	Falling height (m)	Camera location
1	18	Geometric	2.5	Outside the car
2	20	Geometric	2.7	In the car
3	23	Geometric	3	In the car

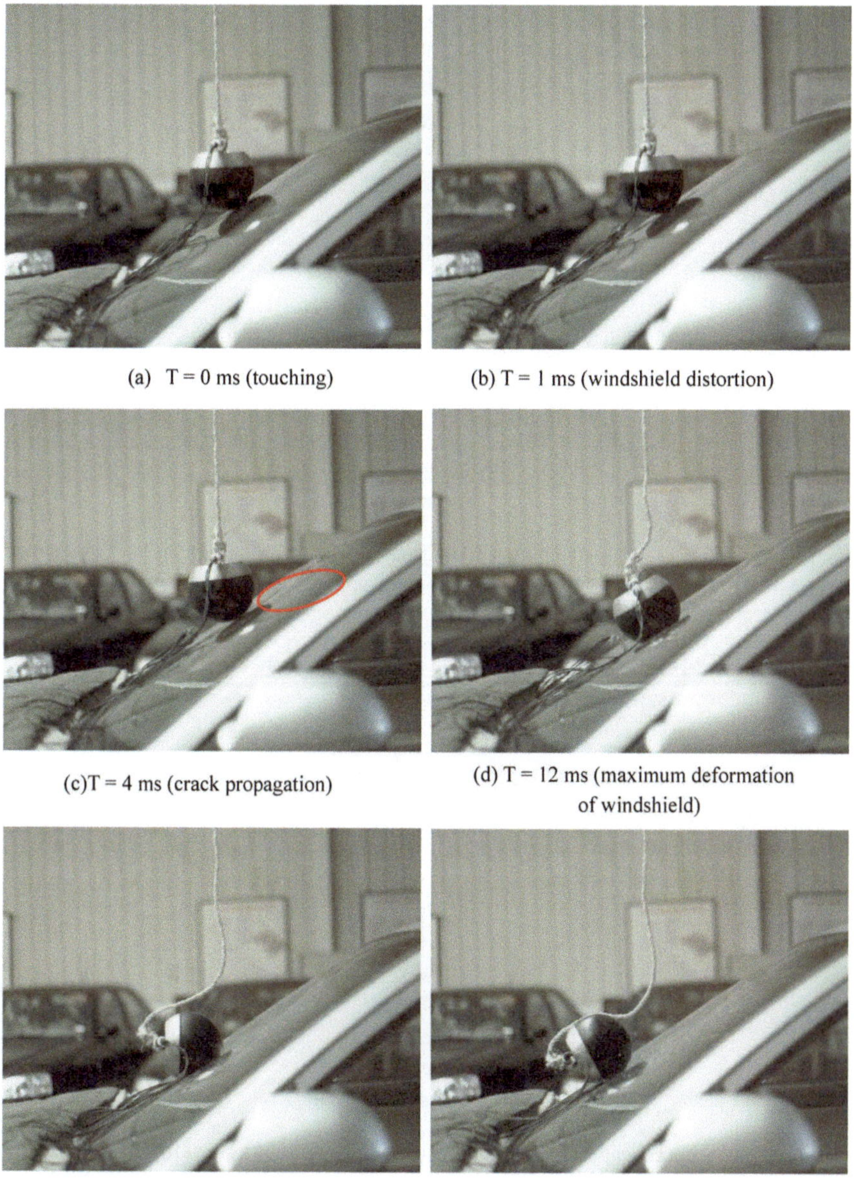

(a) T = 0 ms (touching)

(b) T = 1 ms (windshield distortion)

(c)T = 4 ms (crack propagation)

(d) T = 12 ms (maximum deformation of windshield)

(e) T = 40 ms (headform rolling and windshield rebounding)

(f) T = 44 ms (flying off)

Fig. 8.2 Headform movement upon impacting the windshield at approximately 18 km/h [10]

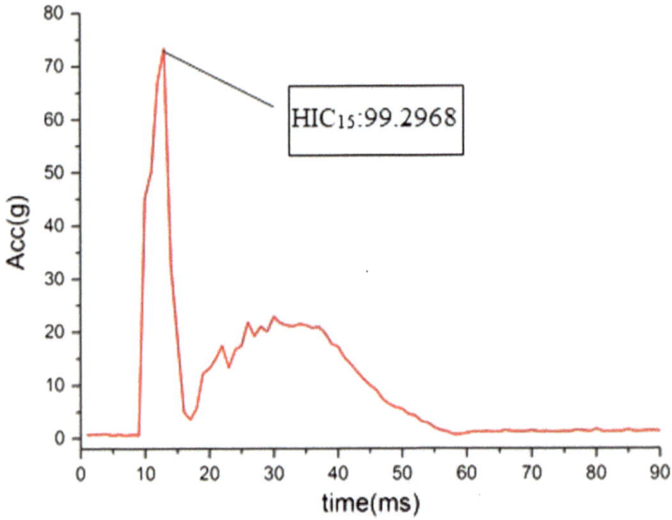

Fig. 8.3 Resultant acceleration of the headform (impact speed is 18 km/h) [10]

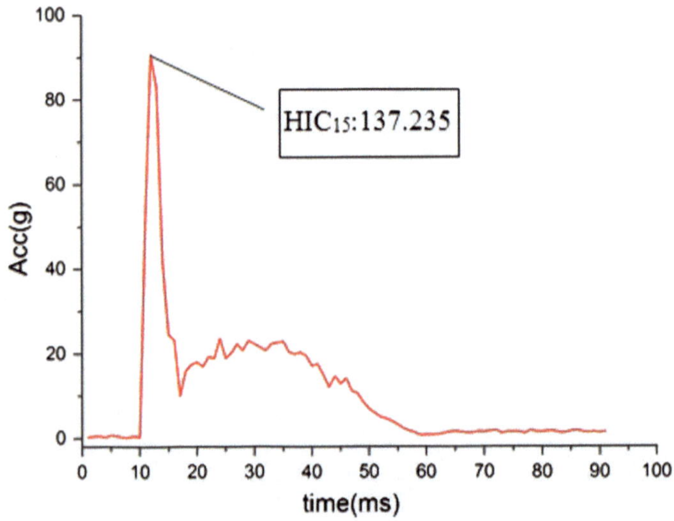

Fig. 8.4 Resultant acceleration of the headform (impact speed is 20 km/h) [10]

The first peak of acceleration shown in Figs. 8.3 and 8.4 is caused by increasing contact force during impact. Then, the crack grows as shown in Fig. 8.2c. The rolling process follows because of the velocity of the headform. Headform acceleration reaches the second peak and then the headform flies away due to the rebound of the windshield, as shown in Fig. 8.2e, f [10].

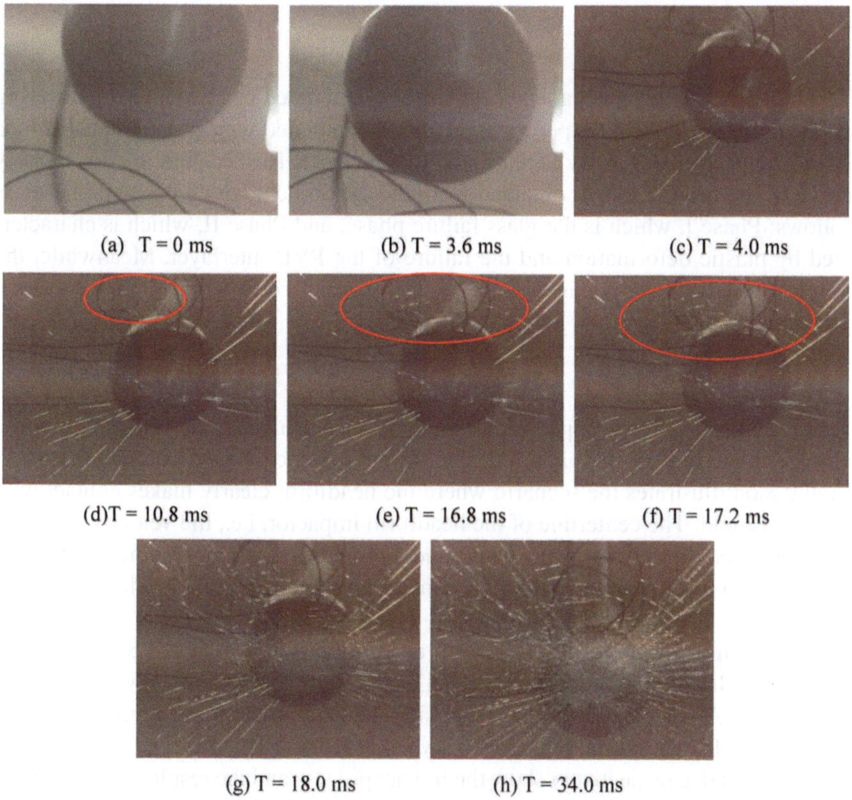

Fig. 8.5 Crack propagation in the impact experiment with a headform impact velocity of 23 km/h [10]

The formula for the head injury criterion (HIC) values is found in Sect. 8.2.3. This formula mainly depends on the first peak of the acceleration curve, as shown in Figs. 8.3 and 8.4, where the first and second peak stages mostly have the same duration and differs mainly in the acceleration curve values of the first peak. When velocity is high, the HIC value is also high. In addition, the radial crack grows first upon impact, followed by the circumferential crack, because the hoop stress field is larger than the radial stress field (Fig. 8.5). As the headform drops lower, the glass near the impact point breaks completely, the crack propagates and generates small scaly cracks to absorb more impact energy [10].

8.1.2 *Experiment 2*

To study the dynamic mechanical behavior of a laminated windshield in engineering applications, a full-scale experiment on a PVB laminated windshield impacted by a headform was conducted by Jaguar Land Rover (JLR). The acceleration time history curve of the headform was obtained from the experiment which has two phases as follows: Phase I, which is the glass failure phase, and Phase II, which is characterized by plastic deformation and the failure of the PVB interlayer. Meanwhile, the acceleration time history curve of the headform can effectively reflect the pedestrian protection capability of the laminated windshield.

(1) Experimental setup

Five tests on headform impact on a laminated windshield were conducted with a cutoff vehicle model for validation. The different velocities are listed in Table 8.2. Figure 8.6a illustrates the scenario where the headform clearly makes contact with the windshield. The centerline of the headform impactor, i.e., the velocity direction line of the headform, was 35° to the ground reference line when it hit the windshield. The target impact point was the first contact point between the headform and the windshield.

As shown in Fig. 8.6b, the headform used in this test weighed 4.8 kg in accordance with the adult headform regulation of the European Enhanced Vehicle-safety Committee. A velocity generator was positioned at the bottom end of the headform, which provided the initial impact velocity, as shown in Fig. 8.6c.

Figure 8.6d, e respectively show the impact process and the result of impact point 2 with cracks on the laminated windshield. A data acquisition kit and a high-speed camera with 1000 frames per second were also used.

As shown in Fig. 8.7a, the headform impact test zone was divided into an adult zone (A) and a child zone (C) by a 1500 mm wraparound line based on the pedestrian testing protocol of Euro NCAP [11]. Each zone was divided into six areas labeled with numbers 1–6 alternately from the left side to the right side of the vehicle. Then, each area was split into a, b, c, and d concrete test areas. For simplicity and efficiency, five representative impact points were selected, as indicated in Fig. 8.6a, c. In addition, Fig. 8.7b shows the geometric dimensions and components of the laminated windshield used in this study.

(2) Experimental results

Figure 8.8 presents the acceleration time history curve of the headform obtained from the test when it impacts on the windshield at location point 3 with an initial velocity

Table 8.2 Individual initial impact velocity for the five tests

	Impact point 1	Impact point 2	Impact point 3	Impact point 4	Impact point 5
Velocity (m/s)	9.770	9.640	9.640	9.690	9.660

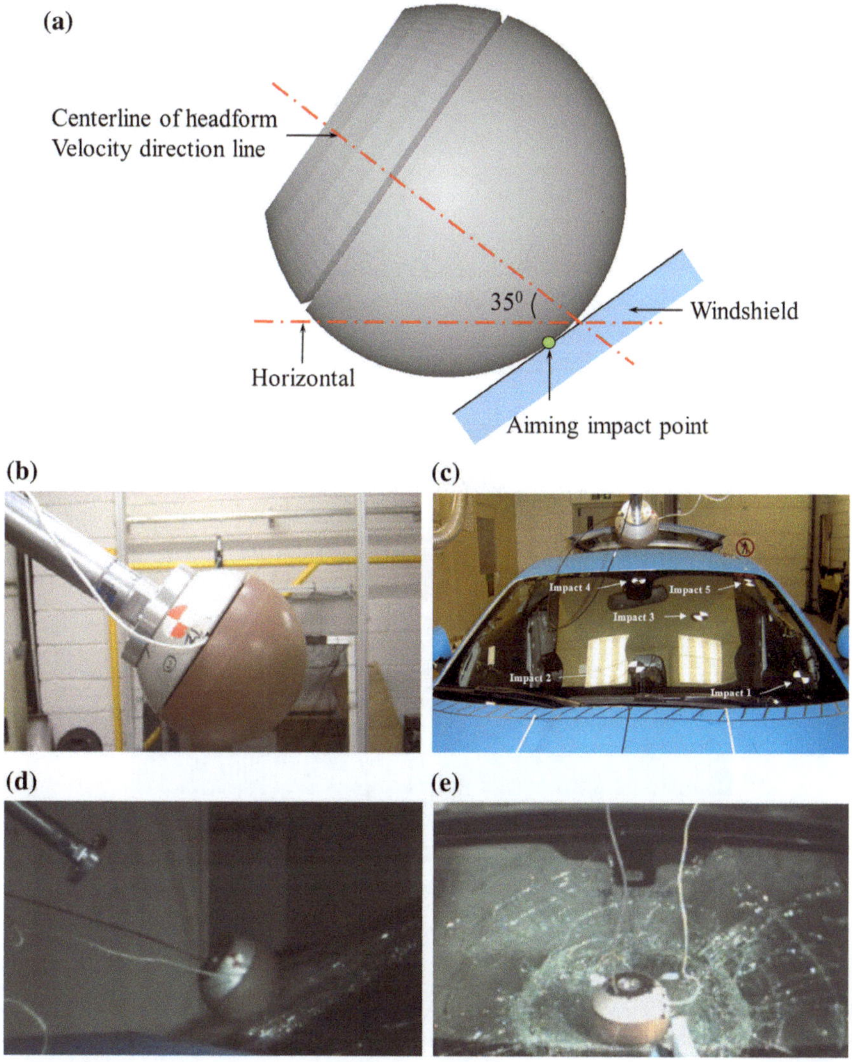

Fig. 8.6 **a** Abbreviated illustration of the contact condition when the headform initially impacts on the windshield. Main equipment, process, and results of the test: **b** headform impactor, **c** test vehicle, **d** impact process, and **e** impact result at impact point 2

of 9.64 m/s. The curve is characterized by the first two high peaks (the first peak is typically higher than the second one) within the first 4 ms (t_2), a significant decrease at approximately 4 ms (t_2), and a buffering phase ranging from 4 to 40 ms.

Reference [12] reported that PVB slightly influences windshield failure during Phase I because of the considerable stiffness difference between the glass and PVB.

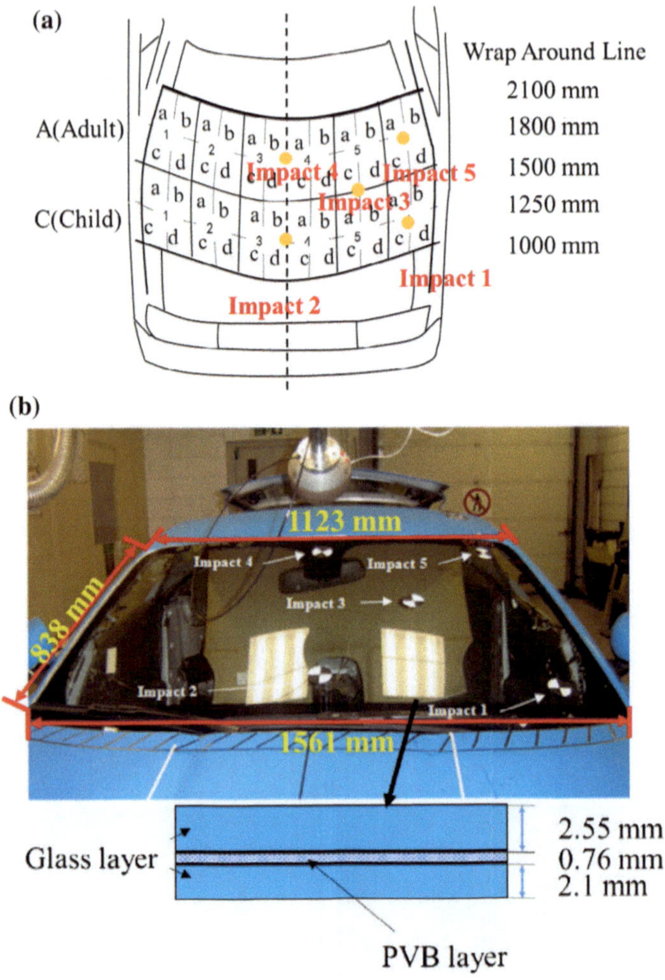

Fig. 8.7 **a** Division of impact zone by Euro NCAP. **b** Selected impact points and geometric dimensions of the windshield used in this test

Thus, Phase I can be called the "glass failure phase." In the beginning, the glass layers bear the main loading and mainly experience bending deformation [12].

Previous works [13, 14] indicated that the outer glass layer is first impacted during the impact test of PVB laminated glass. This impact glass sheet is passed through compression waves because of the deformation of PVB laminated glass and eventually undergoes compression failure. Compression waves are reflected onto tensile waves from the boundaries when they reach the lower glass layer. Brittle materials (e.g., glass) fail earlier under tensile state than under compression state. Thus, the lower glass layer fails first. A radial crack initially occurs on the lower glass because of considerable hoop stress, which is the reason for the failure of the lower glass

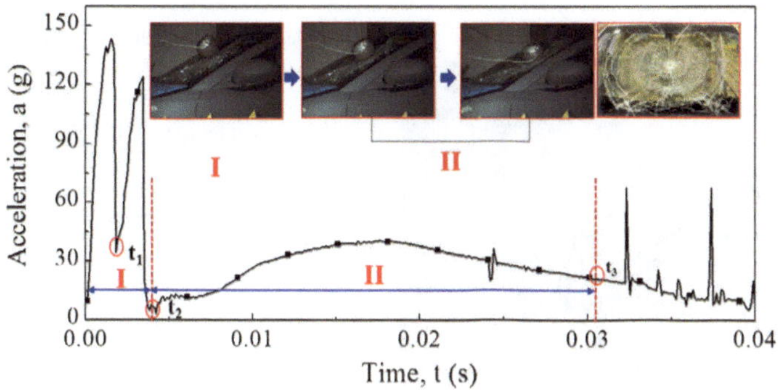

Fig. 8.8 Linear acceleration curve of the headform impactor dynamic response with an initial impact velocity of 9.64 m/s and 35° to the ground reference line

sheet that results in a sudden slight drop at approximately 2 ms (t_1). The outer glass layer mainly bears the compressive force loaded by the headform, which results in the second acceleration peak. The compression failure of this glass sheet occurs at approximately 4 ms (t_2). The load-carrying capacity of the two glass sheets weakens gradually with the evolution of cracks. The counterforce between the outer glass and the headform eventually drops to nearly zero when the glass material within the vicinity of the impact zone fails completely at approximately 4 ms (t_2).

During Phase II, the PVB layer plays a dominant role in bearing loads and in dissipating the impact energy of the headform through a large deformation. During the early stage of Phase II, the hyperelasticity of PVB increases the counterforce between the headform and the laminated windshield with the displacement of the headform.

However, the plasticity of PVB reduces the increase in force in the area where a part of the kinetic energy of the headform dissipates through the plastic deformation and internal energy of PVB because of the friction between the headform and the windshield. PVB fails during the late stage of Phase II; thus, the counterforce of the laminated windshield also declines. In addition, the failure of the glass layer, deformation of the plastic, and failure of the PVB interlayer dissipate most of the kinetic energy of the headform, which characterizes the pedestrian protection capability of the laminated windshield.

8.1.3 *Experiment 3*

(1) Experimental setup

Liu [9, 15] reconstructed real-life head–windshield impact by using a specially designed experimental platform (Fig. 8.9) to investigate the energy absorption mechanism of a PVB laminated windshield. The headform was launched through a pedes-

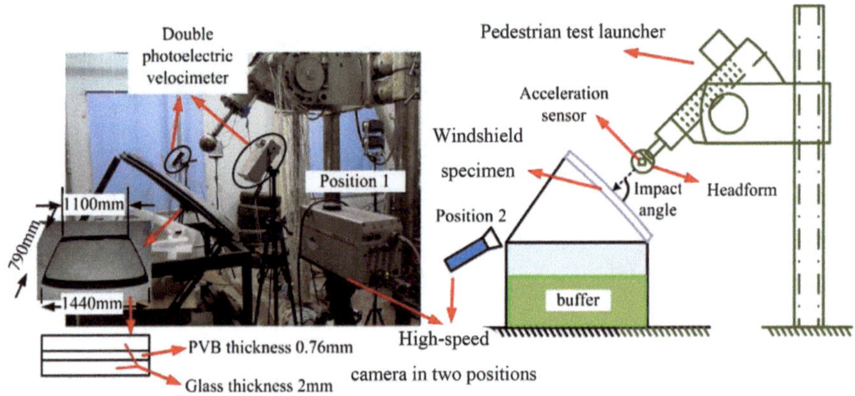

Fig. 8.9 Schematic of the experimental system [9]

trian test [16] at different impact speeds and angles. The impact point is the geometric center of the windshield. To reflect the assembly of real vehicles, the laminated windshield was bonded with a steel frame using glass cement and then fixed onto a mounting bracket. The angle of the windshield to the horizontal plane can be modified to adjust the impact angle between the headform and the windshield. The speed of the headform was measured using a pair of photoelectric velocimeters placed on both sides of the headform; the high-speed cameras in the two positions shown in Fig. 8.7 aimed to record the headform movement and the laminated windshield crack propagation process [9].

(2) Experimental results

In accordance with the experiments results, the dynamic response of the headform and the failure form of the windshield are shown in Figs. 8.10 and 8.11, respectively, where $g = 9.81\ \text{m/s}^2$. The acceleration, velocity, and displacement history of the headform are shown in Fig. 8.10. Meanwhile, Fig. 8.11 presents the crack morphology in the experiment, which includes radial and circular cracks, similar to a real traffic accident.

As shown in Fig. 8.10, the dynamic response of the headform is divided into three stages [9].

Stage I (0–4 ms) is the failure stage of the glass layer, called "glass failure stage," where the acceleration of the headform reaches the peak value. The kinetic energy of the headform decreases during Stage I (velocity decreases from 8.0 to 5.35 m/s) due to the failure of the glass layer. Kinetic energy of headform which is decreases by 83.1 J (from 8.0 to 5.35 m/s) is mainly converted into the fracture surface energy of glass.

Stage II (4–30 ms) is called the "large deformation stage," where the displacement of the headform increases until the maximum value while velocity decreases to zero. The acceleration of the headform increases slowly, which is shaped as an energy absorption plateau in this stage. The kinetic energy of the headform during this stage

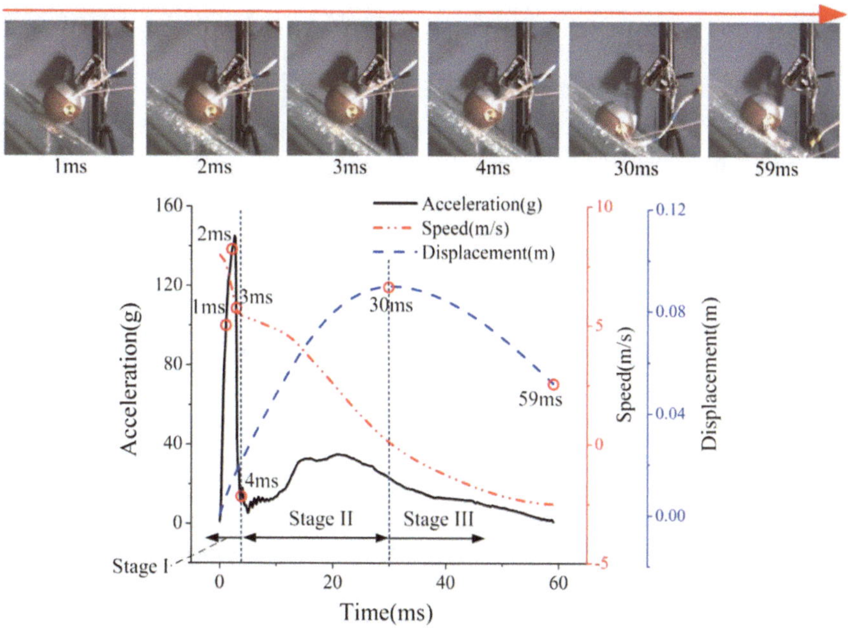

Fig. 8.10 Dynamic response of the headform upon vertical impact at 8.0 m/s [9]

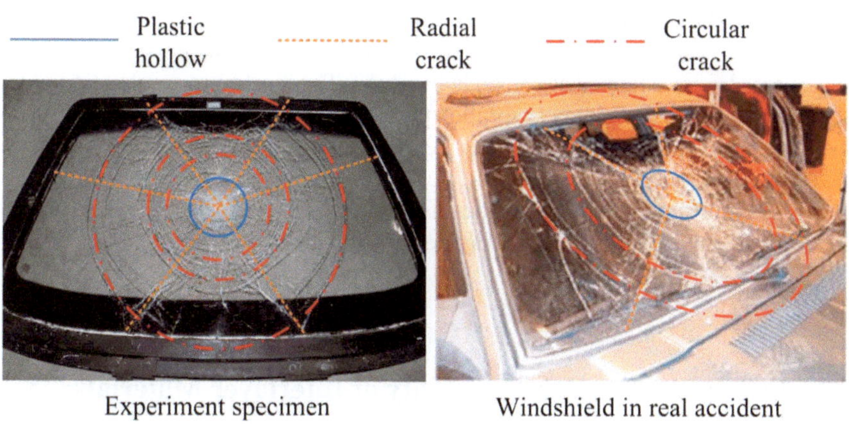

Fig. 8.11 Failure form of the windshield in the experiment (left) and a windshield in a real accident (right) [9]

also decreases (velocity decreases from 5.35 m/s to 0). Kinetic energy of headform which is decreases by 68.7 J (from 5.35 m/s to 0) is mainly converted to the fracture surface energy of glass and PVB internal friction energy (PVB exhibits tension deformation, which mitigates impact energy).

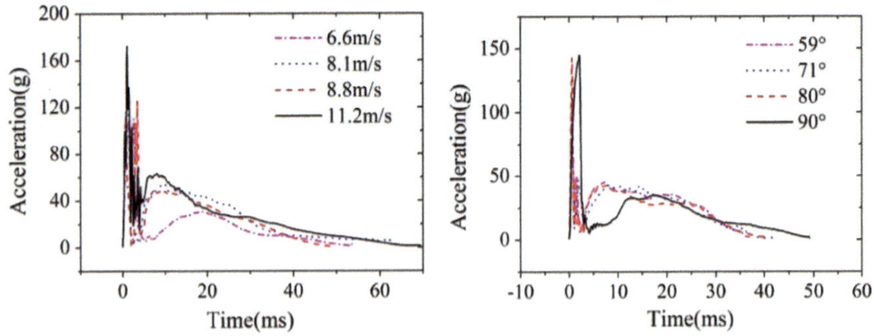

(a) vertical impact speed varying from 6.6 m/s to 11.2 m/s

(b) impact angle varying from 59° to 90° at a speed 8.0 m/s

Fig. 8.12 Acceleration–time curves of the headform at different impact speeds and angles

Stage III (30–59 ms) is called the "rebounding stage." The displacement and acceleration of the headform decrease, while the headform moves backward with velocity increasing from 0 to 2.5 m/s. A limited energy of 15 J is conveyed from the PVB interlayer to the headform due to the viscoelasticity of the PVB interlayer.

The kinetic energy of the headform, which accounts for 89.9% of the initial impact energy, is absorbed by the PVB windshield by presenting failure and deformation according to the analysis of the impact process [9].

The acceleration–time curves of the headform at different impact speeds and angles are also investigated as shown in Fig. 8.12. The maximum acceleration of the headform a_{max} and the energy absorption plateau increase with an increase in impact speed, and the plateau appears earlier. As shown in Fig. 8.12b, a_{max} increases and the energy absorption plateau appears later and decreases slightly with an increase in impact angle. Friction between the headform and the glass surface has a negligible effect on the failure of the glass layer due to the small contact area and low friction coefficient [9].

8.2 Energy Absorption Capability of Interlayer Materials Based on Simulations

Previous studies have mainly focused on the type and mechanism of windshield damage (or fracture) under different impact conditions. For example, Zhao et al. [17–19] suggested a continuum damage mechanics model and simulated the damage and failure of a windshield upon the impact of a featureless headform. Timmel et al. [20] used a smeared model based on an explicit finite element solver to compute the dynamic response of a windshield before and after fracture. Although these results are extremely useful in clarifying the damage mechanism of a windshield, they do not provide sufficient information regarding the energy absorption characteristics of a windshield upon impact with a pedestrian's head.

To provide high impact resistance and high energy absorption capabilities, polymeric foams and honeycomb structures have been widely applied (typically in sandwich structures combined with high-stiffness face sheets) in transportation tools [21]. Pyttel and Weyer [22] tested glassy polymers at different temperatures and loading rates, and verified the established constitutive relation via finite element analysis (FEA). Yasuki and Kojima [23] introduced a new FEA method for aluminum honeycomb structures.

At present, the PVB interlayer material [24], which is widely used in windshields, absorbs a limited amount of energy with a large deformation (whereas the outer glass layer resists impact to a certain extent) [25]. However, the energy absorption capability of traditional foams and honeycomb structures remain low, i.e., within the order of 0.1–10 J/g [26, 27].

The laminated glass interlayer plays a role in energy absorption upon impact; thus, we propose a competitive candidate for laminated glass interlayer, namely, the nanoporous energy absorption system (NEAS), which is a high-performance protection and damping system. NEAS applies the following principle: as a nanoporous material is immersed in a non-wetting second phase (e.g., a liquid or gel-like solution), the second phase can be forced to infiltrate otherwise energetically unfavorable nanopores when the external mechanical pressure exceeds a critical threshold [28–33]. Therefore, the ultra-large specific surface area of a nanoporous material (~1000 m^2/g) can absorb the external mechanical pressure into the interface energy, with an overall energy absorption capability of over 100 J/g. Accordingly, NEAS is a potential choice as an interlayer material for energy-absorbing windshields.

The potential performance of a windshield interlayer made from NEAS should be validated via numerical simulations. In particular, a comparison between the energy absorption characteristics of a conventional PVB interlayer and the new NEAS interlayer may highlight the advantage and possible suitability of NEAS as a potential material for windshield interlayer.

The complete steps of the simulation are as follows. First, we set up a virtual pedestrian–vehicle accident simulation environment in PC-CRASH, in which real-life pedestrian–vehicle accident speed is converted to head–windshield impact speed and equivalent mass. Subsequently, the model problem of a spherical head impacting on a laminated windshield is simulated via FEA. The representative stress–strain relations of the PVB material and NEAS are inputted into the interlayer material model. The contact force of the windshield plate, acceleration and velocity of the headform, and their distributions in time history are comprehensively and parametrically studied. In addition, several explicit formulae for the design variables of a windshield are established to improve understanding of windshield safety design.

8.2.1 Simulation of a Real-Life Accident Between a Pedestrian and a Vehicle

In a pedestrian–vehicle accident, the kinetic movement/trajectory of the pedestrian is complex due to the different front-end shapes of vehicles, contact points, and heights of pedestrians [34]. Nevertheless, the initial impact speed on the windshield, point of impact on the windshield, and equivalent mass of a foreign object are the three major factors that contribute to windshield damage and pedestrian injury [1]. For the first step of this research, we will convert real-life pedestrian–vehicle accident speed (under typical impact conditions) to head–windshield impact speed and equivalent mass (for simplicity, the effect of the point of impact is not discussed in this paper).

8.2.1.1 Virtual Pedestrian–Vehicle Accident Test Field

A virtual pedestrian–vehicle accident field is set up through PC-CRASH [35], a commercial software that is widely used in traffic accident reconstruction, under different conditions, i.e., varying vehicle speeds.

A typical modern passenger car is selected as the model vehicle with mass $M_s =$ 1310 kg and dimensions $L_V \times W_V \times H_V$ = 4480 mm × 1850 mm × 1420 mm. The windshield measures $a \times b \times h$ = 1435 mm × 634 mm × 4.76 mm, where a, b, and c refer to length, width, and thickness, respectively (the curvature of the windshield is included). The pedestrian is modeled via a multi-rigid-body method, including the head, torso, pelvis, upper limbs, lower limbs, and feet, with a typical height H_p =1.75 m and weight H_p=75 kg**. The coefficients of stiffness, friction, and contact are maintain as default values in the software, and the tire–road coefficient is $\mu = 0.8$. The initial condition of the pedestrian is stationary and facing the vehicle, while the vehicle is traveling at a certain speed. The vehicle performs a full brake (i.e., deceleration is set as 0.8 g) as soon as it hits the pedestrian. Figure 8.13 shows the schematic model. In this study, the pedestrian's head hits the center of the windshield normally.

8.2.1.2 Simulation Results

Our simulations show that the impact between the head and the windshield is nearly normal. As the pedestrian–vehicle accident speed v_{p-v} is varied, the head–windshield impact speed v_{h-w} and the equivalent projectile (head) mass M_e are investigated under the current model accident conditions. The results are presented in Fig. 8.14. In accordance with Euro NCAP, the vehicle must move toward a stationary pedestrian at a speed of 40 km/h (2009) in a standard pedestrian protection test. Thus, we select reference accident speed v_0=40 km/h.

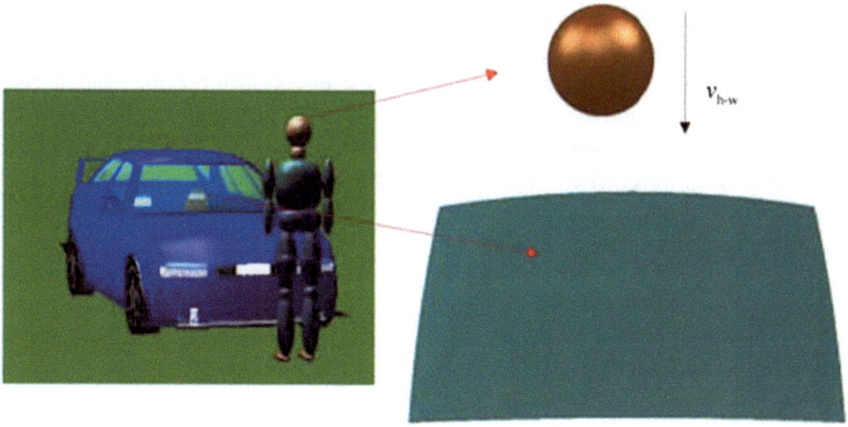

Fig. 8.13 Virtual simulation setup of the pedestrian–vehicle testing field

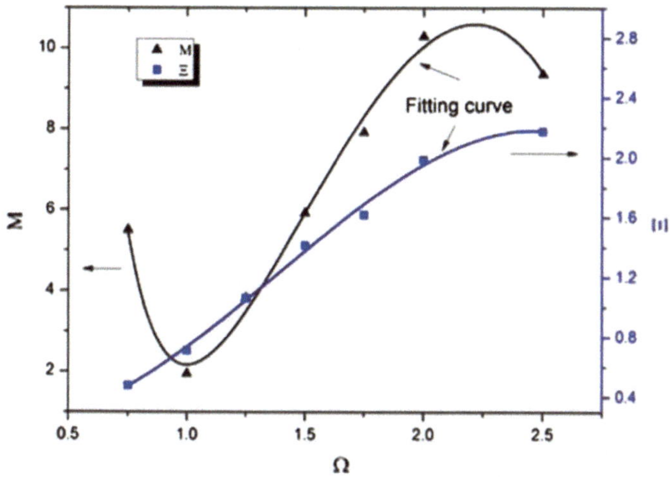

Fig. 8.14 Relationship of pedestrian–vehicle accident speed to head–windshield impact speed and equivalent head mass

With the definition of the normalized variable $\Omega = v_{h-w}/v_0$, $\Xi = v_{p-v}/v_0$, and $M = M_e/4.5$ (standard industrial headform mass $M_0 = 4.5$ kg (2009)), the data in Fig. 8.14 can be fitted as

$$\Omega = -2.712 + 14.98\Xi^{1.5} - 3.719e^{\Xi} - 12.36\ln\Xi/\Xi^2, \tag{8.1}$$

$$M = 0.1618 + 1.309\Xi^{2.5} - 0.6944\Xi^3 - 0.04249\Xi^{0.5}, \tag{8.2}$$

When the trends of the two relations are compared, the initial pedestrian–vehicle accident speed v_{p-v} and the head–windshield impact speed v_{h-w} increase, which agrees with the results of Wood's research [36]. The difference between v_{h-w} and v_{p-v} becomes smaller, and M_e increases except for the cases between 30 and 40 km/h, and above 90 km/h, primarily because body inertia effects differ on the head in cases below 40 km/h and above 90 km/h. During a high-speed accident, the pedestrian generally hits the windshield before the windshield accelerates to the same speed as the vehicle. Therefore, when impact angle and impact point are disregarded, a high accident speed can increase the impact speed between the head and the windshield and the equivalent impact mass.

Determining the relationship between v_{h-w} and v_{p-v} in accidents has long been a challenge to researchers given that no sensor can be installed on a pedestrian's head. The only solution for validation is through sled test within crash laboratories. However, the contact force, position, and duration of head versus windshield in real-life differ considerably from laboratory test results, as evidenced by the on-site accident investigations of Schmucker et al. [37] and Thollon et al. [38].

In the succeeding analysis, impact simulations with the data obtained in previous sections are conducted to investigate the energy absorption capability of interlayers.

8.2.2 Finite Element Modeling of Head–Windshield Impact

A featureless spherical headform is used to model the human head (a rigid ball, the same used in PC-CRASH), which is a common approach for testing and evaluation in the automotive industry; the model head has radius $R = 90$ mm [17]. The Young's modulus of the headform is $E = 6.5$ MPa. The mass of the headform is changeable through a change in density to represent the different equivalent masses of the impactor (the variation in M_e conforms to that of v_{h-w} in Fig. 8.14 for a given v_{p-v}). The windshield dimension is the same as that in PC-CRASH (Fig. 8.13). However, the windshield currently has a tri-layer structure. The outer glass sheets are assigned with typical soda–lime glass properties: density $\rho = 2500$ kg/m^3, Young's modulus $E = 2500$ GPa, and Poisson's ratio $\nu = 0.22$ [39]. Glass is brittle and exhibits a mode I energy release rate when cracking $G_I = 10$ J/m^2 [39]. Thus, the energy dissipated by each fully developed crack (i.e., the longest crack) is approximately $e_I = G_I \cdot l \cdot 2h_{glass} \approx 0.028$ J, where l denotes the crack length and h_{glass} indicates the thickness of the mono glass sheet.

Assume that the maximum number of cracks is 100 (which is considerably more than that in real life). The total amount of energy dissipated by cracking is only approximately 2.8 J, which is significantly less than the total kinetic energy and the energy absorbed by the interlayer and supporting materials. Therefore, the cracking of glass is not considered. The properties are assumed to be rate-independent [40], which is a common practice when dealing with glass. The thickness of the two glass sheets is the same, i.e., 2 mm. The PVB interlayer is modeled as a viscoelastic material with the following parameters: short-term shear modulus $G_0 = 0.33$ GPa, long-term shear modulus $G_\infty = 0.69 \times 10^{-3}$ GPa, bulk modulus $K = 20$ GPa, and decay time $\beta = 12.6$ s^{-1} [18]. The aforementioned material parameters are fully in accord with the experimental data in [25]. The standard thickness of the PVB interlayer used in the automotive industry is 0.76 mm. In this study, thickness is varied to explore its effect on energy absorption.

For the NEAS interlayer, the pressure–volume response measured from the experiment of a typical NEAS is converted to an equivalent stress–strain curve and used as the constitutive relation based on the test results using ZSM-5 [33]. The thickness of the NEAS interlayer is also varied.

The boundary condition of the windshield is assumed to be fixed (clamped), which is qualitatively consistent with the frame setup in practice [41]. The plate and the head are modeled with eight-node linear brick elements, and a mesh convergence study is conducted to determine the suitable mesh density and friction coefficient between the head and the windshield. The size of the mesh density is determined to be 1 mm × 1 mm. The contact between the head and the plate follows Coulomb's friction law with coefficient $f = 0.1$ (a small slip exists between the head and the windshield according to our accident scene investigation). The calculation time step is 0.1 ms.

8.2.3 Results and Discussions

First, four indices are defined to quantitatively evaluate the energy absorption capability of the interlayer:

$F_{\max}$: Maximum contact force that occurs during impact.
α: Force uniformity, i.e., $\alpha = \frac{\int_0^t Fdt}{t_{\max}}/F_{\max}$, where $t_{\max}$ is the time when the head detaches from the windshield.
β: Energy absorption ratio, i.e., $(v_{h-w}^2 - v_{h-w}^{\prime 2})/v_{h-w}^2$, where $v_{h-w}^{\prime 2}$ is the speed at the moment when the head detaches from the windshield plate.
HIC: Head injury criterion, a standard indicator adopted in the automotive industry, i.e., [42]

$$HIC = \max_{(t_1-t_2)}\left\{(t_2-t_1)\left[\frac{1}{t_2-t_1}\int_{t_1}^{t_2} a(t)dt\right]^{2.5}\right\},$$

where a is the acceleration of the head, and t_1 and t_2 are two instances during impact. Here, we choose $t_1 - t_2 = 0.15$ ms.

These indices indicate the energy absorption capability of a structure/material in a comprehensive manner.

8.2.3.1 Parametric Study: Impact Speed

(1) Contact force analysis

Figure 8.15 shows the force–displacement curve when the accident speed v_{p-v} is 10, 20, and 40 m/s, which cover the typical range of pedestrian–vehicle accidents [4]. The windshield quickly deforms after being subjected to the impact of the pedestrian's head via the compression of the interlayer, followed by the bending of the glass sheets (denoted as the "hard contact stage" in Fig. 8.15), thereby leading to a high contact force. NEAS is more ductile; hence, the first stage of interlayer compression is longer than that of the PVB interlayer, such that more impact energy is absorbed. Moreover, NEAS also mitigates the peak contact force during the hard contact stage, which is extremely beneficial for head protection [43]. Meanwhile, if impact speed and impactor mass increase, then contact time becomes shorter. Table 8.3 shows that the contact force in the case with the PVB interlayer is approximately 8.9, 13.05, and 14.98% higher than that of the NEAS interlayer at the three aforementioned accident speeds.

Moreover, the variation in contact force with the NEAS interlayer is also smoother (i.e., the acceleration of the head is lower) than that in the case with the PVB interlayer, which can be further confirmed in Fig. 8.16.

(2) Acceleration analysis

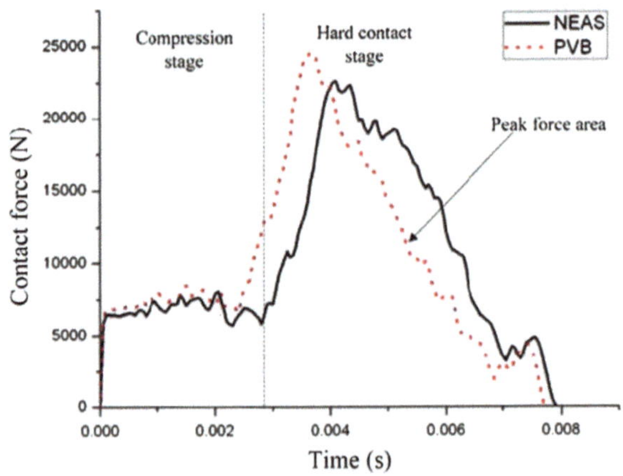

Fig. 8.15 Contact force in the time history curve at an accident speed of 10 m/s

Table 8.3 Maximum force and force uniformity at different accident speeds of the PVB and NEAS interlayers

	$v_{p-v} = 10$ ms		$v_{p-v} = 20$ ms		$v_{p-v} = 40$ ms	
	$F_{\max}$(kN)	α	$F_{\max}$(kN)	α	$F_{\max}$(kN)	α
NEAS	22.6	0.457	68.9	0.460	193	0.398
PVB	24.6	0.417	77.9	0.455	222	0.372

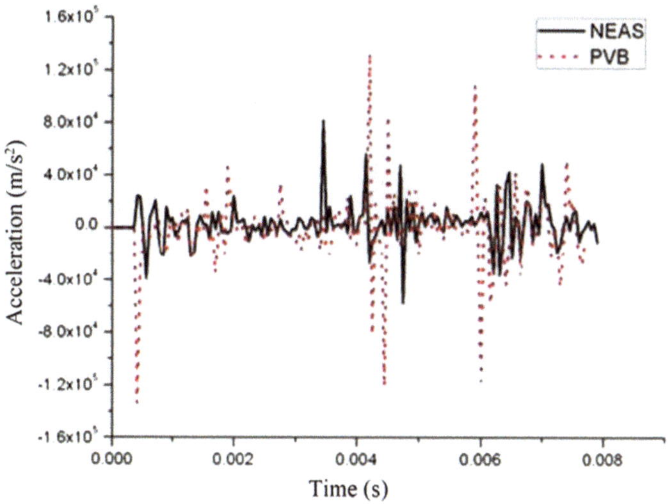

Fig. 8.16 Acceleration in the time history curve at an accident speed of 10 m/s

The time history curve of acceleration shown in Fig. 8.16 indicates that the NEAS interlayer provides gentler and slighter acceleration for the impacting head than the PVB interlayer. The peak acceleration of the head in the NEAS laminated windshield is approximately 50% that of the head in the PVB laminated windshield. The HIC value can be calculated from the acceleration time history curve, which is equal to 1245.76, 4197.26, and 15,977.35 for the PVB laminated windshield when the accident speed is 10, 20, and 40 m/s, respectively. The results are similar to those in [44], and all of the results are above the non-severe head injury threshold of $HIC = 1000$ [42]. If the interlayer is substituted with NEAS, then the HIC values decrease to 891.53, 3416.98, and 3271.45 in the three cases. In real-life accidents, the probability that the pedestrian will die when the accident speed is 50 km/h is generally high [1, 45, 46]. Our analyses show that with the possible use of the NEAS interlayer, the HIC value can be below the threshold when the accident speed is below 40 km/h, thereby considerably enhancing vehicle crash safety. A parametric study is conducted to increase recognition of the relationship between head injury and accident speed v_{p-v}. With the same normalization definition of $\Xi = v_{p-v}/v_0$ and $H = HIC/HIC_0$, where $HIC_0 = 1000$ is the threshold of severe head damage, the results can be fitted as

$$HIC = \begin{cases} 0.447 - 0.0369\Xi^{2.5} + 0.5449\Xi^3 - 0.1037\ln\Xi \ \ \textit{for NEAS interlayer} \\ 0.3730 + 0.4151\Xi^{2.5} + 0.4418\Xi^3 - 0.2642\ln\Xi \ \textit{for PVB interlayer} \end{cases}, \quad (8.3)$$

as shown in Fig. 8.17.

As stated earlier, if the pedestrian's height and vehicle specifications are changed, then the relationship between HIC and v_{p-v} may undergo moderate variations. The

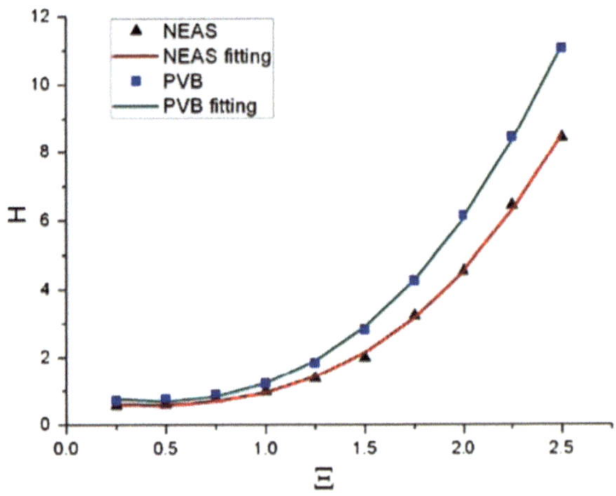

Fig. 8.17 HIC values calculated at different accident speeds for the PVB and NEAS interlayers and their nonlinear fitting curves

results may provide useful insights not only into the estimation of pedestrian head injury during an accident, but also into the establishment of crash test standards or regulations.

(3) Velocity analysis

Figure 8.18 illustrates the velocity change of the head for different interlayers at various accident speeds. Head velocity rapidly decreases during the approaching stage and increases during the detaching stage in the NEAS laminated windshield system. The total energy of the head–windshield system E_{total} is equal to the kinetic energy of the head before impact. It becomes $E_{total} = E_{d-head} + E_{d-windshield} + E_{k-head} + E_{k-windshield}$ after impact, where E_{d-head} and $E_{d-windshield}$ are the deformation (strain) energy of the head and the windshield, and E_{k-head} and $E_{k-windshield}$ are the kinetic energy of the head and the windshield. The deformation of the head and the windshield is small; hence, the kinetic energy of the head is dominant over other energy components (see Fig. 8.19, the case of the PVB interlayer). β is used as a direct indicator for determining the amount of kinetic energy dissipated by the windshield plate, as shown in Fig. 8.20.

The β values for the NEAS laminated windshield are 37.65, 57.10, and 41.25%, whereas those for the PVB laminated windshield are 24.48, 49.13, and 28.75%. The results of the PVB interlayer is consistent with those in [19]. Therefore, NEAS can mitigate the impact energy of foreign objects more efficiently. With the same normalization definition of $\Xi = v_{p-v}/v_0$, the fitting formula can be expressed as follows:

$$\beta = \begin{cases} \frac{0.3467-0.5354\Xi^{0.5}-0.2196\Xi}{1-1.519\Xi^{0.5}+0.6002\Xi} & for\ NEAS\ interlayer \\ \frac{0.2353-0.3664\Xi^{0.5}+0.1509\Xi}{1-1.514\Xi^{0.5}+0.5906\Xi} & for\ PVB\ interlayer \end{cases}, \quad (8.4)$$

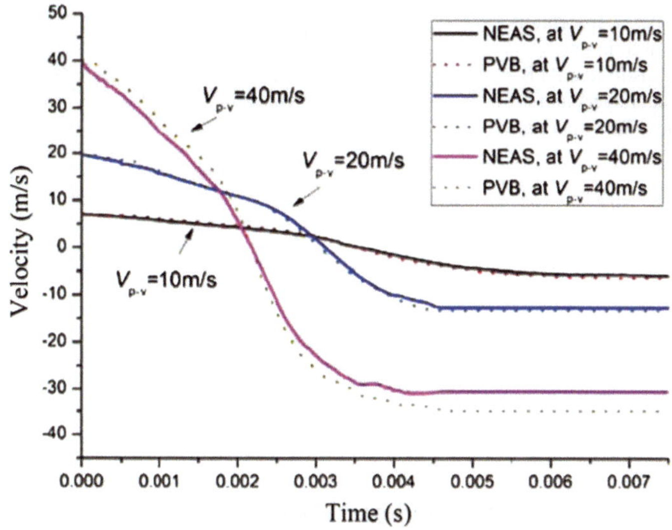

Fig. 8.18 Velocity changes in the time history curve at different accident speeds for the PVB and NEAS interlayers

Fig. 8.19 Comparison of the kinetic energy between the head and the windshield during the impact process at different accident speeds

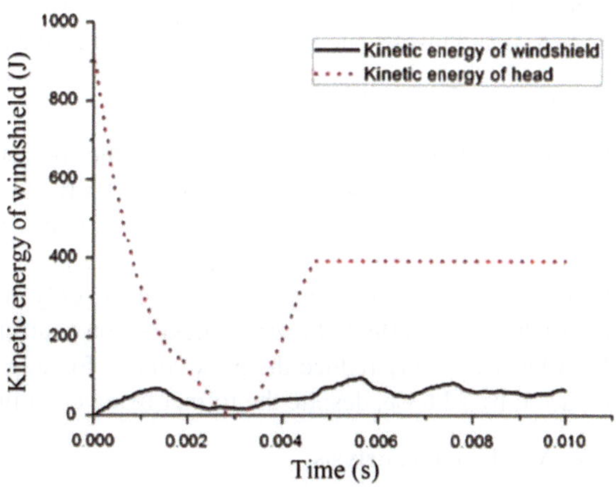

as shown in Fig. 8.20. At an accident speed of approximately 70 km/h, PVB and NEAS can dissipate the highest percentage of incident kinetic energy. If the impact speed is low, then NEAS can perform considerably better than PVB in terms of energy absorption due to its ductility.

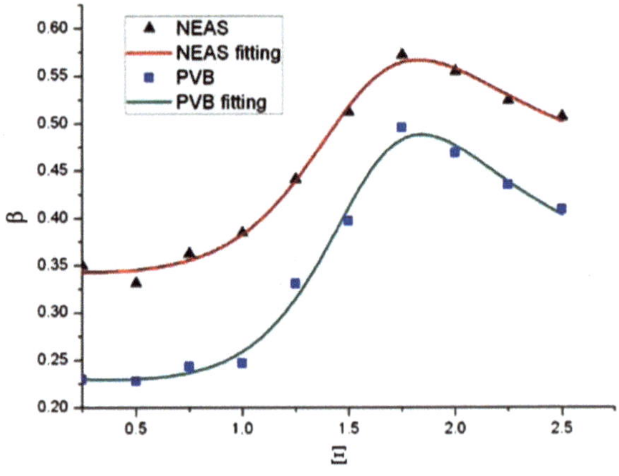

Fig. 8.20 Energy absorption efficiency β at different accident speeds for the PVB and NEAS interlayers and their nonlinear fitting curves

8.2.3.2 Parametric Study: Interlayer Thickness

Windshield interlayer is currently standardized, i.e., the thickness values of PVB interlayers in passenger cars and trucks are 0.76 and 1.52 mm, respectively. In this section, we fix the accident speed at a typical value of $v_{p-v} = 20$ m/s (except in Sect. 8.2) and investigate the effect of interlayer thickness.

(1) Contact force analysis

Figure 8.21 shows the time history curve of the contact force. As interlayer thickness increases, $F_{\max}$ decreases and α increases in PVB and NEAS laminated windshields, thereby indicating that the present design of interlayer thickness is insufficient in terms of pedestrian protection. Impact mostly induces localized deformation (i.e., the energy absorption mechanism is prominent only in a small region near the impact point); hence, the effect of overall thickness on contact force is insignificant. A thick PVB layer can only reduce the peak contact force by 11.78%, whereas NEAS can reduce it by 22.63%, despite the former being eight times thicker than the latter.

(2) Acceleration analysis

Acceleration becomes smooth with an increase in interlayer thickness, which will be reflected in the HIC value. We normalize *HIC* as $H = HIC/HIC_0$. The dimensionless interlayer thickness is $\Gamma = h/h_0$, where $h_0 = 0.76$ mm, which is the standard thickness for cars. As mentioned earlier, accident speed in the standard pedestrian protection test is 40 km/h, which indicates that the pedestrian is generally safe below this accident speed. Here, we choose the accident speed $v_{p\text{-}v} = 50$ km/h to check HIC variation under different interlayer thickness values. Figure 8.22 illustrates the relationship between H and Γ, which may be fitted as

$$H = \begin{cases} 2.214 + 0.0163\Gamma^{1.5} - 0.83051\Gamma^{0.5} + 0.3303\ln\Gamma \; \textit{for NEAS interlayer} \\ 2.795 + 0.01757\Gamma^{1.5} - 0.98361\Gamma^{0.5} + 0.443\ln\Gamma \; \textit{for PVB interlayer} \end{cases}, \tag{8.5}$$

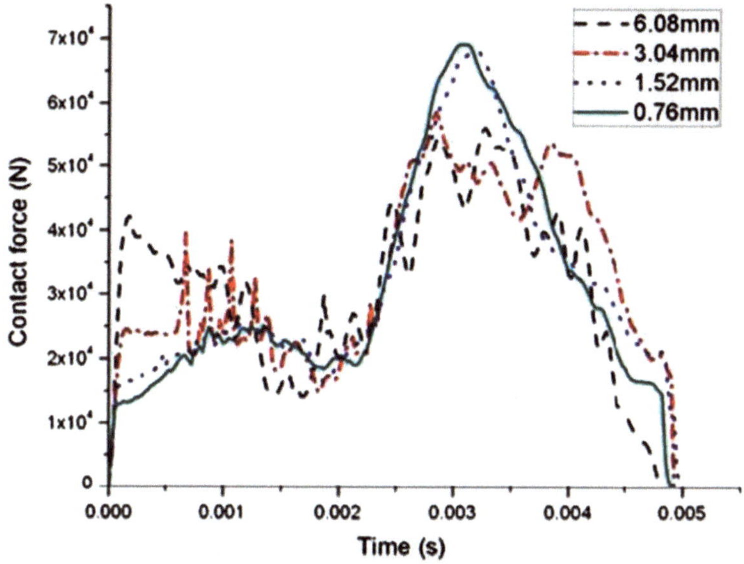

(a) Time history curve of the contact force of the NEAS laminated windshield with different thickness values.

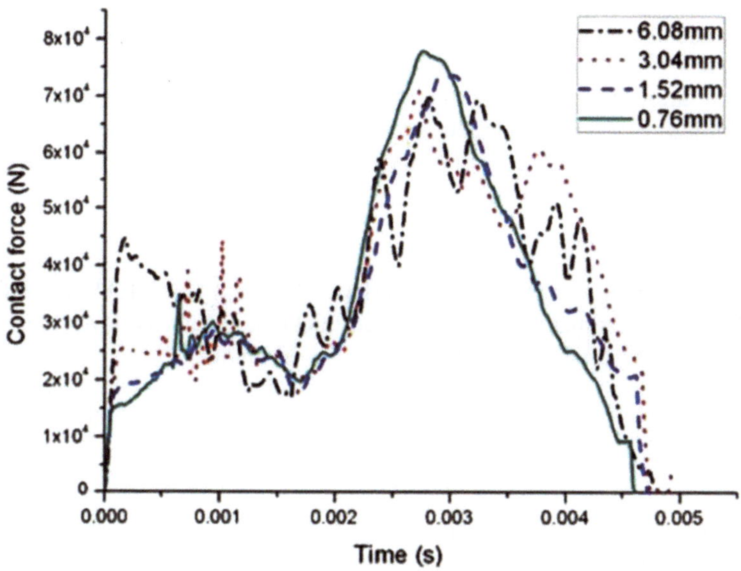

(b) Time history curve of the contact force of the PVB laminated windshield with different thickness values.

Fig. 8.21 Contact force variation during the impact process with different interlayer thickness values at an accident speed of 72 km/h

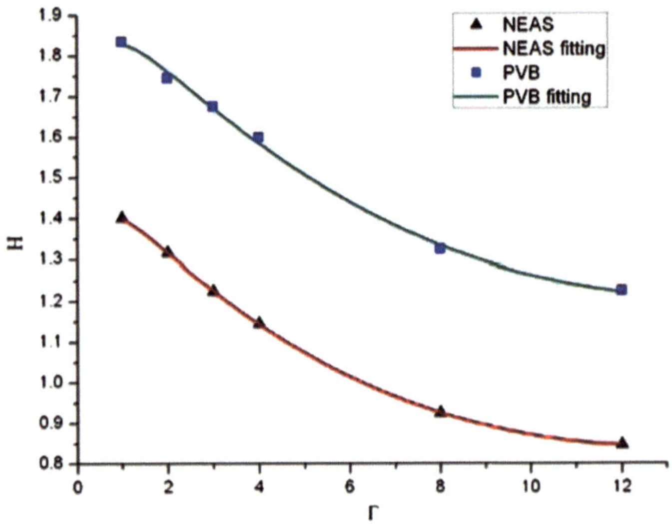

Fig. 8.22 Relationship between the normalized HIC value and interlayer thickness at an accident speed of 72 km/h

Theoretically, the HIC value for the NEAS interlayer can only be kept below 1000 when $h \geq 4.8$ mm according to Eq. (8.5). Meanwhile, keeping the HIC value below the safety threshold is considerably more difficult if the interlayer is made of PVB (for $v_{p\text{-}v} = 50$ km/h). Similar results can be established under other accident speeds and conditions, which will be useful in reducing the critical thickness of the interlayer at which injury is below the threshold.

(3) Velocity analysis

Similarly, β can be correlated with interlayer thickness (for NEAS and PVB) to provide a quick estimation of the required thickness of an interlayer for a given energy mitigation criterion (Fig. 8.23, for $v_{p\text{-}v} = 72$ km/h):

$$\beta = \begin{cases} 0.7232 - 0.1530/\,\Gamma^{0.5}\ for\ NEAS\ interlayer \\ 0.5619 - 0.0773/\,\Gamma^{0.5}\ for\ PVB\ interlayer \end{cases}. \tag{8.6}$$

For other accident speeds, results similar to Eq. (8.6) can be established, which will facilitate the decision-making process of an automotive safety engineer in choosing the suitable thickness for a specific windshield interlayer material.

In addition, the glass sheet is kept the same, thereby providing the NEAS laminated windshield with an equivalent capability to prevent an occupant from being ejected. Consequently, the NEAS laminated windshield exhibits superior pedestrian/occupant energy absorption capability and similar occupant ejection capability compared with the traditional PVB laminated windshield, which indicates its potential function as an interlayer.

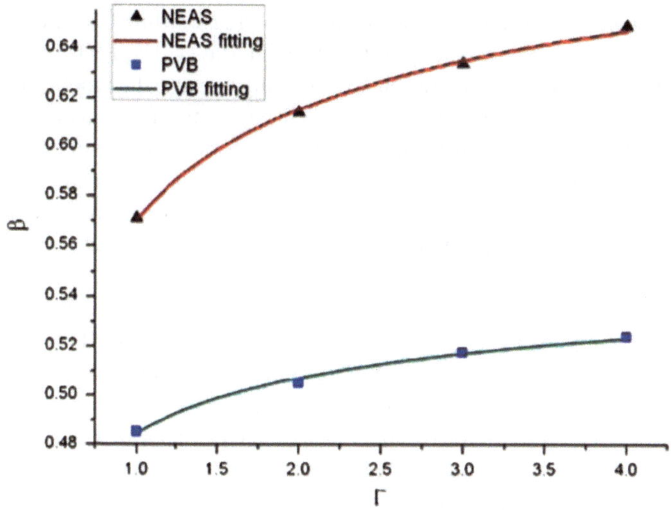

Fig. 8.23 Relationship between energy absorption efficiency β and normalized thickness with their nonlinear fittings

8.3 Summary

Several studies on the pedestrian protection and energy absorption capabilities of PVB laminated glass based on experiments and simulations are presented in this chapter. We also investigate the energy absorption capability of interlayer materials based on simulations, i.e., the performance of a traditional material (PVB) is compared with that of a new material (NEAS).

We present three experiments on PVB laminated windshield impacted by a headform to study the dynamic mechanical behavior of laminated windshields. The results of all the experiments show that the dynamic response of the headform can be divided into three stages: the failure stage of the glass layer, the large deformation stage, and the rebounding stage. During the first stage, the glass layers bear the main load and mostly experience bending deformation. Headform acceleration increases as contact between the headform and the windshield increases. The two acceleration peaks observed in the experiments are attributed to compression waves being reflected onto tensile waves from the boundaries. This phenomenon occurs when compression waves reach the lower glass layer after the outer glass layer is first impacted. The lower glass layer fails earlier under tensile state than under compressive state. The maximum acceleration of the headform increases with increasing impact based on the results of Experiment 3.

During the second stage, when the acceleration peak begins to decline, the PVB layer plays a dominant role in bearing loads and in dissipating the impact energy of the headform through a large deformation. The acceleration of the headform reaches its peak again. After the peak value is reached, the final stage begins. The displacement and acceleration of the headform decrease, while the headform moves backward due to the viscoelasticity of the PVB interlayer. The PVB fails during the last stage while headform acceleration is decreasing because the kinetic energy of the headform and the internal energy of the PVB dissipate through plastic deformation, which is attributed to the friction between the headform and the windshield.

Then, the relationships of pedestrian–automobile accident speed v_{p-v} to human head–windshield impact speed v_{h-w} and equivalent head mass M_e are determined under typical impact conditions using PC-CRASH. The deduced combinations of v_{p-v} and M_e are then used to simulate head–windshield impact using FEA. Simulations focus on several energy absorption characteristics with NEAS or PVB as the interlayer. Interlayer thickness is also varied to explore its effect. Four indices are first defined and used to quantitatively evaluate energy absorption performance. These indices are explicitly related to accident speed, which may provide a quick and effective guideline for automotive safety design and evaluation. The results indicate that NEAS can perform better than PVB in terms of providing pedestrian protection, particularly during high-speed impact accidents, and thus, NEAS can be an ideal candidate for laminated windshield design. In addition, the thickness of the interlayer should be increased for PVB and NEAS to improve safety given that currently available PVB laminated windshields cannot provide sufficient protection to pedestrians against windshield impact.

References

1. J. Xu, Y. Li, Crack analysis in PVB laminated windshield impacted by pedestrian head in traffic accident. Int. J. Crashworthiness **14**, 63–71 (2009)
2. D. Otte, Severity and mechanism of head impacts in car to pedestrian accidents, in *Proceedings of the 1999 International IRCOBI Conference on the Biomechanics of Impact*, 23–24 Sept 1999, Sitges, Spain, 1999
3. C. Oh, Y.S. Kang, Y. Youn, A. Konosu, Development of probabilistic pedestrian fatality model for characterizing pedestrian-vehicle collisions. Int. J. Automot. Technol. **9**, 191–196 (2008)
4. J. Xu, Y.B. Li, Study of damage in windshield glazing subject to impact by a pedestrian's head. Proc. Inst. Mech. Eng. Part D: J. Automobile Eng. **223**, 77–84 (2009)
5. J. Xu, Y. Li, G. Lu, W. Zhou, Reconstruction model of vehicle impact speed in pedestrian–vehicle accident. Int. J. Impact Eng. **36**, 783–788 (2009)
6. S. Shahbeyk, A. Abvabi, A numerical study on the effect of accident configuration on pedestrian lower extremity injuries. Sci. Iranica **16**, 376–387 (2009)
7. M.K. Shin, S.I. Yi, O.T. Kwon, G.J. Park, Structural optimization of the automobile frontal structure for pedestrian protection and the low-speed impact test. Proc. Inst. Mech. Eng. Part D: J. Automobile Eng. **222**, 2373–2387 (2008)
8. J.W. Lee, M.K. Shin, K.H. Yoon, G.J. Park, An orthogonal-array-based design-of-experiments method for designing a vehicle hood and bumper structure. Proc. Inst. Mech. Eng. Part D: J. Automobile Eng. **222**, 161–171 (2008)
9. B. Liu, T. Xu, X. Xu, Y. Wang, Y. Sun, Y. Li, Energy absorption mechanism of polyvinyl butyral laminated windshield subjected to head impact: experiment and numerical simulations. Int. J. Impact Eng. **90**, 26–36 (2016)
10. N. Yang, J. Wang, G. Zhao, Y. Zhong, D. Wang, Experimental study of headform–PVB laminated windshield impact. Int. J. Crashworthiness 1–11 (2016)

11. E. NCAP, European New Car Assessment Programme, Pedestrian Testing Protocol (2014)
12. B. Liu, T. Xu, X. Xu, Y. Wang, Y. Sun, Y. Li, Energy absorption mechanism of polyvinyl butyral laminated windshield subjected to head impact: experiment and numerical simulation. Int. J. Impact Eng. (2015)
13. S. Mukherjee, A. Chawla, P. Mahajan, D. Mohan, N. Mane, M. Singh, Modeling of head impact on laminated glass windshield, in ed. by A. Charpenne. *International IRCOBI Conference on the Biomechanics of Impacts IRCOBI*, Montpellier, France, 2000, pp. 328–336
14. J. Chen, J. Xu, X. Yao, X. Xu, B. Liu, Y. Li, Different driving mechanisms of in-plane cracking on two brittle layers of laminated glass. Int. J. Impact Eng. **69**, 80–85 (2014)
15. B. Liu, Experimental research on energy absorption characteristic of PVB laminated windshield subjected to headform low-speed impact. J. Mech. Eng. **49**, 128 (2013)
16. J. Huang, Q. Liu, B. Nie, Y. Huang, H. Du, Y.A. Xia, Development and evaluation of pedestrian impact test facility based on EEVC/WG17 requirements. Autom. Eng. (2008)
17. S. Zhao, L.R. Dharani, X. Liang, L. Chai, S.D. Barbat, Crack initiation in laminated automotive glazing subjected to simulated head impact. Int. J. Crashworthiness **10**, 229–236 (2005)
18. S. Zhao, L.R. Dharani, L. Chai, S.D. Barbat, Analysis of damage in laminated automotive glazing subjected to simulated head impact. Eng. Fail. Anal. **13**, 582–597 (2006)
19. S. Zhao, L.R. Dharani, L. Chai, S.D. Barbat, Dynamic response of laminated automotive glazing impacted by spherical featureless headform. Int. J. Crashworthiness **11**, 105–114 (2006)
20. M. Timmel, S. Kolling, P. Osterrieder, P.A.D. Bois, A finite element model for impact simulation with laminated glass. Int. J. Impact Eng. **34**, 1465–1478 (2007)
21. G.a.M. Lj, *A Cellular Solids* (Cambridge University Press, Cambridge, 1999)
22. T. Pyttel, S. Weyer, Crash simulation with glassy polymers – constitutive model and application. Int. J. Crashworthiness **8**, 433–442 (2003)
23. T. Yasuki, S. Kojima, Application of aluminium honeycomb model using shell elements to offset deformable barrier model. Int. J. Crashworthiness **14**, 449–456 (2009)
24. T.S. Valera, N.R. Demarquette, Polymer toughening using residue of recycled windshields: PVB film as impact modifier. Eur. Polym. J. **44**, 755–768 (2008)
25. J. Xu, Y. Li, B. Liu, M. Zhu, D. Ge, Experimental study on mechanical behavior of PVB laminated glass under quasi-static and dynamic loadings. Compos. B Eng. **42**, 302–308 (2011)
26. H.C. Jo, Y.E. Kim, A study on the influence of the seat and head restraint foam stiffnesses on neck injury in low speed offset rear impacts. Int. J. Precis. Eng. Manuf. **10**, 105–110 (2009)
27. K. Siringam, N. Juntawong, S. Cha-Um, C. Kirdmanee, Aluminium foam-polymer composites: processing and characteristics. J. Mater. Sci. **44**, 1506–1511 (2009)
28. X. Chen, F.B. Surani, X. Kong, V.K. Punyamurtula, Energy absorption performance of steel tubes enhanced by a nanoporous material functionalized liquid. Appl. Phys. Lett. **89**, 331 (2006)
29. L. Liu, Y. Qiao, X. Chen, Pressure-driven water infiltration into carbon nanotube: the effect of applied charges. Appl. Phys. Lett. **92**, 2604 (2008)
30. Y. Qiao, L. Liu, X. Chen, Pressurized liquid in nanopores: a modified Laplace-Young equation. Nano Lett. **9**, 984 (2009)
31. L. Liu, X. Chen, W. Lu, A. Han, Y. Qiao, Infiltration of electrolytes in molecular-sized nanopores. Phys. Rev. Lett. **102**, 184501 (2009)
32. X. Chen, G. Cao, A. Han, V.K. Punyamurtula, L. Liu, P.J. Culligan, T. Kim, Y. Qiao, Nanoscale fluid transport: size and rate effects. Nano Lett. **8**, 2988 (2008)
33. J. Zhao, P.J. Culligan, J.T. Germaine, X. Chen, Experimental study on energy dissipation of electrolytes in nanopores. Langmuir ACS J. Surf. Colloids **25**, 12687 (2009)
34. L.Y.U. Xu Jun, An overview on the technique of accident reconstruction for pedestrian-vehicle crash. Autom. Eng. **31**, 1029–1033 (2009)
35. H.G. Xu, Y.H. Fan, Simulation research on form and kinematics law of contact process for automobile-pedestrian collision based on the coupling of PC-crash and MADYMO, in *International Conference on Transportation Engineering*, 2007, pp. 2235–2240
36. D.P. Wood, Impact and movement of pedestrians in frontal collisions with vehicles. ARCHIVE Proc. Inst. Mech. Eng. Part D Transp. Eng. 1984–1988 **198–202**(2), 101–110 (1988)

37. U. Schmucker, M. Beirau, M. Frank, D. Stengel, G. Matthes, A. Ekkernkamp, J. Seifert, Real-world car-to-pedestrian-crash data from an urban center. J. Trauma Manage. Outcomes 1–6 (2010)
38. L. Thollon, C. Jammes, M. Behr, P.J. Arnoux, C. Cavallero, C. Brunet, How to decrease pedestrian injuries: conceptual evolutions starting from 137 crash tests. J. Trauma **62**, 512–519 (2007)
39. J. Menčík, *Strength and Fracture of Glass and Ceramics* (Elsevier, 1992)
40. W. Holand, G.H. Beall, *Glass Ceramic Technology* (2012)
41. X. Yuan, S.M. Li, Analysis of rectangular thin plate vibration under different support conditions. *Aeroengine* **31**, 39–43 (2005)
42. C.W. Gadd, *Use of a Weighted-Impulse Criterion for Estimating Injury Hazard* (1966)
43. D. Marjoux, D. Baumgartner, C. Deck, R. Willinger, Head injury prediction capability of the HIC, HIP, SIMon and ULP criteria. Accid. Anal. Prev. **40**, 1135 (2008)
44. X.U. Jun, B.H. Liu, G.E. Dong-Yun, L.I. Yi-Bing, Research on dynamic response of PVB laminated windshield subjected to low velocity impact. Acta Armamentarii (2010)
45. V. Giavotto, Compatibility of vehicles with safety barriers: head ejection through side windows (2004)
46. T. Maki, J. Kajzer, K. Mizuno, Y. Sekine, Comparative analysis of vehicle–bicyclist and vehicle–pedestrian accidents in Japan. Accid. Anal. Prev. **35**, 927–940 (2003)

Chapter 9
Accident Reconstruction Based on Laminated Windshield Cracking

Traffic accidents are becoming increasingly frequent worldwide due to various complex factors. Pedestrian–vehicle accidents comprise the most common type of traffic accidents. Considerable efforts have been exerted to improve the protection of vulnerable road users, particularly pedestrians, against injuries and deaths. Evidently, traffic accident reconstruction techniques, which play an important role in accident investigations, should be studied.

China has been consistently ranked as one of the countries with a high pedestrian fatality rate. The Road Traffic Accident Annual Census Report of China [1] indicated that over 89,455 individuals died in at least 378,781 accidents in 2006. Among which, pedestrians accounted for 26.01%, which is the highest proportion of all traffic fatalities. On average, a pedestrian is injured every 5 min in China and one is killed every 17 min. Even in a country where traffic management is comparatively well organized, e.g., the US, pedestrian safety is also the focus of public safety. In 1999, 4907 pedestrians were killed in the US, accounting for 12% of all traffic fatalities [2].

Accident reconstruction and assessment must have stringent requirements given the high pedestrian death rate during accidents. Thus, accident reconstruction methods should be investigated.

9.1 Reconstruction Method Based on Information Left on the Road

Although the age and state of health of the pedestrian, nature of impact, and vehicle shape affect the outcome of an injury, the primary factor in injury/fatality risk is vehicle impact speed [3–5]. Vehicle impact speed is the primary focus of accident investigators.

J. Xu and Y. Li, *Impact Behavior and Pedestrian Protection of Automotive Laminated Windshield*, https://doi.org/10.1007/978-981-13-2441-3_9

9.1.1 Tire Marks

The moment speed of a vehicle at the starting point mark can be obtained based on the energy conservation law.

In the beginning, the reconstruction model used in vehicle speed estimation is based on the energy conservation law. First, a coefficient is determined based on the skid mark pattern according to previous experience. Then, the coefficient and length of the skid mark are applied to determine the initial velocity at the beginning of the mark [6–8], as shown in Eq. (9.1):

$$v_c = \sqrt{2\mu g s}, \tag{9.1}$$

where μ is the tire-to-road coefficient, s is the length of the skid mark, g is the acceleration due to gravity, and v_c refers to the velocity of the vehicle.

The obtained friction coefficient can be relatively accurate because the vehicle brake test can simulate the behavior of the tires during a vehicle accident.

9.1.2 Pedestrian Throw Distance

However, many pedestrian–vehicle accidents do not leave skid marks. The anti-lock braking system (ABS) is widely used at present; hence, few marks are left when a vehicle breaks. In addition, a road surface will have no mark left on it under certain weather conditions, such as snow and rain. To solve this problem, pedestrian throw distance can be considered in calculating vehicle impact speed.

A collision accident has three phases: contact between vehicle and pedestrian, pedestrian is thrown, and pedestrian slides on the road [9]. The impact velocity and geometric configuration of the vehicle front end affect vehicle–pedestrian contact. The influence factor for the pedestrian throw phase is mainly vehicle brake strength. For the pedestrian sliding phase, the influencing factors are moment speed and direction and road condition when the pedestrian separates from the vehicle.

Several researchers have investigated the dynamic response of vehicle–pedestrian accidents during the 20th century. A number of empirical formulas based on the projectile motion formula have been proposed based on the experiment results presented in Table 9.1, where v_c denotes the vehicle impact speed, and x_L indicates pedestrian rolling distance. μ is the friction coefficient; g is the acceleration due to gravity; and d_t and d_0 are the pedestrian throw distance and sliding distance, respectively. h_p is the pedestrian throw height, and a is the average deceleration. m_c and m_p are the vehicle mass and pedestrian mass, respectively. The formulas are suitable for projectile-type accidents. Several researchers have also developed a speed calculation formula for coiling-type accidents (Table 9.2) due to the occurrence of a second impact in such type of accidents.

Table 9.1 Summary of speed–pedestrian throw distance formulas

References	Model
[18]	$v_c = \mu\sqrt{2g}\left(\sqrt{h_p + \frac{x_L}{\mu}} - \sqrt{h_p}\right)$
[19]	$v_c = \sqrt{\mu^2 h_p + 2\mu g d_t} - \mu h_p$
[19]	$v_c = 101.6 \times \left(-\frac{\sqrt{h_p}}{7.97} \pm \sqrt{\frac{h_p}{63.52} - 0.02d_t}\right)$
[19]	$v_c = \sqrt{2\mu g d_t}(\mu = 0.66{-}1.2)$
[19]	$d_t = \frac{v_c^2}{2a} + 0.03av_c$
[19]	$v_c = 2a\left(-\frac{\sqrt{h_p}}{g} \pm \sqrt{\frac{h_p}{g^2} - 2d_t}\right)$
[19]	$v_c = \frac{\sqrt{(d_t-\mu-d_0)2\mu g}}{m_c^2} \times (m_c + m_p)$
[19]	$v_c = \sqrt{2.6 + 3.5d_t} - 4.7$

Table 9.2 Summary of coiling-type speed–pedestrian throw distance formulas

References	Model
[6]	$v_c = \sqrt{\frac{2\mu g}{\mu^2 \sin^2\theta + \mu \sin 2\theta + \cos^2\theta}} \times \left(\frac{m_c+m_p}{am_c}\right)\sqrt{d_t - (x_L + \mu h_p)}$
[20]	$v_c = 11.3\sqrt{d_t} - 0.3$
[21]	$v_c = 9.84d_t^{0.57}$
[12]	$v_c = 8.3604d_t^{0.6046}$
[22]	$v_c = 12.7d_t^{0.5} - 2.6$

In addition, a pedestrian throw distance model for estimating initial vehicle impact speed based on kinematics law was developed by Schmidt–Nageld [10]. Several passenger cars with different vehicle masses, dimensions, and initial impact velocities were tested. The following empirical formula was obtained through data fitting:

$$\mathrm{X} = 0.0178\mu g v_c + 0.027v_c^2/\mu g, \tag{9.2}$$

where X is the throw distance.

Numerous authors have conducted tests to determine the coefficients in the throw distance model and implemented several changes to the parameters [11–17]. However, another problem arises with this model, i.e., determining pedestrian throw distance is a challenging task because traces or marks that indicate vehicle–pedestrian contact point are difficult to obtain in an accident scene.

Hence, the following issues arise when calculating speed based on pedestrian throw distance:

(1) Contact position of the vehicle, pedestrian, and road.

(2) The second contact between the pedestrian and vehicle can lead to a considerable change in throw distance.
(3) Grinding and rolling accidents have not been investigated by experiments, and modeling an analysis based on dynamics is difficult.

9.1.3 Road Scattered Materials

Road scattered materials provide important information in the accident reconstruction process because of the difficulty in acquiring road marks. Researchers have found that vehicle impact speed can be determined by calculating the relationship between the fragment distribution of windows (windshield or side windows) and vehicle impact velocity. In [23, 24], the generalized throw distance formula for glass fragments has been developed as follows:

$$L_{\Sigma} = A_3\left[v^2 + (2gH)\right]^{A_4} + A_1\left[v\sqrt{2H/g}\right]^{A_2} + A_5\left[v^2 + (2gH)\right]/g\exp[-A_6(v/H)], \tag{9.3}$$

where $A_1 \sim A_6$ have different values because L_{Σ} and H have various combinations. However, this formula is seldom used in accident reconstruction because numerous parameters have to be inputted and calculation results are poor.

9.2 Reconstruction Method Based on the Laminated Windshield

Braun and Strobl [14, 25] used the throw distance of spinning fragments from the windshield and headlamp to estimate the impact speed of a vehicle. On the basis of their work, Xu [8, 26–28] proposed a generalized fragment throw distance model to calculate the impact speed of a vehicle by using the dimensions of the glass fragment field based on kinematics law.

The exact values of the parameters that describe glass fragments are difficult to determine based on the authors' accident investigation experiences. Windshield glazing on modern vehicles consists of two soda–lime glass plies fused with a polymer interlayer. PVB, a widely used windshield interlayer, has two major advantages over monolithic glass, namely, energy absorption and fragment glutinosity, which indicates that less fragments will spin on the ground. Therefore, the aforementioned methods are limited for investigation and reconstruction.

Pedestrian–vehicle accidents have three phases according to the motion of the pedestrian. First, the vehicle bumper impacts with the pedestrian's leg (defined as the "contact phase"). Then, the pedestrian's head impacts with the windshield, engine hood, or A-pillar (called the "impact phase"). Finally, the pedestrian slides down onto the ground (called the "fall-over phase"). A total of 237 pedestrian–vehicle accidents were recorded by NTADTU. Among which, head impact on the windshield accounted for 81.02% of all vehicle parts that came in contact with the pedestrian's head. Evidently, the windshield of a vehicle can provide considerable information about accidents.

Accordingly, a suitable impact dynamics model is presented in the current study to describe the impact between the pedestrian head and the windshield. A mathematical model that characterizes the deflection of the impact point and the impact speed of the pedestrian head is also proposed. After combining the two models, a complete reconstruction model for vehicle impact speed calculation is developed. Finally, to demonstrate the validity of the model, 10 actual accidents are selected to compare the results.

9.2.1 Accident Reconstruction Model of Vehicle Impact Speed

9.2.1.1 Abstract Physical Model

As mentioned earlier, a typical composite PVB windshield glazing has two pieces of glass with an interlayer between them. When the head crashes onto the windshield plate, a clear deflection of the impact point is observed on the glazing, as shown in Fig. 9.1. The deflection is considerably larger than the thickness of the entire glazing, thereby indicating that it is a large deformation problem. Thus, we consider that no sliding occurs between the layers for simplicity. The shape of the windshield glazing can be regarded as a rectangle, and the entire windshield glazing is considered a composite thin plate. The comparative directions of pedestrian motion to that of the vehicle vary. To simplify the problem, we disregard protuberances on the head, e.g., eyes, nose, and ears. That is, the head is regarded as a sphere headform. Windshield glazing is typically supported by a rubber bar made of polyurethane, a type of elastic material. Without losing generality, one border ($y = 0$) is selected as an example in such boundary conditions. The equations that describe the boundary condition are [29]

$$[V_y]_{y=0} = D\left[\frac{\partial^3 w}{\partial y^3} + \frac{(2-v)\partial^3 w}{\partial \chi^2 \partial y}\right]_{y=0} = -K_{11}[w]_y = 0, \tag{9.4}$$

$$[M_y]_{y=0} = D\left[\frac{\partial^2 w}{\partial y^2} + \frac{\mu \partial^2 w}{\partial \chi^2}\right]_{y=0} = 0, \tag{9.5}$$

where V_y is the shear force of the unit length, $D = Eh^3/12(1-\mu^2)$, E is the Young's modulus of the plate, h is the thickness of the plate, w is the deflection of the plate, (x, y) is the displacement of the point on the plate in the global coordinate, n is the Poisson's ratio, K_{11} is the spring stiffness of the elastic foundation, and M_y is the

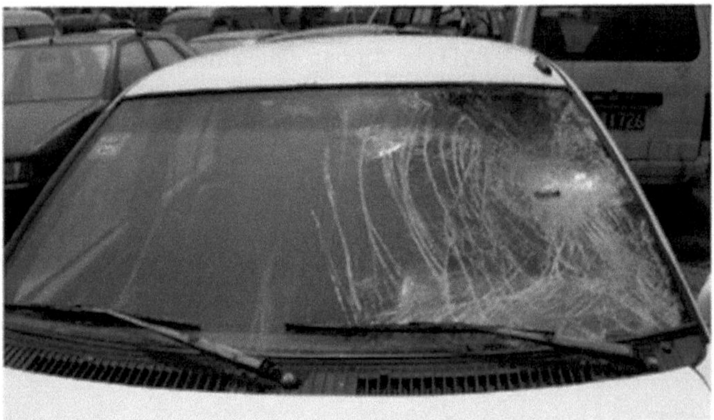

Fig. 9.1 Windshield glazing upon the impact of a pedestrian head

rotation moment of the unit length. Evidently, $K_{11} = 0$ represents a free boundary, whereas $K_{11} = \infty$ refers to a simply supported boundary.

Then, we conduct FEA to show the difference in plate deflection under different boundary conditions. A rectangular plate made of a composite laminated glass material is studied via FEA. Figure 9.2 indicates that only approximately 3% difference exist between the two boundary conditions. The plate cannot bear much deflection and bending moment due to the brittleness of the glass material. Consequently, the effect of the boundary conditions on plate deflection is minimal. For simplicity, we consider the support boundary as a simply supported one. Thus, an abstract physical model of the impact between the pedestrian head and windshield glazing is developed, as shown in Fig. 9.3.

The aforementioned typical three-layered windshield glazing is considered. t_i denotes the thickness of a certain layer. The subscripts "g" and "PVB" refer to glass and PVB film, respectively. R is the radius of the average pedestrian head (impactor/indenter). The head crashes into the windshield glazing at a velocity of v_0. We define the glass layer that will come in contact with the head as the inner glass layer, whereas the other glass layer is called the outer glass layer.

9.2.1.2 Material Model

The stress–strain relations of PVB are influenced by time because PVB is a viscoelastic material. As pointed out in [30], the time of impact between the pedestrian head and windshield glazing is comparatively short, i.e., approximately 0.1 s. In such a short period, the shear modulus $G(\mathrm{t})$ of PVB will not change significantly. Thus, we consider PVB as a linear elastic material during impact. The inner and outer glass layers are common float glasses; thus, the relations among shear modulus $G(\mathrm{t})$, bulk modulus K, Young's modulus E, and Poisson's ratio n are [31]

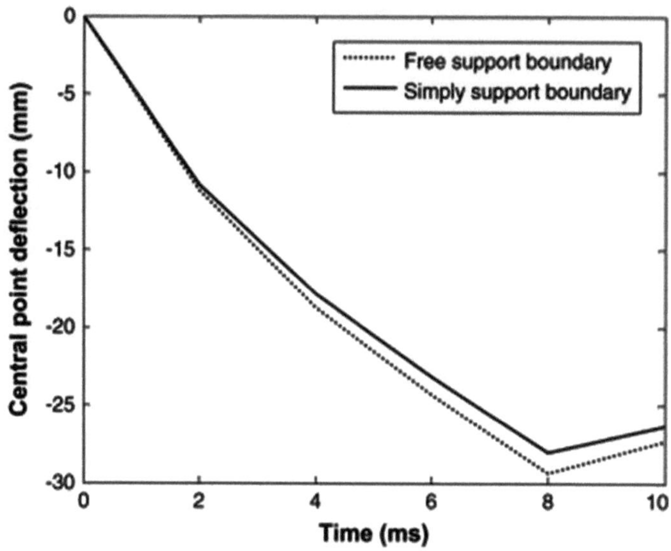

Fig. 9.2 Comparison between central point deflections with a free support boundary and a simply supported boundary

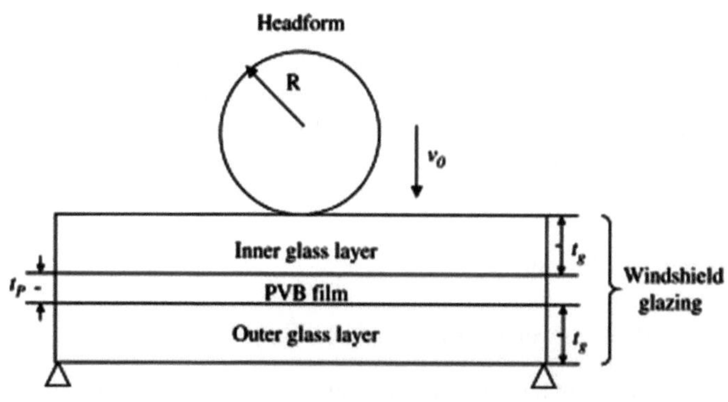

Fig. 9.3 Schematic of a spherical pedestrian headform impacting on wind-shield glazing

$$G = \frac{E}{2(1+v)}, \tag{9.6}$$

$$K = \frac{E}{3(1-2v)}. \tag{9.7}$$

The same formulae can be used to calculate the PVB material. Glass and PVB films can be regarded as orthotropic materials according to their physical and material properties. Therefore, we assume that each layer of the composite windshield is orthotropic with respect to its material symmetry lines and follows Hooke's law.

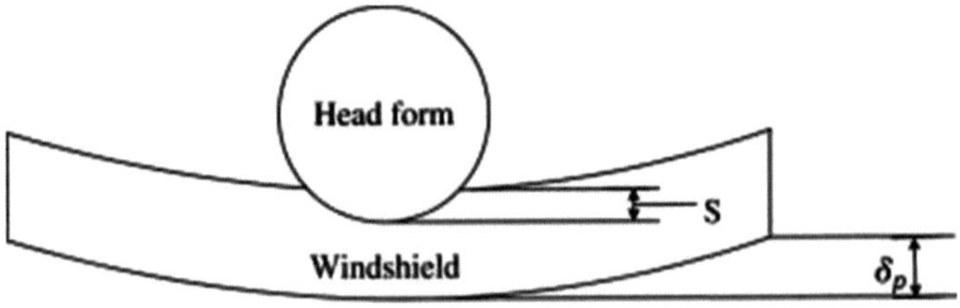

Fig. 9.4 Headform impact on the windshield glazing

Accordingly, considering the integrity of the composite windshield becomes reasonable. Consequently, an equivalent elastic modulus and Poisson's ratio should be introduced as follows [31]:

$$\bar{E} = \frac{2t_g E_g + t_p E_p}{2t_g + t_p} \tag{9.8}$$

$$\bar{\nu} = \frac{2t_g \nu_g + t_p \nu_p}{2t_g + t_p} \tag{9.9}$$

where $\bar{E}$ and $\bar{\nu}$ are equivalent Young's modulus and equivalent Poisson's ratio, respectively, of windshield glazing.

We are concerned with the dynamic response of windshields in this study. Thus, investigating various details of the pedestrian head is unnecessary. Only the values of the physical parameters from references are substituted into the featureless spherical headform used in this study.

9.2.1.3 Dynamics Model

(1) Model of deflection versus pedestrian head impact speed

In general, the pedestrian head will intrude into the windshield and cause the glazing plate to curve or even break, shown in Fig. 9.4.

The time during impact is denoted as t, then the impact force of the pedestrian head is

$$F = -M\frac{dv_0}{dt}, \tag{9.10}$$

where M and v_0 is the mass and initial speed of the headform, respectively.

If we denote the distance in which the headform approaches the windshield by s, then evidently

$$\dot{s} = v_0 \tag{9.11}$$

Greszczuk [32] showed that if the contact duration between the impactor and the target is extremely long compared with their natural periods, then the vibrations of the system can be neglected. The impact force according to the Hertz contact law can be described as follows:

$$F_c = ns^{2/3}, \tag{9.12}$$

where n is the contact stiffness defined by the shape of the impactor. The headform is a spherical object, and n can be written as

$$n = \frac{4\sqrt{R}}{3\pi(k_1 + k_2)}, \tag{9.13}$$

where

$$k_1 = \frac{1 - v_1^2}{\pi E_1} = \frac{1 - \bar{v}^2}{\pi \bar{E}}, \tag{9.14}$$

$$k_2 = \frac{1 - v_2^2}{\pi E_2}. \tag{9.15}$$

When studying the contact behavior of a composite material, Sankar [33] indicated that Eq. (9.15) can be rewritten as follows:

$$k_2' = \frac{1}{\pi E_2}. \tag{9.16}$$

The impact force and the corresponding deflection exhibit a certain relation in an impact problem:

$$F_p = K_p \delta_p. \tag{9.17}$$

where F_p represents the contact force, K_p is the elastic property of the windshield glazing (depending on the material of the glazing), and δ_p is the curvature of the glazing.

The governing equation can be derived according to the system energy conservation law as follows:

$$\frac{1}{2}Mv_0^2 = \int_0^{\delta_{\max}} F_p d\delta_p + \int_0^{s} F_c ds \tag{9.18}$$

The substitution of Eqs. (9.12) and (9.17) into Eq. (9.18) yields

$$\frac{1}{2}Mv_0^2 = \frac{1}{2}K_p\delta_p^2 + \frac{2}{5}ns^{2/5}. \tag{9.19}$$

When $F_c = F_p = F$ is considered, we can rewrite Eq. (9.19) as follows:

$$\frac{1}{2}Mv_0^2 = \frac{1}{2}\left(\frac{F^2}{K_p}\right) + \frac{2}{5}\left(\frac{F^{3/5}}{n^{2/3}}\right). \tag{9.20}$$

Young [34] suggested that the aforementioned elastic property can be determined when a rectangular plate is simply supported:

$$K_p = \frac{E_2 h_w^3}{ab^2} \tag{9.21}$$

where h_w is the thickness of the plate, and a and b are the lengths of the two sides.

The substitution of Eqs. (9.13) and (9.21) into Eq. (9.20) yields:

$$v_0 = \sqrt{\frac{2}{M}\left(F^2\left(\frac{ab^2}{2E_2h_w^3}\right) + F^{3/5}\left(\frac{2}{5}\left(\frac{3\pi(k_1+k_2)^{2/3}}{16R}\right)\right)\right)}. \tag{9.22}$$

Thus, on the basis of Navier's method [35], the deflection of the windshield glazing (rectangular plate) under the impact of a pedestrian's head is expressed as

$$w(x, y) = \frac{4F}{\pi^4 E_2 I ab}\sum_{m=1}^{\infty}\sum_{n=1}^{\infty}\frac{\sin\frac{m\pi\xi}{a}\sin\frac{n\pi\eta}{b}}{\left(\left(\frac{m}{a}\right)^2+\left(\frac{n}{b}\right)^2\right)^2}\sin\frac{m\pi x}{a}\sin\frac{n\pi y}{b}, \tag{9.23}$$

where E_2I is the flexural rigidity of the windshield, ξ and η are the distances to the nearest side of windshield, and I is the largest moment of inertia of the windshield.

Therefore, we can obtain

$$F = \frac{w(x, y)\pi^4 E_2 I ab}{4\sum_{m=1}^{\infty}\sum_{n=1}^{\infty}\frac{\sin\frac{m\pi\xi}{a}\sin\frac{n\pi\eta}{b}}{\left(\left(\frac{m}{a}\right)^2+\left(\frac{n}{b}\right)^2\right)^2}\sin\frac{m\pi x}{a}\sin\frac{n\pi y}{b}}. \tag{9.24}$$

When Eq. (9.24) is substituted into Eq. (9.22), the first half of the entire model that describes the relations between deflection and head impact velocity can be developed. In this case, we can easily calculate the impact velocity of the pedestrian's head if we obtain the exact measurement.

(2) Model of pedestrian head impact speed versus vehicle impact speed

Mini-cars; small, mid-size, and large sedans; sports and specialty vehicles; and wagons and sports utility vehicles (SUVs) can be grouped as bonnet-type vehicles according to the shape of their front end [36]. Vehicle impact speed is not the same as pedestrian head impact speed for bonnet-type vehicles due to the revolution of the pedestrian upon impact. Such vehicles are the most involved in pedestrian–vehicle accidents.

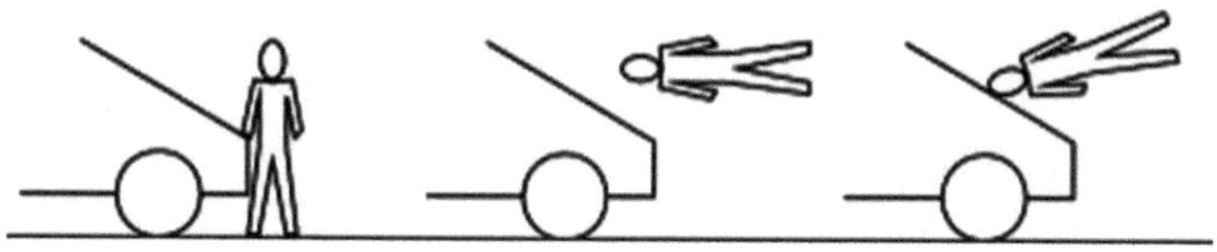

Fig. 9.5 Schematic of contact phase and impact phase

Two phases occur between the first contact of the vehicle with the pedestrian and pedestrian head impact with the windshield: contact phase and impact phase (Fig. 9.5).

A pedestrian is regarded as a single rigid body with initial momentum I and rotation radius R_p. Hence, the momentum of a pedestrian is

$$Fh = I\alpha, \tag{9.25}$$

where F is the contact force, h denotes the height of the contact point, and α is the angular acceleration of the pedestrian.

During impact time t_i, h and I are constant. When Eq. (9.25) is integrated, we derive

$$h\int\limits_0^{t_i} Fdt = I\int\limits_0^{t_i} \alpha dt. \tag{9.26}$$

In accordance with the momentum conservation law, we can also obtain

$$\int\limits_0^{t_i} Fdt = m_p(v_{p1} - 0), \tag{9.27}$$

where v_{p1} and m_p are the velocity and mass of the pedestrian, respectively.

Moreover, the following relation is widely recognized:

$$\int\limits_0^{t_i} \alpha dt = \omega_1 - 0, \tag{9.28}$$

where ω_1 is the angular velocity of the pedestrian.

The combination of Eqs. (9.27) and (9.28) and their substitution into Eq. (9.26) yields [11]

$$m_p v_{p1} h = I\omega_1. \tag{9.29}$$

$I = m_p R_p^2$; thus, the following equation can be derived [11]:

$$v_{p1}h = \omega_1 R_p^2. \tag{9.30}$$

The vehicle impact speed v_0 can be expressed as [11]

$$v_0 = v_{p1} + \omega_1 h. \tag{9.31}$$

As a result of horizontal momentum conservation, the kinetic momentum of a pedestrian–vehicle system is [11]

$$m_v v_0 = m_p v_{p1} + m_v\left(\omega_1 h + v_{p1}\right), \tag{9.32}$$

where m_v is the mass of the vehicle.

When Eq. (9.29) is substituted into Eq. (9.30) and then into Eq. (9.32), the pedestrian impact speed v_i is [11]

$$v_0 = \frac{m_v v_{p1} h^2}{\left(m_v + m_p\right)R_p^2 + m_v h^2}. \tag{9.33}$$

The combination of Eqs. (9.23), (9.25), and (9.33) yields

$$v_0 = \frac{\sqrt{\frac{2}{M}\left(F^2\left(\frac{ab^2}{2E_2h_w}\right) + F^{3/5}\left(\frac{2}{5}\left(\frac{3\pi(k_1+k_2)^{2/3}}{16R}\right)\right)\right)}\left(m_v + m_p\right)R_p^2 + m_v h^2}{m_v h^2}, \tag{9.34}$$

where

$$F = \frac{w(x, y)\pi^4 E_2 I ab}{4\sum_{m=1}^{\infty}\sum_{n=1}^{\infty}\frac{\sin\frac{m\pi\xi}{a}\sin\frac{n\pi\eta}{b}}{\left(\left(\frac{m}{a}\right)^2+\left(\frac{n}{b}\right)^2\right)^2}\sin\frac{m\pi x}{a}\sin\frac{n\pi y}{b}}.$$

9.2.1.4 Model Summary

The entire model is illustrated in Fig. 9.6 to summarize the model for convenience. Inputting numerous values of different variables will not be burdensome as some of the variables have constant values.

9.2.1.5 Results and Discussion

(1) Determining the parameters in the model

In real-life accident reconstruction, the stature and weight of the pedestrian can be easily obtained from police investigation files. The masses, center of masses, inertias, and distance between inertias of each body part can be acquired from [37]. Parallel

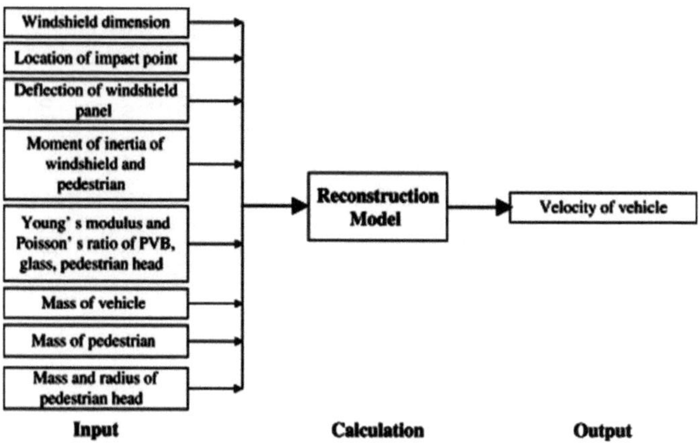

Fig. 9.6 Simplified illustration of the model structure

axis theorem is applied to calculate the rotation radius and inertia of the pedestrian. The PVB material properties of windshield glazing are nearly the same. In accident reconstruction, we consider $\nu_{\mathrm{PVB}} = \mathrm{const} = 0.49$ for simplicity. The parameters of PVB and glass are determined from Table 3.3 in Chap. 3.

(2) Actual accidents for comparison

Exact impact velocity cannot be obtained in real-life traffic accidents; thus, we have to select actual pedestrian–vehicle accidents with skid marks and regard the speed calculated based on skid marks as the exact speed. We choose 10 real-life accidents from NTADTU that have skid marks and windshield deflection. The results after the comparison are provided in Table 9.3.

(3) Results and discussion

First, the differences, with the maximum value being 10.78% and the average value being 7.56%, demonstrate that the newly developed model is sufficiently accurate and reliable for use in pedestrian–vehicle accidents without skid marks or pedestrian throw distances. The application of this model to further verify the results of accident reconstruction can enhance the accuracy and reliability of reconstruction results. This model provides a new and convenient method for investigating pedestrian–vehicle accidents.

Second, the 10 speed values computed from the model are slightly lower than the exact speed values. One reason for this result is the simplicity of the supporting boundary. Exact deflection is larger than the deflection that occurs with a simply supported boundary plate, thereby leading to the lower results of the model. If a more exact elastic condition is adopted, then the results will be improved. Another reason is that the windshield is typically broken in high-speed accidents because of the large deformation in most cases. The measurements of deflection are insufficiently accurate (they are usually smaller than the exact measurements). Therefore, the results will

Table 9.3 Comparison between exact speed and speed calculated using the model

	Case I	Case II	Case III	Case IV	Case V	Case VI	Case VII	Case VIII	Case IX	Case X
Length of skid mark (m)	11.23	7.80	9.30	15.40	23.00	8.40	15.30	21.10	19.80	24.00
Road–tire adhesion coefficient	0.841	0.790	0.690	0.740	0.862	0.766	0.760	0.812	0.820	0.760
Deflection of windshield on the impact point (mm)	37	23	19	42	49	27	41	47	45	48
Location of impact point (x mm, h mm)	(−145, −650)	(350, 600)	(−230, 100)	(250, −400)	(240, 470)	(135, 460)	(735, 160)	(−135, 695)	(−635, 395)	(715, 480)
Dimensions of windshield (a mm, b mm)	(1320, 620)	(1310, 730)	(1350, 730)	(1280, 630)	(1200, 680)	(1310, 670)	(1400, 710)	(1350, 660)	(1295, 625)	(1380, 690)
Height of bumper (mm)	505	495	525	495	505	499	530	512	501	510
Weight of pedestrian (kg)	46	53	75	65	78	69	79	76	76	69
Mass of vehicle (kg)	1325	1880	1225	1480	1290	1340	2105	1890	1450	1655
Stature of pedestrian (cm)	175	168	172	166	173	158	176	173	166	169
Exact speed (km/h)	48.98	39.56	40.37	53.80	70.97	40.42	54.34	65.97	64.22	68.07
Result of the model (km/h)	45.94	37.42	38.18	50.20	63.72	37.31	51.29	60.99	57.81	60.73
Difference (%)	6.20	5.40	5.42	6.70	10.22	7.69	5.61	7.55	9.98	10.78

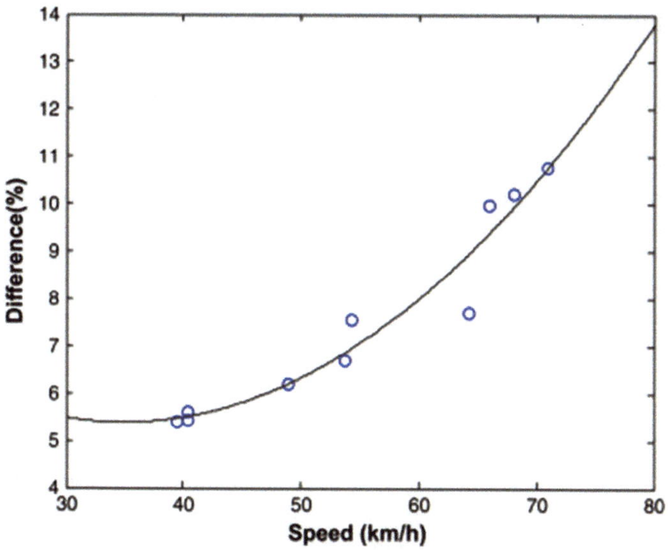

Fig. 9.7 Differences increase as vehicle impact speed increases

be smaller to a certain extent. Moreover, the effective mass of the head will increase considerably if the pedestrian is impacted by a vehicle at high speed and the pedestrian rotates several times. In conclusion, this model may be unsuitable for reconstructing high-speed pedestrian–vehicle accidents.

Third, the model does not consider the curvature and thickness of the windshield and the windshield angle. In most cases, the curvature of a windshield is small, and thus, we regard a windshield as a plate and not a shell. Most windshields of modern passenger vehicles have the same thickness according to the automotive industry standard. In fact, windshield angle may vary from one vehicle to another. Thus, the impact speed of the pedestrian head will decrease slightly according to the exact angle, thereby affecting the accuracy of the model.

In addition, the comparison results show that the differences increase as speed increases. Figure 9.7 illustrates that the relation between speed and difference is speed = difference α. This relation suggests that the proposed vehicle impact speed reconstruction model may be unsuitable for high-speed accidents. The deflection of a windshield is difficult to measure, and such difficulty in high-speed accidents suggests that the model is also unsuitable for such cases.

Lastly, the shoulder of the pedestrian, and not only the head, also impact on the windshield glazing in some cases. In such scenario, the shoulder's impact on the windshield will add to the deflection of the windshield, which is difficult to distinguish. The proposed model should not be adopted under such circumstances.

9.3 Summary

In this chapter, we introduced current accident reconstruction methods and then developed a model of vehicle impact speed versus deflection of windshield glazing.

First, current accident reconstruction methods were introduced. These methods include those based on information left on the road: tire marks, pedestrian throw distance, and road scattered materials.

Then, accident reconstruction based on a laminated windshield was developed given that an investigation would be nearly impossible if marks or throw distances cannot be obtained for accident reconstruction. We first constructed an abstract physical model that describes the dynamic process of the pedestrian head impacting on the composite PVB windshield glazing. Several necessary simplifications were implemented. A material model was selected. The most important part was the development of the dynamic model of vehicle impact speed versus windshield deflection. The entire process of a pedestrian–vehicle accident, excluding the sliding down phase, was discussed based on impact dynamics and rectangular plate theory. The results of the new model exhibit good agreement with those calculated from skid marks, which are considered the exact speed in 10 actual accidents. This finding demonstrates the accuracy and reliability of the proposed model. Therefore, a new and reliable method for accident reconstruction was established

However, the new model may be unsuitable for calculating vehicle impact velocity in high-speed collision accidents between a pedestrian and a vehicle because the deflection of the windshield is difficult to measure, particularly in high-speed accidents. In addition, the impact point should be checked to ensure that the head (and not the shoulder) has impacted on the windshield before the new calculation method is used.

References

1. Traffic management bureau of Police Ministry. Road traffic accident annual census report of China, Beijing (2007)
2. M.F. Ballesteros, P.C. Dischinger, P. Langenberg, Pedestrian injuries and vehicle type in Maryland, 1995–1999. Accid. Anal. Prev. **36**, 73–81 (2004)
3. F.H. Waiz, M. Hoefliger, W. Fehlmann, Speed limit reduction from 60 to 50 km/h and pedestrian injuries, in *Stapp Car Crash Conference with IRCOBI and Child Injury and Restraint Conference with IRCOBI*, 1983, pp. 311–318
4. M. Howard, A. Thomas, W. Koch, J. Watson, R. Hardy, Validation and application of a finite element pedestrian humanoid model for use in pedestrian accident simulations, in *Proceedings of the 2000 International Ircobi Conference on the Biomechanics of Impact*, 20–22 Sept 2000, Montpellier, France, 2000
5. D.P. Wood, C.K. Simms, D.G. Walsh, Vehicle-pedestrian collisions: validated models for pedestrian impact and projection. Proc. Inst. Mech. Eng. Part D: J. Automob. Eng. **219**, 183–195 (2005)
6. I. Han, R.M. Brach, Impact throw model for vehicle-pedestrian collision reconstruction. Proc. Inst. Mech. Eng. Part D: J. Automob. Eng. **217**, 69–71 (2002)

7. L. Yibing, Study on the road traffic accident reconstruct analysis theory, in *Automotive Engineering* (Tsinghua University, Beijing, 2004)
8. X. Hg, *Automotive Accident Engineering* (People's Traffic Press, Beijing, 2004)
9. R. Limpert, *Motor Vehicle Accident Reconstruction and Cause Analysis* (Accident Investigation, 1984)
10. D.N. Schmidt, D.A. Nagel, Pedestrian impact case study, in *American Association for Auto Medicine Conference Proceedings*, 1970
11. D.P. Wood, D.P. Wood, Impact and movement of pedestrians in frontal collisions with vehicles. ARCHIVE Proc. Inst. Mech. Eng. Part D: Transp. Eng. **202**, 101–110 (1988)
12. T.F. Fugger Jr, B.C. Randles, J.J. Eubanks, Comparison of pedestrian accident reconstruction models to experimental test data for wrap trajectories, in *Vehicle Safety 2000—Proceedings of an International Conference held* 7–9 June 2000 (Institution of Mechanical Engineers, London, UK, 2000)
13. D. Wood, C. Simms, A hybrid model for pedestrian impact and projection. Int. J. Crashworthiness **5**, 393–404 (2000)
14. H. Braun, Splitterwurfweiten. Eine Experimentelle UntersuchungG. Verkehrsunfall **18** (1980)
15. D.M. Severy, Auto-pedestrian impact experiments (1963)
16. G.S. Hill, Calculations of vehicle speed from pedestrian throw. Impact (1994)
17. The Institute of Traffic Accident Investigator, Impact results of pedestrian crash tests, in *The first conference of ITAI*, Hull, UK, 1993
18. Y. Lin, Y. huang, *The Practical Automobile Accident Identification* (China Communications Press, 2001)
19. X. Hongguo, *Automobile Accident Engineering* (China Communications Press, 2004)
20. A. Toor, M. Araszewski, R. Johal, R. Overgaard, A. Happer, Revision and Validation of Vehicle/Pedestrian Collision Analysis Method, in *SAE 2002 World Congress & Exhibition*, 2002, pp. 1014–1024
21. A. Toor, M. Araszewski, Theoretical vs. empirical solutions for vehicle/pedestrian collisions, in *SAE 2003 World Congress & Exhibition*, 2008
22. A. Happer, M. Araszewski, A. Toor, R. Overgaard, R. Johal, Comprehensive analysis method for vehicle/pedestrian collisions, in *SAE 2000 World Congress*, 2000
23. H. Xu, Y. Gao, L. Chen, The theory and application of the traffic accident collision velocity based on glass debris. China Road J. 75–81 (1994)
24. H. Xu, Y. Gao, L. Chen, Research on the theoretical model of glass fragment used on the calculation of collision velocity. Automot. Eng. 246–251 (1995)
25. H. Strobl, Betrachung zum Thema Steinschlag (1996)
26. X.U. Hongguo, Y. Ren, L. Wang, Determination on mechanics parameters of colliding on roadway for throwing object of traffic accident. J. Highw. Transp. Res. Dev. (2003)
27. H. Xu, W. Gao, J. Su, Y. Gao, L. Chen, Multi-value problem of collision velocity in motor vehicle accident. China J. Highw. Transp. **9** (1996)
28. H.G. Xu, Y.L. Gao, L.F. Chen, H. Apeil, China J. Highw. Transp. **7**(2), 75–81 (1994)
29. W.L. Li, Vibration analysis of rectangular plates with general elastic boundary supports. J. Sound Vib. **273**, 619–635 (2010)
30. L.R. Dharani, S.R. Mettu, S. Zhao, S.D. Barbat, L. Chai, Modeling Fracture in Laminated Automotive Glazing Impacted by Spherical Featureless Headform, Sae Technical Papers (2003)
31. S. Zhao, L.R. Dharani, L. Chai, S.D. Barbat, Analysis of damage in laminated automotive glazing subjected to simulated head impact. Eng. Fail. Anal. **13**, 582–597 (2006)
32. N. Nicholas. J.A. Zukas, H.F. Swift, L.B. Greszczuk, in *Impact Dynamics*, ed. by D.R. Curran (Wiley, New York, 1982)

33. B.V. Sankar, Low-velocity impact response and damage in composite materials. Key Eng. Mater. **120–121**, 549–582 (1996)
34. W.C. Young, R.G. Budynas, *Roarks's Formulas for Stress and Strain* (McGraw-Hill, 2003)
35. E. Ventsel, T. Krauthammer, E. Carrera, Thin plates and shells: theory, analysis, and applications. Appl. Mech. Rev. **55**, 1813–1831 (2002)
36. T. Maki, J. Kajzer, K. Mizuno, Y. Sekine, Comparative analysis of vehicle–bicyclist and vehicle–pedestrian accidents in Japan. Accid. Anal. Prev. **35**, 927–940 (2003)
37. Q. Cao, *Man-Machine Design Engineering* (Southwest Jiaotong University, Chengdu, 1988)

MIX
Papier aus verantwortungsvollen Quellen
Paper from responsible sources
FSC® C105338

If you have any concerns about our products,
you can contact us on
ProductSafety@springernature.com

In case Publisher is established outside the EU,
the EU authorized representative is:
Springer Nature Customer Service Center GmbH
Europaplatz 3, 69115 Heidelberg, Germany

Printed by Libri Plureos GmbH
in Hamburg, Germany